Cancer Metastasis – Related Genes

Cancer Metastasis – Biology and Treatment

VOLUME 3

Series Editors

Advisory Editorial Board

Titles published in this Series are:

Volume 1: Cancer Metastasis, Molecular and Cellular Mechanisms and Clinical Intervention.

Editors: Wen G. Jiang and Robert E. Mansel.
ISBN 0-7923-6395-7

Volume 2: Growth Factors and Receptors in Cancer Metastasis.

Editors: Wen G. Jiang, Kunio Matsumoto, and Toshikazu Nakamura.
ISBN 0-7923-7141-0

Cancer Metastasis – Related Genes

Edited by

Danny R. Welch
Jake Gittlen Cancer Research Institute,
The Pennsylvania State University College of Medicine,
Pennsylvania, The United States of America

KLUWER ACADEMIC PUBLISHERS
DORDRECHT / BOSTON / LONDON

A C.I.P. Catalogue record for this book is available from the Library of Congress.

ISBN 1-4020-0522-9

Published by Kluwer Academic Publishers,
P.O. Box 17, 3300 AA Dordrecht, The Netherlands.

Sold and distributed in North, Central and South America
by Kluwer Academic Publishers,
101 Philip Drive, Norwell, MA 02061, U.S.A.

In all other countries, sold and distributed
by Kluwer Academic Publishers,
P.O. Box 322, 3300 AH Dordrecht, The Netherlands.

Printed on acid-free paper

Printed in the Netherlands.

TABLE OF CONTENTS

Preface ... vii
Danny R. Welch

List of Contributors ... xi

Chapter 1 ... 1
Metastasis-Suppressor Genes: A Review and Perspective on an Emerging Field
Barbara A. Yoshida, Zita Dubauskas, Mitchell H. Sokoloff, Danny R. Welch, Carrie W. Rinker-Schaeffer

Chapter 2 ... 35
The Roles of Map Kinases in Controlling Cancer Metastasis
Alessandro Alessandrini

Chapter 3 ... 51
Tumor Metastasis-associated Human Mta1 Gene: Role in Epithelial Cancer Cell Proliferation and Regulation
Garth L. Nicolson, Akihiro Nawa, Yasushi Toh, Shigeki Taniguchi, Katsuhiko Nishimori

Chapter 4 ... 65
Cooperative Integrin Interactions in the Regulation of Tumor Metastasis
Peter C. Brooks

Chapter 5 ... 89
Brain Metastasis Associated Genes
Dario Marchetti

Chapter 6 ... 109
Autocrine Motility Factor and its Receptor as Regulators of Metastasis
Yasuharu Onishi, Arayo Haga, Avraham Raz

Chapter 7 ... 123
nm23 Metastasis Suppressor Gene
Patricia S. Steeg, Taoufik Ouatas, Michael Mair, Susan E. Clare, Melanie T. Hartsough

Chapter 8 145
Gene Regulation in Melanoma Metastasis
Menashe Bar-Eli

Chapter 9 169
Heterochromatin-associated Protein 1, HP1$^{Hs\alpha}$, in Breast Cancer Invasion and Metastasis
Dawn A. Kirschmann, Mary J.C. Hendrix

Chapter 10 191
Inhibition of Invasion and Metastasis During Specific Amino Acid Restriction Associated with Metastasis Suppressor and Other Gene Changes
Gary G. Meadows, Xiaokang Ge, Hui Zhang, Daniel R. Oros, Ya-Min Fu

Chapter 11 209
Role of BRMS1 in Breast Carcinoma Metastasis
Rajeev S. Samant, Lalita R. Shevde, Danny R. Welch

Chapter 12 219
The Role of KISS1 in Melanoma Metastasis Suppression
John F. Harms, Danny R. Welch

Chapter 13 231
Osteopontin: a Ras-regulated Gene That Contributes to Tumor Metastasis
Ann F. Chambers and Alan B. Tuck

Chapter 14 247
The Emerging Role for the mRNA Cap-binding Protein, EIF-4E, in Metastatic Progression
Stephen G. Zimmer and Jeremy R. Graff

Index 267

PREFACE

Being diagnosed with cancer is devastating. But when the cancer cells have to spread to form secondary colonies, the prognosis for the patient is worse. If meaningful improvements in survival are to occur, then control of metastasis will be a foundation. Relatively little is known about the control of the metastatic process at the molecular level. This volume begins to explore our current knowledge regarding the underlying molecular and biochemical mechanisms controlling the metastatic phenotype. While all of the authors attempted to put their findings into a context for translation to the clinical situation, the state-of-the-art does not fully allow this. Nonetheless, we write these summaries of our work as an early effort toward that end. I am grateful to all of the authors who have contributed generously of their time and energies to make this volume a reality.

To metastasize, neoplastic cells dissociate from the primary tumor, enter a circulatory compartment (typically lymphatics or blood vasculature), survive transport, arrest, exit the circulation and finally proliferate at a discontinuous site in response to local growth factors. Unless cells accomplish *every* step of the metastatic cascade, metastases cannot develop. The process is highly inefficient, i.e., $<0.1\%$ of cells entering the vasculature form clinically detectable secondary tumors. At each step of the metastatic cascade, multiple genes and proteins are involved. Because inappropriate movement of cells with subsequent colonization of secondary sites implies that some of those genes are either mutated or aberrantly regulated, it follows that identifying and manipulating metastasis-regulatory genes could lead to decreased efficiency of the metastatic process and better systemic control of neoplasia. Moreover it must be emphasized that each of the genetic defects responsible for developing metastatic potential is superimposed over those already involved in the genesis of a tumor. A paradigm describing the genetics of progression toward metastasis may be modeled after the oncogene and tumor suppressor gene paradigm in the development of carcinomas. Analogous to the role of oncogenes in tumorigenesis, metastasis-promoting genes drive conversion from nonmetastatic to metastatic. Similar to tumor suppressor genes, metastasis suppressors would inhibit the metastatic process. In the case of negative regulators, the distinction between tumor suppressors and metastasis suppressors is critical! Tumor suppressors, by definition also block metastasis since tumor formation is a prerequisite to metastasis. However, using this functional definition, metastasis suppressors only block spread to distant sites. They do not reduce tumor formation.

To date, only a limited number of genes have been shown to functionally regulate the metastatic cascade; but, fortunately, the number of genes identified is growing rapidly. Our criteria for claiming a role in metastasis requires *in vivo* validation. Simply put, *in vitro* surrogates of component steps of the metastatic cascade are inadequate to measure a complex, multistep, multigenic phenotype like metastasis.

The ability to metastasize can be due to inherent deficiencies within tumor cells themselves (i.e., genetics) or to defective responses to the host environment (i.e., epigenetics). The relative contribution of each has yet to be fully determined. But because of the currently available technologies, the former will be the focus of this volume. Within this volume, several candidate metastasis-regulatory genes are described in detail. The chapters are organized into loose sections. Because the field of metastasis genetics is still in its infancy, the clusters are somewhat arbitrary and artificial. However, they provide one attempt to overview this rapidly expanding area of research.

Yoshida and colleagues review the emerging field of cancer metastasis genetics, highlighting the context under which the genes were discovered and how they fit into a larger picture. This chapter is followed by discussion of genes which promote tumor progression (i.e., metastasis-promoting genes). Identification of metastasis-promoting genes is notoriously difficult because of the inherent nature of the metastatic process. Since only one step needs to be blocked in order to prevent metastasis, introduction of a *bona fide* metastasis-promoting gene into a cell would not necessarily enhance metastasis if that cell were defective for another step. Hence, the model system from which one starts is critical. Alessandro Alessandrini highlights components of the MAP kinase signaling cascade and how they confer tumorigenic and metastatic potential upon NIH3T3 cells. Garth Nicolson and colleagues describe a recently discovered gene mta1, which promotes metastasis. mta1 appears to be involved in regulation of gene expression, perhaps downstream of such signaling pathways.

Peter Brooks summarizes the environmental milieu in which tumor cells reside and the importance of surface adhesion molecules in mediating the metastatic phenotype. Outside-in signaling is critical in the efficiency of metastasis. Coupling the surface molecules to intracellular events (as highlighted above) will be an area of fruitful future research. In a similar vein, Dario Marchetti describes an example for organ-specific melanoma metastasis to brain. Identification of neurotrophins represents a class of exogenous signals which contribute to metastasis. They remain some of the few well-defined molecules which can explain why some cells colonize certain organs while others do not. While tumor cell behavior can be modulated by environment, they sometimes carry the necessary machinery themselves. Onishi and colleagues describe roles for autocrine signals of motility as contributors to metastasis.

The majority of chapters herein describe roles for metastasis-suppressor genes. Since the discovery of Nm23 by Patricia Steeg in 1984, the list of metastasis suppressors has grown significantly to include AP2, KiSS1, BRMS1, and MKK4 among others. Menashe Bar-Eli describes a role for the transcriptional regulator AP2 in melanoma metastasis, including how this gene regulates the expression of other metastasis-associated genes. Dawn Kirschmann and Mary Hendrix provide evidence that heterochromatin associated protein ($HP1^{HS\alpha}$) might regulate gene expression critical for metastasis. Both of these genes make logical candidates for metastasis regulation since each may be involved in coordinated regulation of expression of multiple critical genes. Likewise, Gary Meadows and colleagues present some intriguing data which shows that diet can modulate metastatic potential. Specific amino acid deprivation can markedly inhibit metastasis. They hypothesize that amino acid response elements may be controlling families of metastasis-regulatory genes.

Carrie Rinker-Schaeffer's laboratory has described a metastasis suppressor effect with a member of the stress-activated MAP kinase family. Like the MAP kinase family and the promotion of metastasis, the downstream effectors of MKK4 are not yet known. As these become more finely defined, crosstalk and feedback mechanisms will likely emerge. These pathways will be important as the field begins to dissect whether there are metastasis suppressors which function universally (i.e., for all tumor types) or whether there are metastasis suppressor genes which act only upon one tumor type (e.g., breast cancer, but not prostate cancer or melanoma).

Thus far, most metastasis suppressor genes have been identified and tested using only a limited number of models. Two examples from the Welch laboratory are described. BRMS1 was isolated from breast carcinoma cells inhibited for metastasis following introduction of human chromosome 11. The mechanism of action is speculated upon by Rajeev Samant, but thus far appears to be involved in transcriptional regulation of other genes. KiSS1 was discovered in human melanoma cells suppressed for metastasis following introduction of chromosome 6. John Harms summarizes the current knowledge of the KiSS1 gene and the steps of the metastatic process inhibited in the cells from which this gene was derived. The chromosome 6 hybrid cells complete every step of the metastatic cascade, except growth at the secondary site. The mechanisms of action for all of the metastasis regulatory genes remain largely enigmatic, but this result provides one small example of the types of progress being made. Ann Chambers and colleagues have utilized the powerful technique of intravital video microscopy to challenge the notion that metastatic cells must extravasate in order to colonize an organ. While it is far from certain whether intravascular or perivascular proliferation of metastases is more prevalent, the implications regarding the genetic control of the metastatic phenotype must take both possibilities into account.

Finally, Jeremy Graff and Steve Zimmer review an understudied area with regard to metastasis regulation B translation. They provide some compelling data highlighting correlations between translational efficiency and metastasis.

Together these chapters provide a glimpse into the complex world of metastasis genetics. Certainly, there remains a great deal of work to be done. Yet, the future looks bright for translating the types of research represented herein to the clinic.

LIST OF CONTRIBUTORS

Allessandrini, Allessandro. Medical Services, Massachusetts General Hospital, Department of Medicine, Harvard Medical School, U.S.A.

Bar-Eli, Menashe. University of Texas, M.D. Anderson Cancer Center, Department of Cancer Biology, Houston, Texas, U.S.A.

Brooks, Peter C. New York University School of Medicine, Kaplan Cancer Center, Departments of Radiation Oncology and Cell Biology, New York, NY 10016, U.S.A.

Chambers, Ann F. London Regional Cancer Centre, London, Ontario N6A 4L6, Canada.

Clare, Susan E. Women's Cancers Section, Laboratory of Pathology, National Cancer Institute, Bethesda, MD; Division of Surgical Oncology, Department of Surgery, Northwestern University, Chicago, IL, U.S.A.

Dubauskas, Zita. The University of Chicago, Section of Urology and Genitourinary Oncology Research Program, Chicago, Illinois, U.S.A.

Fu, Ya-Min. Cancer Prevention and Research Center, Department of Pharmaceutical Sciences, College of Pharmacy, Washington State University, Pullman, WA 99164-6510, U.S.A.

Ge, Xiaokang. Cancer Prevention and Research Center, Department of Pharmaceutical Sciences, College of Pharmacy, Washington State University, Pullman, WA 99164-6510, U.S.A.

Graff, Jeremy R. Lilly Research Laboratories, Cancer Research Division, Eli Lilly and Company, Indianapolis, IN, U.S.A.

Haga, Arayo. Karmanos Cancer Institute, Wayne State University, Detroit, MI, U.S.A.

Harms, John F. Jake Gittlen Cancer Research Institute, The Pennsylvania State University College of Medicine, Hershey, Pennsylvania, U.S.A.

Hartsough, Melanie T. Women's Cancers Section, Laboratory of Pathology, National Cancer Institute, Bethesda, MD; Division of Surgical Oncology, Department of Surgery, Northwestern University, Chicago, IL, U.S.A.

Hendrix, Mary J.C. Department of Anatomy & Cell Biology, Holden Cancer Center at the University of Iowa, College of Medicine, Iowa City, Iowa, 52242-1109, U.S.A.

Kirschmann, Dawn A. Department of Anatomy & Cell Biology, Holden Cancer Center at the University of Iowa, College of Medicine, Iowa City, Iowa, 52242-1109, U.S.A.

Mair, Michael. Women's Cancers Section, Laboratory of Pathology, National Cancer Institute, Bethesda, MD; Division of Surgical Oncology, Department of Surgery, Northwestern University, Chicago, IL, U.S.A.

Marchetti, Dario. Department of Neurosurgery, The University of Texas-Houston Health Science Center, Houston, TX, U.S.A.

Meadows, Gary G. Cancer Prevention and Research Center, Department of Pharmaceutical Sciences, College of Pharmacy, Washington State University, Pullman, WA 99164-6510, U.S.A.

Nawa, Akihiro. Department of Obstetrics and Gynecology, Nagoya University School of Medicine, Nagoya 466, Japan.

Nicolson, Garth L. The Institute of Molecular Medicine, Huntington Beach, California 92649, U.S.A.

Nishimori, Katsuhiko. Department of Obstetrics and Gynecology, Nagoya University School of Medicine, Nagoya 466, Japan.

Onishi, Yasuharu. Karmanos Cancer Institute, Wayne State University, Detroit, MI, U.S.A.

Oros, Daniel R. Cancer Prevention and Research Center, Department of Pharmaceutical Sciences, College of Pharmacy, Washington State University, Pullman, WA 99164-6510, U.S.A.

Ouatas, Taoufik. Women's Cancers Section, Laboratory of Pathology, National Cancer Institute, Bethesda, MD; Division of Surgical Oncology, Department of Surgery, Northwestern University, Chicago, IL, U.S.A.

Raz, Avraham. Karmanos Cancer Institute, Wayne State University, Detroit, MI, U.S.A.

Rinker-Schaeffer, Carrie W. The University of Chicago, Section of Urology and Genitourinary Oncology Research Program, Chicago, Illinois, U.S.A.

Samant, Rajeev S. Jake Gittlen Cancer Research Institute, The Pennsylvania State University College of Medicine, Hershey, Pennsylvania, U.S.A.

Shevde, Lalita R. Jake Gittlen Cancer Research Institute, The Pennsylvania State University College of Medicine, Hershey, Pennsylvania, U.S.A.

Sokoloff, Mitchell H. The University of Chicago, Section of Urology and Genitourinary Oncology Research Program, Chicago, Illinois, U.S.A.

Steeg, Patricia S. Women's Cancers Section, Laboratory of Pathology, National Cancer Institute, Bethesda, MD; Division of Surgical Oncology, Department of Surgery, Northwestern University, Chicago, IL, U.S.A.

Taniguchi, Shigeki. Laboratory of Molecular Biology, Graduate School of Agricultural Science, Tohoku University, Sendai 981-8555, Japan.

Toh, Yasushi. Department of Gastroenterorogic Surgery, National Kyushu Cancer Center, 811-1395 Japan.

Tuck, Alan B. London Regional Cancer Centre, London, Ontario N6A 4LS, Canada.

Welch, Danny R. Jake Gittlen Cancer Research Institute, Pennsylvania State University College of Medicine, Hershey, Pennsylvania, U.S.A.

Yoshida, Barbara A. The University of Chicago, Section of Urology and Genitourinary Oncology Research Program, Chicago, Illinois, U.S.A.

Zhang, Hui. Cancer Prevention and Research Center, Department of Pharmaceutical Sciences, College of Pharmacy, Washington State University, Pullman, WA 99164-6510, U.S.A.

Zimmer, Stephen G. L.P. Markey Cancer Center, University of Kentucky, Department of Microbiology and Immunology, Lexington, U.S.A.

Chapter 1

METASTASIS-SUPPRESSOR GENES: A REVIEW AND PERSPECTIVE ON AN EMERGING FIELD

Barbara A. Yoshida[1], Zita Dubauskas[1], Mitchell H. Sokoloff[1], Danny R. Welch[2], Carrie W. Rinker-Schaeffer[1]

[1]*The University of Chicago, Section of Urology and Genitourinary Oncology Research Program, Chicago, Illinois* [2]*Pennsylvania State University College of Medicine, Hershey, Pennsylvania.*

Abstract
Metastasis is the most lethal attribute of a cancer. There is a critical need for markers that will distinguish accurately those histologic lesions and disseminated cells with a high probability of causing clinically important metastatic disease from those that will remain indolent. While the development of new diagnostic markers of metastasis was the initial motivation for many studies, the biologic approach used to identify metastasis-suppressor genes has provided surprising insights into the *in vivo* mechanisms regulating the formation of metastases. This chapter reviews the evolving view of the mechanisms that regulate metastasis and the importance of metastasis-suppressor genes in this process. The known metastasis-suppressor proteins or genes and the microcell-mediated chromosomal transfer strategy used to identify many of them are reviewed. New evidence for the role of these metastasis-suppressor activities (genes) in regulating the growth of disseminated cancer cells at the secondary site, the potential for the identification of novel therapeutic targets, and the multidisciplinary approach needed to translate this information into clinical tools for the treatment of metastatic disease are discussed.

1. THE CLINICAL PROBLEM: PREDICTING METASTATIC PROPENSITY

Our ability to detect and successfully treat localized cancers has improved appreciably in recent years. However, metastatic disease presents a continuing therapeutic challenge and is the most common cause of cancer-related death. Thus, there is an emphasis on the diagnosis of cancers at an early stage, when they are localized and most likely to be curable. Although screening for early stage disease is logical, its utility is limited by the inability of conventional diagnostic and histologic parameters to predict accurately the true extent and prognosis of a substantial proportion of clinically localized cancers (1-3). This

D.R. Welch (ed.), Cancer Metastasis – Related Genes, 1–33.

limitation is due, in part, to the inherent limitations and subjectivity of current grading and staging systems (4, 5).

The incidence of disease recurrence in surgical patients treated for prostatic and breast cancer illustrates this problem particularly well. Although we have a wealth of clinical and biologic information on these diseases, a large percentage of apparently resectable and theoretically curable lesions are found to be more advanced at the time of resection than envisaged, resulting in a substantial failure rate after attempted curative surgery (6-8). In studies of prostate cancer patients, even when patient selection excludes men with factors predicting poor prognosis (e.g. poorly differentiated histology, high prostate specific antigen [PSA] levels, clinical suspicion of local invasion) the relapse rate after radical retroperitoneal prostatectomy has approached 20%-30% (9-11). Similarly, one-third of surgical patients with node-negative breast cancer will develop metastases, while the other two-thirds, despite receiving no chemotherapy, do not (12). Even in patients with small tumors and tumor-negative lymph nodes (T1N0), there is a 15 to 25% likelihood of distant metastases (8).

Since the current staging systems for breast and prostate cancers do not accurately identify those patients curable by regional treatment alone, the evaluation of additional parameters associated with the metastatic phenotype will be very important for the differentiation of patients curable by surgery alone from those requiring systemic therapy. For instance, men at high risk for relapse of prostate cancer can be identified (e.g. serum PSA > 10 ng/ml, clinical stage T_1 or T_2 with greater than 50% of tissue at Gleason grade 4 (3, 4) on biopsy or clinical stage T_3 prostate cancer) and would be immediate candidates for adjuvant anti-metastatic therapies if they existed (10, 11, 13-16). Likewise, breast cancer patients with particularly poor prognoses can be identified by the detection of high microvessel counts concurrent with low expression of Nm23 and/or E-cadherin in the primary tumor (12-17). In fact, these parameters are better prognostic biomarkers than the conventional analysis of tumor size and grade. The information obtained from the simultaneous evaluation of biomarkers such as these have the potential to lead to a reduction in the morbidity in those patients not requiring chemotherapy, and possibly identify those patients requiring more aggressive therapies than indicated by current methods.

Overall, it is clear that there is a critical need for markers that will distinguish accurately those histologic lesions and disseminated cells that have a high probability of causing clinically important metastatic disease from those that will remain indolent (5, 15). Concerns have been raised that "metastasis" has often occurred by the time of diagnosis of the primary tumor, the implication being that it is then too late for anti-metastatic therapy to be of use (18). However, the mere spread of cancer cells into the vasculature or to a secondary site does not constitute metastasis. Development of clinically significant metastases requires that a cancer cell complete a series of well-defined steps, generally referred to as the metastatic cascade (13). If a cell fails to complete any one of these steps, overt metastases will not develop (13-15).

The clinical importance of disseminated cancer cells (detected by sensitive methods such as reverse transcriptase polymerase chain reaction [RT-PCR]) has become an issue of considerable interest (19). Several such studies have reported the detection of tumor-derived cells in the circulation and bone marrow without future development of disease (16, 20). Other reports have demonstrated an increased risk of disease recurrence in patients with bone marrow micrometastases both for prostate cancer (by the detection of mRNA transcripts for prostate specific antigen [21]), and breast cancer (by the detection of cytokeratin-positive cells [22]). Even in these later studies, however, the majority of patients with tumor cell-positive bone marrow samples did not actually develop recurrent disease, although the proportion with recurrence could increase given extended time for patient follow-up. The discrepancy regarding the clinical importance of disseminated cells is likely due to differences in the experimental approaches used to identify cells (i.e. RT-PCR vs. immunohistochemical detection).

Tumor-cell growth at the site of metastasis is an important clinical target since cells must survive and proliferate in order to grow into overt, macroscopic metastases. The first step toward developing effective therapies to inhibit such growth is to identify the genes/proteins that regulate metastatic colonization. To this end, a growing number of laboratories are focusing translational research efforts on the discovery of genes that specifically regulate the metastatic ability of cancer cells. For example, several metastasis-promoting genes – including *WDNM-1*, *WDNM-2*, *MMP11* (stromelysin-3), *MTA1*, and *ERBB2* – have been identified in association with the development of metastatic breast cancer (23-27). One must keep in mind, however, that it takes the coordinated expression of many genes to allow the development of metastases (28, 29). Thus, while it is relatively easy to demonstrate an association for a given gene with metastatic ability, it is difficult to prove that a particular gene is essential. On the other hand, it only takes one gene to block metastasis since inability to complete any step of the metastatic cascade renders a cell nonmetastatic. Metastasis-suppressor genes suppress the formation of spontaneous, macroscopic metastases *without affecting the growth rate of the primary tumor*. It has now been more than ten years since the discovery of the first metastasis-suppressor gene *nm23* (NME1) (30). Since then, both *in vitro* and *in vivo* (eg. animal) studies have documented the important role of the loss of metastasis-suppressor gene function in the acquisition of metastatic ability (15, 30-32).

While the initial motivation for these studies was the development of new diagnostic markers of metastasis, the biologic approach used to identify metastasis-suppressor genes has provided surprising insights into the *in vivo* mechanisms regulating the formation of metastases. We anticipate that identifying the molecular pathways that regulate metastatic colonization and growth control at the secondary site will provide additional, potentially novel therapeutic targets for the treatment of metastatic disease. The purpose of this chapter is to:

- Present the evolving view of the mechanisms that regulate metastasis
- Describe the functional strategy used to identify metastasis-suppressor genes and discuss important principles learned from these studies
- Document the known metastasis-suppressor genes and report new evidence that supports their role in the regulation of growth control at the secondary site
- Discuss the multidisciplinary approach needed to translate metastasis-suppressor genes into clinical tools

1.1 Regulation of Metastatic Propensity – Evolving Paradigms

Metastasis is defined as the formation of progressively growing secondary tumor foci at sites discontinuous from the primary lesion (15). This process is illustrated by the spontaneous hematogenous metastasis of tumor cells to the lung (Figure 1, Panel A). The formation of a primary tumor requires a cadre of molecular and cellular alterations that enable a cell(s) to circumvent normal growth control mechanisms, as well as, to manipulate its local environment (14). These changes include the development of a blood supply once the focus of transformed cells grows beyond a size that can be nourished by nutrient or metabolite diffusion (33, 34). Tumor progression and the acquisition of metastatic competence requires additional changes in gene expression (e.g. protein degrading enzymes, adhesion molecules) that culminate in a malignant phenotype. Following invasion into adjacent tissues, tumor cells disseminate via blood vasculature or lymphatics and travel individually or as emboli comprised of tumor cells or tumor and host cells. At the secondary site, cells or emboli arrest either because of their physical size or by binding to specific molecules in particular organs or tissues (15, 35). In order for disseminated cells to grow into overt metastases, they must survive and proliferate in the vasculature or in the surrounding tissue after extravasation. The formation of clinically important metastases depends upon the completion of *every step of this cascade*, the last of which is metastatic colonization (Figure 1) (14).

The presence of isolated cells at a secondary site represents a risk to the patient. Cells getting to the secondary site certainly have the potential to colonize, and therefore, it is crucial not to ignore the presence of neoplastic cells anywhere. On the other hand, as we will show, the mere presence of cells does not necessarily mean that metastatic colonization will occur. The challenge is to determine how to discriminate between disseminated cells that will form overt metastases from those that will not.

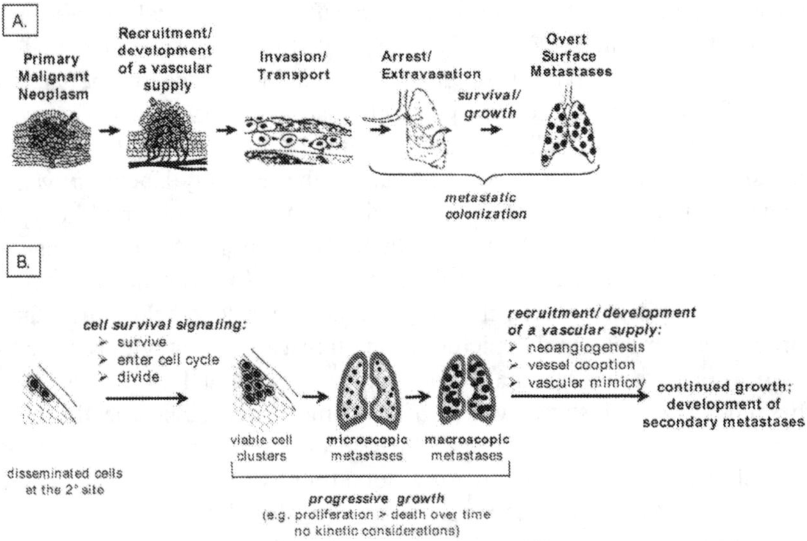

Figure 1. Development of spontaneous hematogenous metastases. Panel A. The development of spontaneous hematogenous metastases requires cancer cells to complete a well-defined series of steps. This figure is adapted from reference 13. Panel B In order to form overt metastases, disseminated cells must complete additional steps at the metastatic site(s).

Cancer metastasis, both clinically and experimentally, is known to be inefficient (36). In experimental models, less than 0.1% of cells injected into the circulation go on to form secondary tumors (15, 37). While many factors contribute to the observed inefficiency of metastasis formation, those considered to be most important include low survival rates of cells in the circulation and the low percentage of cells that successfully escape from the vasculature into surrounding tissues (18). At this time, there is some question as to whether post extravasational growth control or growth within a vessel are more predominant (38). This process has, for the most part, been studied using assays in which the number and kind of cells injected are known and the numbers and sizes of metastases formed are assessed (18, 39). The processes that are responsible for metastatic efficiency *in vivo* remain hidden, thus mechanistic paradigms have largely been based upon logical inference rather than direct observation. The development of new technologies has enabled researchers to test the possibility that cancer cell dissemination, arrest (nonspecific arrest and/or specific adhesion events), and growth at the secondary site are critical determinants in metastasis formation.

The ability to observe single cells *in vivo* has been greatly enhanced by improvements in intravital microscopy and the utilization of vital fluorescent dyes like green fluorescent protein (GFP) (18, 40). Studies that couple these two powerful techniques have added greatly to our knowledge of the metastatic processes following tumor cell entry into circulatory compartments. The use of *in vivo* video microscopy allows for the direct observation of experimental metastasis over time (39). Cancer cells can be fluorescently labeled *in vitro* and then injected into an animal. The cells can then be viewed at different time points, by both fluorescence and oblique transillumination, in thin tissues or superficial ($\leq$50 μm) regions of thick tissues *in vivo* (39). Experiments utilizing this technology have demonstrated that, in contrast to the long-held belief, the vast majority of cancer cells in the microcirculation manage not only to survive there, but also to extravasate into the surrounding tissue within 1 to 2 days (41, 42). Such studies have translated well into the clinical arena. Specifically, the vast majority of clinical studies using RT-PCR to detect prostate tumor cells in the peripheral circulation and bone marrow found no association between detection of disseminated cells and treatment failure (16, 20). Interestingly, in a recent study of breast cancer patients, detection of cytokeratin-positive cancer cells in the bone marrow was associated with the development of overt metastases and death (22). The apparent difference between these two findings may be due to differences in study design (e.g. detection methods, markers used, etc.) or in factors that influence the growth of disseminated cancer cells at the metastatic site. Additional studies will be necessary to distinguish between these possibilities. Taken together, the clinical and experimental evidence supports the observation that dissemination from the primary tumor site is a frequent event. Furthermore, these independent and complementary studies strongly suggest that growth control of individual disseminated cells determines the efficiency of metastatic colonization.

1.2 Metastatic Colonization

Metastatic colonization is the lodging and subsequent growth of disseminated cancer cells to form clinically significant metastases (Figure 1, Panel B). In order to proliferate, surviving disseminated tumor cell(s) must be able to initiate cell appropriate context-dependent signaling cascades, which enable them to survive, enter the cell cycle, and divide. While disseminated cells are likely to be present in numerous organs, only certain environment(s) appear to allow their survival and subsequent growth (37, 43, 44). Intercellular interactions with the stroma and with other tumor cells are critical for tumor cell survival and involve the activation of adhesion-dependent survival pathways such as those described for e-cadherin (45, 46) and integrin molecules (47). Clusters of proliferating cells grow into lesions consisting of a few hundred that can be detected reliably by histological methods. Cells within such microscopic lesions can receive oxygen and nutrients by diffusion. Progressive growth of microscopic lesions into overt or macroscopic metastases (greater than 1 mm in diameter) requires that the fraction of proliferating cells exceed the fraction that are quiescent or apoptotic. In recent literature, this transition from microscopic to macroscopic metastasis has often been referred to as the switch to an angiogenic phenotype or the angiogenic switch (48). This terminology implies that microscopic metastases exist in one of two states; either the lesion is angiogenic (forming new blood vessels) or it is not. However, the progression from a "microscopic lesion" to an overt metastasis is more accurately described in terms of growth control. Indeed, the interchangeable use of "angiogenesis" and "growth" has been a source of confusion. This progression may occur over a period of months or even years and is not necessarily dependent upon new blood vessel formation. Vascularization is in fact a late step in metastatic colonization (49). Recent studies have shown that, in addition to the induction of classical neovascularization via endothelial cell recruitment, tumor cell masses can develop a blood supply by alternative means such as the cooption of preexisting host vessels (49) or by formation of tumor channels, a process referred to as vascular mimicry (50). As we will describe in the following paragraphs, recent data from our laboratories suggest that a subset of metastasis-suppressor genes inhibit early steps in metastatic colonization, prior to the need for development or recruitment of vessels.

1.3 Identification of Metastasis-Suppressor Activity: A Functional Approach

Metastasis-suppressor genes suppress the formation of (spontaneous) macroscopic metastases. As their name implies, these genes are distinct from *oncogenes*, which promote cellular transformation, and *tumor-suppressor genes*, which suppress tumor growth. While the first metastasis-suppressor gene, *nm23,* was identified by a cDNA subtraction approach, the majority of metastasis-suppressor activities identified to date, have been discovered using microcell-mediated chromosomal transfer (MMCT; Table 1). The choice of the MMCT strategy was logical since the existence of metastasis-suppressor genes was originally implicated by the results of somatic cell fusion studies, the precursos of MMCT (51-54). The techniques for the generation of genetically stable somatic cell hybrids were developed in the early studies of Barski et al. (55). In most instances, fusion between malignant cells and normal cells results in hybrid cells that are suppressed in their tumorigenic capacity (56). Ichikawa et al. were the first researches to identify specific chromosomal losses associated with the reacquisition of metastatic ability (57). In their study, fusion of nonmetastatic with highly metastatic Dunning rat prostatic cancer cells resulted in nometasttatic hybrid. Importantly, the tumorigenecity (e.g. tumor formation and latency period) and *in vivo* growth rates of the primary tumors of hybrid clones containing a full complement of rat chromosomes were not affected. At the experimental endpoint, none of the animals bearing hybrid tumors developed distant metastases. However, when the nonmetastatic primary tumors were serially passaged *in vivo*, animals occasionally developed distant metastases. Cytogenetic analysis of these metastatic revertants revealed a consistent loss of a copy of rat chromosome 2. This critical study suggested that the loss of specific chromosomes could increase the metastatic potential of prostate cancer cells without affecting growth rate or tumorigenicity.

Table 1. Chromosomal region identified by microcell-mediated chromosomal transfer that suppress metastasis *in vivo*.

Chromosomal Location	Tumor Types [reference]	Cell Lines Tested (species of origin)	*In Vitro* Phenotype†	*In Vivo* Phenotype
Chromosome 1	Melanoma [72]	MelJuSo (human)	ND	↓ spontaneous mets. ↓ experimental mets.
6q16.3-q23	Melanoma [73, 74]	C8161 (human)	↓ motility	↓ spontaneous mets.
chromosome 6	Breast[75]	MelJuSo (human)	ND ND	↓ experimental mets. occasional single cells (detected by GFP tagging) which are growth suppressed but viable
		MDA-MB-435 (human)		↓ spontaneous mets.
7q21-22 and/or 7q31.2-32	Prostate[76]	AT6.3 (rat)	ND	↓ spontaneous mets. ↓ experimental mets.
8p21-p12	Prostate [77, 78]	AT6.2 (rat)	↓ invasion	↓ spontaneous mets. NE experimental mets.
10cen-10q23	Prostate[79]	AT6.3 (rat)	ND	↓ spontaneous mets.
11q13.1-13.2	Breast [75, 80]	MDA-MB-435 (human)	ND ND	↓ spontaneous mets.
11pter-q14		R1564 (rat)		NE spontaneous mets.
11p11.2-13	Prostate[68]	AT6.1 (rat) AT3.1 (rat)	ND ND	↓ spontaneous mets. ↓ spontaneous mets.
12qcen-q13 and/or 12q24-ter	Prostate[64]	AT6.1 (rat)	NE motility‡ NE invasion‡	↓ spontaneous mets. no micrometastases observed at the experimental endpoint
16q24.2	Prostate[82]	AT6.1 (rat)	ND	↓ spontaneous mets.
17p12-11.2 and/or 17cen-q12	Prostate[63]	AT6.1 (rat)	NE motility‡ NE invasion‡	↓ spontaneous mets. Micrometastases observed at the experimental endpoint

ND = Not Determined, NE = Not Examined, GFP = Green Fluorescent Protein mets.=metastases
†Motility was measured by micropipet motility assay or by migration toward a chemoattractant in Boyden chambers. Invasion was measured by migration through Matrigel.
‡Rinker-Schaeffer, C.W., unpublished results. ND = Not Determined, NE = Not Examined, GFP = Green Fluorescent Protein mets. = metastases

The observation of a metastasis-suppression activity being associated with a specific chromosome coincided with the development of MMCT as a technique for the study of genes encoded by individual human chromosomes (51-54, 58-61). In this approach, summarized in Figure 2, well-characterized donor cells, carrying a single human chromosome tagged with a selectable marker or markers (e.g. neomycin phosphotransferase, etc.) are used to transfer the chromosome of interest into recipient cells (62). Briefly, donor cells are sequentially treated with colcemid, to depolymerize microtubules, and cytochalasin-B, to depolymerize actin bundles. The treated cells are centrifuged and the resulting pellet contains the microcells. Microcells are, in effect, micelles that contain single or multiple chromosomes. To enrich for those containing single chromosomes, the microcells are size-fractionated by sequential filtration through polycarbonate membranes of decreasing pore size. Microcells become attached to recipient cells in the presence of phytohemagglutinin and then become fused with the addition of polyethylene glycol. Recipient cells containing human chromosomes are selected in G418-containing media and then characterized by molecular and cytogenetic methods such as sequence tagged site PCR, karyotyping, and fluorescence *in situ* hybridization (63, 64). The complete characterization of the hybrids under study is critical, as it provides information on the addition and /or deletion of donor and recipient chromosomal material, as well as any rearrangements that may have occurred during MMCT.

Several laboratories have employed the technique of MMCT to test the functional significance of chromosomal alterations, such as loss of heterozygosity (LOH), observed in clinical samples. In addition, the use of MMCT, in combination with positional or expression based cloning techniques, has allowed the functional identification of genes conveying phenotypes such as senescence or tumor and metastasis-suppression (15, 65, 66). A review of the literature shows that transfer of a given chromosome can have different phenotypic effects that are dependent on the characteristics of the recipient cell line. For example, the transfer of human chromosome 7 by MMCT into immortalized SUSM-1 fibroblast cells induces senescence (67), while transfer of the same chromosome into choriocarcinoma cells results in suppression of tumor growth *in vivo* (68). Such results have enabled the definition of complementation groups for particular chromosome functions. The potential outcomes of transferring a particular chromosome into highly metastatic cells are summarized in Figure 2.

Studies using highly metastatic Dunning rat prostatic cells as the recipients for chromosomal transfer showed that chromosomes 12 and 17 specifically suppressed the metastatic ability of these cells (63, 64). The observed metastasis suppression had no effect on tumor growth rate. Interestingly, in analogous studies of human prostate cancer cell lines, transfer of these chromosomes suppressed the cells' tumorigenicity (69, 70). These findings could result from at least three alternative mechanisms. First, a given chromosomal region may encode a number of different genes, one or more of which may be active as a tumor-suppressor gene in human prostate cancer cells but be inactive or not

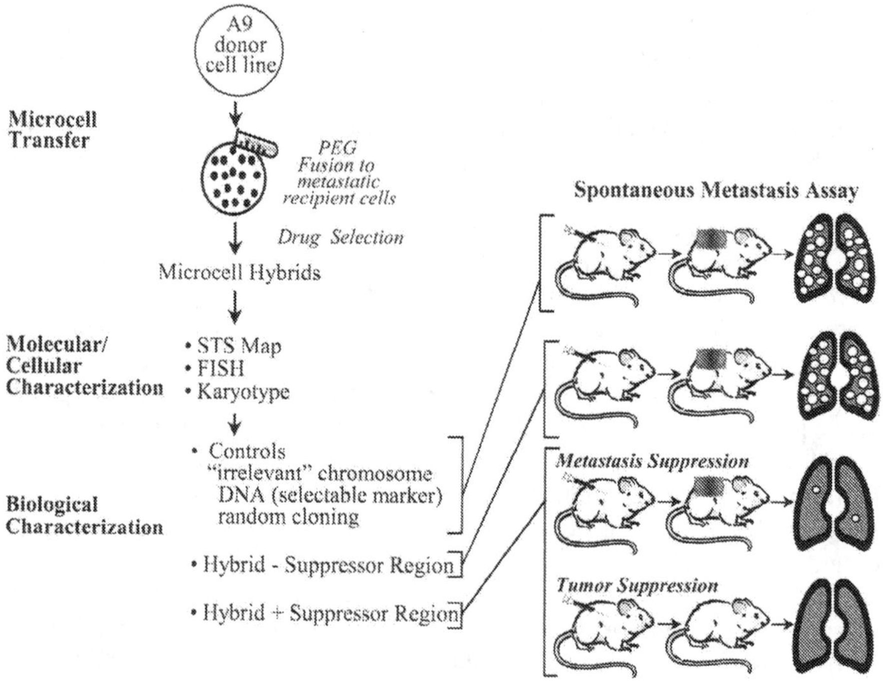

Figure 2. Identification of metastasis-suppressor activities by use of a microcell-mediated chromosomal transfer. A9 donor cells containing a single human chromosome are used for the preparation of microcells that will be transferred to the recipient cell lines. Stable microcell hybrids are selected and characterized by molecular and cellular methods. To screen the microcell hybrids for the minimal metastasis-suppressor region, one employs the following technique: fluorescence in situ hybridization (FISH) karyotyping and PCR amplification for human specific sequences using STS markers. Ultimately, the presence or absence of a suppressor region is determined by subcutaneous injection of the hybrids into the flanks of immunodeficient mice (63). Inclusion of a variety of controls is critical for the definition of metastasis-suppressor activity in vivo. The potential outcomes of in vivo studies using control, suppressed and unsuppressed hybrids are illustrated at left. STS = sequence tagged site; FISH = fluorescence in situ hybridization; PEG = polyethylene glycol. Figure reproduced from ref. 13.

expressed in rat prostate cancer cells. Second, genes may function as metastasis-suppressor genes when expressed in rat prostate cancer cells but may be inactive or not expressed in human prostate cancer cell lines. Third, gene(s) that lie in the same chromosomal region may have different functions depending on the context (i.e. cell type) in which they are expressed.

In the third scenario, the effect of the gene product may be limited or determined by the recipient cells. We refer to this scenario as the "cellular hard-wiring" hypothesis[1]. For example, human prostate cancer cell lines, compared to Dunning rat prostatic cancer cell lines, are weakly metastatic in spontaneous metastasis assays (62). These differences in their *in vivo* biologic activities could be the result of genetic differences between the tumor cells, or they could result from an epigenetic mechanism such as differential tumor–stromal interactions. The nature of cellular interactions with the extracellular matrix can regulate tissue-specific gene expression as cells form an elaborate three-dimensional network composed of the nuclear, cytoskeletal, and extracellular matrices (27, 71). Thus, the differential effects of a given chromosome transferred into different cell types can be the result of differential expression of the genes on the chromosome as determined by the way a cell responds to its environment.

During the past decade, several human chromosomes have been functionally tested through the use of MMCT, and metastasis-suppressor activities have been reported on chromosomes 1, 6, 7, 8, 10, 11, 12, 16 and 17 (63, 64, 72-82) (Table 1). Such functional studies, combined with positional and expression based gene cloning techniques, have enabled the identification of KAI1, KISS-1, MKK4/SEK1 and BRMS1 as metastasis-suppressor genes.

1.4 Metastasis-Suppressor Genes

As discussed in the introduction, metastasis-suppressor genes suppress the formation of spontaneous, macroscopic metastases *without affecting the growth rate of the primary tumor*. To date, five genes, *nm23*/NME1, KAI1, KISS1, BRMS1, and *MKK4*/MAP2K4, have been shown to meet the criteria of a metastasis-suppressor gene [Summarized in Table 2; (31, 32, 83-119)]. The role of other genes, such as CD44 and *maspin*/PI5, in metastasis suppression is less well defined (102, 120-131). The potential mechanism of action of all of these genes has been inferred by analogy to other family members and observations in model systems. How these genes, and their protein products, function to suppress metastasis *in vivo* is the subject of enthusiastic study. Decreased expression of the suppressor gene is the key parameter determining metastatic potential and may occur by a variety of mechanisms, not necessarily loss of heterozygosity (32, 91). To date, *nm23*/NME1 and KAI1 are the best-characterized metastasis-suppressor genes.

---

[1] In our work, the concept of cellular hardwiring refers to the work of Pienta and Coffey (71, 132).

1.4.1 nm23/*NME1*

The prototypical metastasis-suppressor gene, *nm23*, was identified in the murine K1735 melanoma using subtractive hybridization (a method to identify genes differentially expressed between two cell lines), and six human homologs have been identified (90). Loss of Nm23-H1 expression is associated with metastatic potential in many, but not all, late-stage tumors (91). Transfection of *nm23*-H1 cDNA into highly metastatic murine melanoma, rat mammary adenocarcinoma, and human breast cancer and melanoma cells reduces their invasiveness and metastatic ability *in vivo* (91). In cancers, such as lung, colon, prostate, etc. [reviewed in reference (87)], where no alterations in the expression pattern of Nm23-H1 are evident, it is possible that the biologic function of Nm23-H1 does not influence malignant progression in these cell types. Alternatively, its effects may be inhibited by alternate mechanisms. The mechanism of action for metastasis suppression by Nm23 still remains unknown; however, recent evidence suggests that it is phosphorylated and may be involved in a novel signaling pathway that, in turn, controls cell motility (84, 87).

1.4.2 KAI1

The localization of metastasis-suppressor activity to rat chromosome 2 in the cell fusion experiments by Ichikawa *et al.* prompted the search for homologous metastasis-suppressor genes for human prostate cancer. The first of such genes identified was *KAI1*. MMCT was used to transfer human chromosome 11 into Dunning AT6.1 and AT3.1 rat prostate cancer cells, and the resulting microcell hybrids were assayed for metastasis suppression in immunodeficient mice (81). These studies led to the identification of the metastasis-suppressor gene *KAI1,* which maps to 11p11.2-p13 (101). The metastasis-suppressor activity of *KAI1* was subsequently demonstrated by transfecting it into AT6.1 cells and assaying the metastatic ability of individual transfected control cell lines in severe combined immunodeficient (SCID) mice (101). Recent reports suggest that expression of KAI1 decreases both the invasiveness and motility of cells *in vitro* (101, 110). Additional studies show that *KAI1* transfectants exhibit enhanced Ca^{++}-independent aggregation, indicating that KAI1 expression alters cell-cell interactions (109). The metastasis-suppressor activity of *KAI1* was subsequently demonstrated by its transfection into AT6.1 cells and assaying the metastatic ability of individual transfected control cell lines in severe combined immunodeficient (SCID) mice (101). Lowered expression of KAI1 has also been associated with progression in a wide variety of cancers including pancreatic, hepatocellular, bladder, breast, and non-small cell lung cancers (31, 133-136), as well as esophageal cell carcinomas (137) and squamous and lymphoid neoplasms (138). These data suggest that KAI1 has a conserved metastasis-suppressor function. Further, these studies demonstrate that metastasis-suppressor genes can be developed as clinical markers *even before their biochemical mechanism of action has been elucidated.*

Table 2. Summary of metastasis-suppressor genes identified*

Metastasis Suppressor Gene	Discovery Method	Tumor Types	Cell Lines Transfected†	*In Vitro* Phenotype†	*In Vivo* Phenotype	Status in Clinical Disease	Reported Mechanisms of Action
nm23‡ (NME1) (17q21.3) [83]	cDNA subtraction	**Melanoma** [84-89]	K-1735 (mouse)	↓ Motility ↓ Colony formation ↓ Proliferation (TGFβ)	↓ Exp. mets.	Inverse correlation between Nm23 expression and metastatic potential	Nucleotide diphosphate kinase Signal transduction Transcriptional activation [90,91]
			B16 F10 (mouse)	↓ Invasiveness ↑ Cell-cell adhesion ↑ Immuno-sensitivity	↓ Exp. mets.		
			B16 FE7 (mouse)	ND	↓ Exp. mets.		
			MelJuSo (human)	ND	↓ Exp. mets.		
		Breast [84, 87, 92-95]	MDA-MB-435 (human)	↓ Motility ↓ Colony formation	↓ Spont. mets.	Inverse correlation between Nm23 expression and metastatic potential	
			MTLn3 (rat)	ND	↓ Spont. mets.	No trend observed	
		Prostate [87, 96]	DU145 (human)	↓ Colony formation ↓ Invasiveness ↓ Adhesion to extracellular matrix components	ND		

Table 2. Summary of metastasis-suppressor genes identified*

Metastasis Suppressor Gene	Discovery Method	Tumor Types	Cell Lines Transfected†	*In Vitro* Phenotype†	*In Vivo* Phenotype	Status in Clinical Disease	Reported Mechanisms of Action
nm23 cont.		**Colon** [87, 97]	HD3 (human) (AS-oligo study)	↓ Adhesion to tissue culture dish ↓ Growth arrest ↓ Differentia-tion	ND	Aggressive colorectal cancers have high expression of mutated Nm23	
			U9 (human) (AS-oligo study)	NE	ND		
		Oral [98, 99]	LMF4	ND	↓ Exp. mets. ↑ Differentiation	Inverse correlation between Nm23 expression and metastatic potential	
KAI1 ǁ (11p11.2) also known as CD82 [100]	MMCT/ Alu -specific PCR/ hybridization of cDNA library	**Prostate** [32, 100-103]	AT6.1 (rat) AT3.1 (rat) AT6.3 (rat)	↓ Invasiveness ND ND	↓ Spont.mets. NE Spont. mets. ↓ Spont. mets.	Inverse correlation between protein expression and metastatic potential	Integrin signaling Cell adhesion Motility [104-106]
		Breast [31, 75, 107, 108]	MDA-MB-435 (human)	NE	↓ Spont. mets.	Inverse correlation between protein expression and metastatic potential	
			ch 11 MMCT		Protein Expression / modification in the 1° tumors and mets.		
			KAI1 cDNA transfection	ND	↓ Spont. mets.		

Table 2. Summary of metastasis-suppressor genes identified*

Metastasis Suppressor Gene	Discovery Method	Tumor Types	Cell Lines Transfected†	*In Vitro* Phenotype†	*In Vivo* Phenotype	Status in Clinical Disease	Reported Mechanisms of Action
KAI1 cont.		**Melanoma** [109]	MelJuSo (human)	↑ Cell aggregation ↓ Motility ↓ Invasiveness	↓ Exp. mets.	ND	
			B16-BL6 (mouse)	↑ Cell aggregation ↓ Motility	↓ Exp. mets.	Inverse correlation between protein expression and metastatic potential	
		Colon [110-111]	BM314 (human)	↓ Invasiveness	ND		
			DLD-1 (human)	↑ Cell aggregation ↓ Motility ↓ Invasiveness	ND		
KiSS1 (1q32) [112]	MMCT/ cDNA subtraction	**Melanoma** [89, 113, 114]	C8161	NE adhesion to ECM NE invasion ↓ Colony formation ↑ Spread on collagen type IV	↓ Exp. mets. ↓ Spont. mets.	*In situ* hybridization	Signal transduction [113]
			MelJuSo	NE invasion	↓ Exp. mets. ↓ Spont. mets.	ND	
		Breast [114]	MDA-MB-435	NE motility	↓ Spont. mets.	ND	
BrMS1	MMCT/	**Breast**	MDA-MB-435	ND	↓ Spont. mets.	ND	Cell

Table 2. Summary of metastasis-suppressor genes identified*

Metastasis Suppressor Gene	Discovery Method	Tumor Types	Cell Lines Transfected†	*In Vitro* Phenotype†	*In Vivo* Phenotype	Status in Clinical Disease	Reported Mechanisms of Action
(11q13.1-2) [115]	differential display	[115]	(human)				Communication Motility [115]
			MDA-MB-231 (human)	ND	↓ Exp. mets.		
MKK4 (MAP2K4) (17p11.2) [116]	MMCT/ positional EST identification	**Prostate** [117]	AT6.1 (rat)	ND	↓ Spont. mets.	ND	Cytokine/stress-induced signal transduction [118-119]
CD44 (11p13) [120]	MMCT	**Prostate** [102, 121-126]	AT3.1 (rat)	ND	↓ Spont. mets.	Decreased expression of CD44 correlates with higher tumor grade, aneuploidy, and presence of distant metastases.	Receptor for both hyaluronic acid and osteopontin. Cell adhesion [127]
Maspin (PI5) (18q21.3) [128]	Differential display	**Breast** [129-130]	MDA-MB-435 (human)	↓ Invasiveness ↓ Motility	↓ Primary tumor growth	[no cohort studies, although weak expression in malignant cells of invasive breast carcinomas has been reported]	Serine protease inhibitor Modulation of integrin expression [130] Inhibit angiogenesis
		Prostate [131]	AT3.1 (rat)	ND	NE Primary tumor growth NE Spont. mets.	ND	

Table 2. Summary of metastasis-suppressor genes identified*

Metastasis Suppressor Gene	**Discovery Method**	**Tumor Types**	**Cell Lines Transfected†**	***In Vitro* Phenotype†**	***In Vivo* Phenotype**	**Status in Clinical Disease**	**Reported Mechanisms of Action**

* ND = Not Determined, NE= Not Observed, GFP= Green Fluorescent Protein Mets.=metastases † Motility was measured by micropipet motility assay or by migration toward a chemoattractant in Boyden chambers. Invasion was measured by migration through Matrigel.
*TGFβ=transforming growth factor-beta, ND=Not Determined , MMCT= microcell-mediated chromosomal transfer, DNA= complementary DNA, NE=Not Observed, Exp. Mets.=experimental metastases, Spont. Mets.=spontaneous metastases. †Cell motility was determined in chemotaxis assays using Boyden chambers, in phagokinetic track assays on cover slips, or by cinematography studies. Invasion was measured by migration through Matrigel or reconstituted basement membranes in Boyden chambers. Colony formation was evaluated in soft agar. Cell proliferation was measured by counting viable cells using a hemocytometer. Cell adhesion was evaluated by the ability of cells to form conjugates with lymphokine activated killer cells (LAK), the ability to adhere to tissue culture plates coated with laminin, fibronectin, collagen type I, or collagen type IV in the absence of fetal bovine serum (FBS), or by the ability to remain adherent to tissue culture plates after the removal of FBS and the addition of oligonucleotides and transforming growth factor beta (TGFβ). Immunosensitivity was determined in chromium release assays with LAK cells. Cell aggregation was examined by culturing single cell suspensions in Puck's saline + 0.8% fetal bovine serum. Cell spreading over extracellular matrix substrates was monitored over time by photography. ‡Additional clinical studies have examined the expression of Nm23 in hepatocellular, gastric, ovarian, and cervical carcinomas (87). §HD3 and U9 are sublines of the human colon carcinoma line, HT29, and differ in their responses to TGFβ. ||An inverse correlation between KAI1 protein and/or mRNA expression and malignant potential have been observed in pancreatic, nonsmall cell lung, bladder, hepatocellular, and oesophageal squamous cell carcinomas.

2. EMERGING ROLE OF METASTASIS-SUPPRESSOR GENES IN THE REGULATION OF METASTATIC GROWTH

While it is tempting to speculate on the mechanism of action of genes listed in Table 2, examination of how genes such as *MKK4* or *BRMS1* suppress metastasis will require construction of appropriate biochemical constructs and identification of *in vitro* conditions that will enable us to conduct meaningful biochemical and molecular studies. As a first step to accomplishing this goal, our laboratories have initiated studies designed to examine the step in the metastatic cascade inhibited by a chromosome or gene of interest. As an example of these studies, we will present recent work on the metastasis-suppressor activity encoded by chromosomes 17 and 6. These studies have brought us closer to defining mechanisms of metastasis suppression.

2.1 Chromosome 17 Mediates Suppression of Growth at the Secondary Site

We have recently reported the identification of discontinuous portions of human chromosome 17 (D17S952→D17S805, D17S930→ D17S797, and D17S944→ qter) that together suppress the metastatic ability of AT6.1 Dunning rat prostatic cancer cells when introduced via MMCT (63, 80). PCR and Southern blot analyses demonstrated that three of the four markers on 17p13, including *HIC1* and *TP53*, and 12 of the 13 markers in 17q21-23, including *BRCA1* and the metastasis-suppressor gene *NME1* (*nm23*), were not retained in this region (63). AT6.1 microcell hybrids containing this portion of chromosome 17 were tested *in vivo* in spontaneous metastasis assays. Spontaneous metastasis is measured by the ability of tumor cells to form a locally growing tumor at the site of injection and disseminate to secondary sites thereafter.

At the experimental endpoint, the number of overt surface metastases observed in the lungs from mice with AT6.1-17 tumors was reduced 15 to 30-fold compared to lungs from mice bearing parental AT6.1 tumors (63). This suppression could be due to the inhibition of any step within the metastatic cascade. We reasoned that examination of the biology of metastasis suppression would provide clues to the identity of genes responsible for suppression of metastatic growth. A series of *in vivo* experiments were conducted and no evidence was found to suggest that there is a decrease in the number and/or viability of tumor cells colonizing the lung (80).

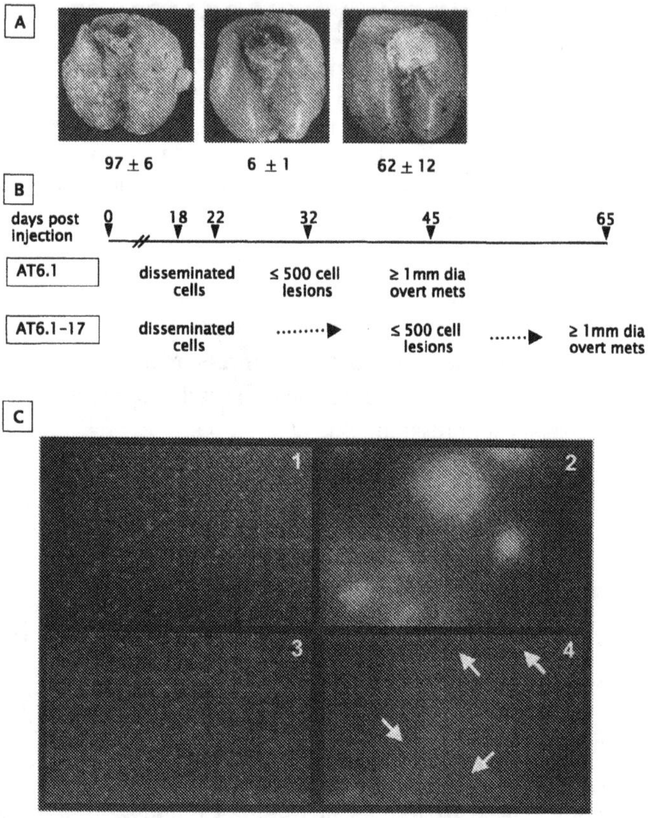

Figure 2. Examination of the mechanism of metastasis suppression by chromosome 17 and 6. Panel A. Quantification of overt surface metastases and micrometastases. AT6.1 cells are highly metastatic rat prostate cancer cells. AT6.1-17-Tβgal cells contain the metastasis-suppressor region of human chromosome 17 and are tagged with a β-galactosidase reporter gene enabling the sensitive detection of microscopic metastases. The numbers of overt and microscopic metastases were determined using Bouin's fixation and β-galactosidase activity, respectively. At the experimental endpoint, lungs were removed from tumor bearing animals. Left, lung from AT6.1 tumor-bearing animal stained with Bouin's solution; Middle, lung from AT6.1-17-Tβgal tumor-bearing animal stained with Bouin's solution; and Right, lung from AT6.1-17-Tβgal tumor-bearing animal stained for βgalactosidase activity. The average number of overt or microscopic metastases and SE are shown below the panels. This figure is adapted from reference 139. Panel B. A combination of techniques has been used to examine the timecourse of cancer cell dissemination and growth in suppressed AT6.1-17 cells as compared to metastatic AT6.1 parental cells. These data indicate that genes encoded by chromosome 17 inhibit a step in metastatic colonization. Panel

C. Photomicrographs of mouse lung following intravenous injection of GFP-tagged C8161 and metastasis-suppressed neo6/C8161 cells (Panel C3) are present. At one month, however, C8161 cells have proliferated to form macroscopic lung lesions (Panel C2); but most neo6/C8161 cells have been cleared. Occasional single cells (Panel C4, arrows) can be found in the lungs, but fail to proliferate. These results imply that chromosome 6 suppresses metastasis by inhibiting the ability of C8161 cells to grow in the lung at an early stage of colonization. Photographs were taken at 300X magnification. Data adapted from reference 13.

Based upon these findings we hypothesized that a gene or genes encoded by the suppressor region of chromosome 17 functions by inhibiting the growth of metastases *in the lung* (139). To test this possibility, AT6.1-17 cells were transduced with a β-galactosidase reporter gene construct (AT6.1-17Tβgal cells) and tested in spontaneous metastasis assays (2). At the experimental endpoint, animals were sacrificed and the excised lungs were stained for β-galactosidase expression. This approach allowed the visualization of microscopic AT6.1-17βgal surface metastases. Subcutaneous injection of AT6.1 parental cells resulted in the formation of a mean number of 97 overt surface metastases (detected using Bouin's fixation) per lung (Fig. 3, A, *left*). As expected, the number of overt macroscopic metastases after the subcutaneous injection of AT6.1-17-Tβgal cells was greatly reduced (Fig. 3, A, *middle*). In contrast, when lungs removed from mice carrying AT6.1-17-Tβgal tumors were stained for β-galactosidase activity, numerous blue-staining microscopic metastases were observed (Fig. 3, A, *right*). Interestingly, the mean number of AT6.1-17-Tβgal micrometastases (i.e. 62 ± 12 SE) detected by this method is on the same order of magnitude as the number of macroscopic AT6.1 metastases (i.e. 97 ± 6 SE). These results demonstrate that AT6.1-17 cells do escape from the primary tumor and arrive in the lungs, but do not form large metastatic foci (139). Development of overt metastases was associated with loss of the metastasis-suppressor region of chromosome 17 (139).

Because of the similarity between our findings to the angiostatin-mediated dormancy reported by Holmgren *et al.* (140), we investigated the possibility that AT6.1-17 primary tumors secrete a substance that suppresses the growth of its own metastases (139). For this experiment, 2×10^5 AT6.1-17 cells were injected subcutaneously into the flanks of severe combined immunodeficient mice, which were then divided into two experimental groups. Once the tumors reached a volume of 1 cm^3, the tumors were surgically removed from the mice in the first group while those in the second group were left intact, although a contralateral sham surgery was performed. It was anticipated that if the AT6.1-17 primary tumor secreted a substance like angiostatin, which suppresses the growth of its own metastases, then a substantial increase in the number of overt metastases should develop in the lungs of mice in which the primary tumors had been removed. However, after approximately 65 days post-injection, the animals were sacrificed and examination of the lungs from both groups showed no difference in the numbers of overt macrometastases (139). Thus, these studies found no evidence for an antiangiogenic mechanism in this model.

Taken together, our data suggested that AT6.1-17 cells escape from the primary tumor but are growth inhibited at the secondary site (139). If this is an early event, we predicted that viable, disseminated AT6.1 and AT6.1-17 cells should be present in the lung at very early time points. We found that viable cells could be harvested from the lungs of both AT6.1- and AT6.1-17-tumor bearers as early as 18 days post-injection (Figure 3, Panel B). Our preliminary time-course data show that AT6.1-17 cells disseminate and lodge in the lungs, but have an extended latency period as compared to AT6.1 parental cells.

2.2 Chromosome 6

Based on the high incidence of chromosome 6 abnormalities in late-stage human melanoma (143), we introduced an intact chromosome 6 into the highly metastatic C8161 human melanoma cells by MMCT. Parental cells formed tumors in every mouse injected intradermally with 1 x 10^6 cells, and more than 90% of the mice developed regional lymph node and lung metastases. In contrast, chromosome 6-C8161 hybrids (neo6/C8161) were still tumorigenic but completely suppressed for metastasis (141). Intravenous injection of neo6/C8161 cells also did not produce metastases. In a recent study, introduction of a version of a chromosome 6 with deletions on the long arm allowed refinement of the metastasis-suppressor locus to a 40 Mb region represented by chromosomal bands 6q16.3-q23 (142).

The mechanism of action for the metastasis-suppressor protein from the gene on chromosome 6 was studied using a variety of *in vitro* and *in vivo* techniques. The neo6/C8161 cells were still locally invasive and cells were even detected in efferent vessels. This finding implied that the step(s) in the metastatic cascade inhibited by introduction of chromosome 6 occurred subsequent to intravasation. The identity of those steps was not further elucidated using *in vitro* assays mimicking adhesion, invasion, motility, or growth. No important differences between the metastatic and nonmetastatic cells were observed using the many *in vitro* assays (72, 73, 141, 143).

To better define the step(s) in metastasis blocked by addition of chromosome 6, cells that constitutively express green fluorescent protein (GFP) were engineered. GFP-tagged C8161 and neo6/C8161 cells were injected intravenously into athymic mice. C8161, as expected, formed overt metastases, but neo6/C8161 cells did not. Microscopic metastases (single cells or clusters of <10 cells) were observed in the lungs following neo6/C8161 cell injection, suggesting that these cells lodged in the lungs but failed to proliferate (144). To determine whether the fluorescing cells were viable, they were isolated from lung up to 60 days post-injection and grown in culture. Upon injection into the skin of athymic mice, the neo6/C8161 cells isolated from the lung grew at rates similar to previously injected neo6/C8161 cells. This result implies that the gene or genes on chromosome 6 interfere specifically with growth regulatory responses in the lung, but not in the skin.

3. FROM GENE DISCOVERY TO CLINICAL UTILITY

This chapter has focused on the identification and development of metastasis-suppressor genes as new additions to our molecular armamentarium. As translational researchers, our immediate goals are: (1) to improve the ability of the pathologist to distinguish unambiguously malignant from indolent lesions and (2) to help the clinician differentiate tumors that are highly likely to metastasize from those that are not. The practical question, therefore, is how can we use these genes, or the pathways that they regulate, to improve patient management? When the search for metastasis-suppressor genes was initiated in the late 1980s, the major challenge was the identification of candidate genes. Recently, however, there has been an explosion in the genetic information that is instantly available. Furthermore, due to the efforts of independent laboratories and cooperative efforts, such as the Cancer Genome Anatomy Project of the NCI, cancer transcriptomes and proteomes will soon be available (145, 146). New technologies will continue to increase our ability to dissect molecular pathways in individual cells within human cancers. While this wealth of information will no doubt be of use, work from the groups of Bissell, Cunha, and Chung has clearly demonstrated that tissue structure determines, or at least greatly influences, gene expression and function (147-154). Thus, it may be extremely difficult to predict the importance of genes expressed in individual microdissected cancer cells to the biology of the intact tumor, the behavior of which is determined by complex interactions among a population of cells. The present challenge is to identify the genes that are *functionally important* in the acquisition of metastatic ability. Achieving this goal will require the use of well-characterized, *in vivo* (animal) models coupled with clinical correlative studies. It must be emphasized that *in vitro* models do not accurately reflect *in vivo* metastasis (155). *Indeed, none of the metastasis-suppressor genes described herein could have been identified using traditional* in vitro *assays*. Given the inherent variability and nonlinear behaviors of biologic systems, it is probable that no one model will prove adequate to separate out the contributions of the multiplicity of genes involved in the development of metastases. Thus, it is more advantageous to focus studies on a particular model and tease out important cellular pathways modulated by a particular gene of interest in that model, and then to test and verify the importance of the target pathway in clinical disease, as well as, in additional model systems.

Technological advances are enabling us to examine the metastatic process and the genes that regulate it in new ways. This ability has led us to reevaluate fundamental concepts concerning the determinants of metastatic propensity. In the past, the escape of cells from the primary site was viewed as the rate-limiting step for the development of metastases. The clinical implication was that disseminated cancer cells were destined to grow into lethal metastases, thus they were not a target for therapeutic intervention (18). Findings from clinical studies and basic research from several independent laboratories have shown that survival and subsequent growth of extravasated cancer cells at the secondary site may determine metastatic efficacy. These observations are driving our laboratories and others to reconsider the role of endothelial cell–tumor cell interactions in survival, signaling, and growth control cascades in order to develop new strategies for controlling the growth of disseminated cancer cells (39, 45, 156).

As metastasis researchers, we find ourselves in the midst of a revolution. In preparing this chapter, we considered the parallels between recent developments in our field and the development of the field of molecular biology. Much of early molecular biology was pursued by individuals who were not trained as biologists, but as physicists, such as Max Delbrück (157). We are respectful of the observations of Erwin Schrödinger, the father of statistical mechanics, who observed that, "all of the physical and chemical laws that are known to play an important part in the life of organisms are of the statistical kind. The behavior of such systems depends entirely on a large number of molecules that cooperate to form the observed function or phenotype (158)." Although this comment was made in regard to normal biologic processes, it is equally applicable to the multiple genetic changes that are required for the acquisition of metastatic ability. Metastasis is a complex, multigenic phenotype. As such, multiple markers will be needed for the accurate assessment of the metastatic ability of tumors and tumor cells. This need is highlighted by the tremendous impact of seemingly trivial experimental manipulations on the outcome of metastasis assays (155). Recently, parallels have been drawn between the behavior of cancer cells and complex adaptive systems (159, 160). As such, very small changes in initial conditions may produce an outcome of such great diversity as to appear random (159). Ultimately, we believe that, in order to translate our molecular findings into meaningful markers, we will have to go beyond our traditional areas of expertise and work with mathematicians, computational biologists and others to take this revolution from bench to bedside.

ACKNOWLEDGEMENTS

We wish to thank Mr. Edwin F. Schaeffer, III for his editorial efforts. We also wish to thank Dr. Charles Brendler of the University of Chicago for his enthusiastic support of this research.

REFERENCES

1. Mackillop WJ, O'Sullivan B, Gospodarowicz M. The role of cancer staging in evidence-based medicine. Cancer Prev Control 1998; 2:269-77.
2. Gospodarowicz M, Benedet L, Hutter RV, Fleming I, Henson DE, Sobin LH. History and international developments in cancer staging. Cancer Prev Control 1998; 2:262-8.
3. Bostwick DG. Staging prostate cancer--1997: current methods and limitations. Eur Urol 1997; 32:2-14.
4. Carter H, Partin, AW. Diagnosis and staging of prostate cancer. In: Walsh PC RA, Vaughn ED, Wein AJ, ed. Campbell's Urology. Philadelphia: W.B.Saunders Company, 1998:2519-2538.
5. Isaacs JT. Molecular markers for prostate cancer metastasis. Developing diagnostic methods for predicting the aggressiveness of prostate cancer. Am J Pathol 1997; 150:1511-21.
6. Schellhammer PF. Radical prostatectomy. Patterns of local failure and survival in 67 patients. Urology 1988; 31:191-7.
7. Lerner SP, Seale-Hawkins C, Carlton CE, Jr., Scardino PT. The risk of dying of prostate cancer in patients with clinically localized disease. J Urol 1991; 146:1040-5.
8. Heimann R, Hellman S. Clinical progression of breast cancer malignant behavior: what to expect and when to expect it. J Clin Oncol 2000; 18:591-9.
9. Catalona WJ, Dresner SM. Nerve-sparing radical prostatectomy: extraprostatic tumor extension and preservation of erectile function. J Urol 1985; 134:1149-51.
10. Lu-Yao GL, McLerran D, Wasson J, Wennberg JE. An assessment of radical prostatectomy. Time trends, geographic variation, and outcomes. The Prostate Patient Outcomes Research Team [see comments]. Jama 1993; 269:2633-6.
11. Voges GE, McNeal JE, Redwine EA, Freiha FS, Stamey TA. Morphologic analysis of surgical margins with positive findings in prostatectomy for adenocarcinoma of the prostate. Cancer 1992; 69:520-6.
12. Heimann R, Lan F, McBride R, Hellman S. Separating favorable from unfavorable prognostic markers in breast cancer: the role of E-cadherin. Cancer Res 2000; 60:298-304.
13. Yoshida, B. A., Sokoloff, M.A., Welch, D.R., Rinker-Schaeffer, C.W. Metastasis-Suppressor Genes: a Review and Perspective on an Emerging Field, Journal of the National Cancer Institute. *92:* 1717-1730, 2000.
14. MacDonald NJ, Steeg PS. Molecular basis of tumour metastasis. Cancer Surv 1993; 16:175-99.
15. Welch DR, Rinker-Schaeffer CW. What defines a useful marker of metastasis in human cancer? J Natl Cancer Inst 1999; 91:1351-3.
16. Melchior SW, Corey E, Ellis WJ, et al. Early tumor cell dissemination in patients with clinically localized carcinoma of the prostate. Clin Cancer Res 1997; 3:249-56.

17. Heimann R, Ferguson DJ, Hellman S. The relationship between nm23, angiogenesis, and the metastatic proclivity of node-negative breast cancer. Cancer Res 1998; 58:2766-71.
18. Chambers AF. The metastatic process: basic research and clinical implications. Oncol Res 1999; 11:161-8.
19. Gomella LG, Raj GV, Moreno JG. Reverse transcriptase polymerase chain reaction for prostate specific antigen in the management of prostate cancer [see comments]. J Urol 1997; 158:326-37.
20. Christiano AP, Yoshida BA, Dubauskas Z, Sokoloff M, Rinker-Schaeffer CW. Development of markers of prostate cancer metastasis: review and perspective. Urol Oncol (in press).
21. Wood DP, Jr., Banerjee M. Presence of circulating prostate cells in the bone marrow of patients undergoing radical prostatectomy is predictive of disease-free survival. J Clin Oncol 1997; 15:3451-7.
22. Braun S, Pantel K, Muller P, et al. Cytokeratin-positive cells in the bone marrow and survival of patients with stage I, II, or III breast cancer [see comments]. N Engl J Med 2000; 342:525-33.
23. Ebralidze A, Tulchinsky E, Grigorian M, et al. Isolation and characterization of a gene specifically expressed in different metastatic cells and whose deduced gene product has a high degree of homology to a Ca2+-binding protein family. Genes Dev 1989; 3:1086-93.
24. Dear TN, Ramshaw IA, Kefford RF. Differential expression of a novel gene, WDNM1, in nonmetastatic rat mammary adenocarcinoma cells. Cancer Res 1988; 48:5203-9.
25. Dear TN, McDonald DA, Kefford RF. Transcriptional down-regulation of a rat gene, WDNM2, in metastatic DMBA-8 cells [see comments]. Cancer Res 1989; 49:5323-8.
26. Basset P, Bellocq JP, Wolf C, et al. A novel metalloproteinase gene specifically expressed in stromal cells of breast carcinomas. Nature 1990; 348:699-704.
27. Fujita M, Spray DC, Choi H, et al. Extracellular matrix regulation of cell-cell communication and tissue-specific gene expression in primary liver cultures. Prog Clin Biol Res 1986; 226:333-60.
28. Fidler IJ, Radinsky R. Search for genes that suppress cancer metastasis [editorial; comment]. J Natl Cancer Inst 1996; 88:1700-3.
29. Fidler IJ, Radinsky R. Genetic control of cancer metastasis [editorial]. J Natl Cancer Inst 1990; 82:166-8.
30. Steeg PS, Bevilacqua G, Kopper L, et al. Evidence for a novel gene associated with low tumor metastatic potential. J Natl Cancer Inst 1988; 80:200-4.
31. Yang X, Welch DR, Phillips KK, Weissman BE, Wei LL. KAI1, a putative marker for metastatic potential in human breast cancer. Cancer Lett 1997; 119:149-55.
32. Dong JT, Suzuki H, Pin SS, et al. Down-regulation of the KAI1 metastasis suppressor gene during the progression of human prostatic cancer infrequently involves gene mutation or allelic loss. Cancer Res 1996; 56:4387-90.
33. Folkman J. Angiogenesis and angiogenesis inhibition: an overview. Exs 1997; 79:1-8.
34. Folkman J. The role of angiogenesis in tumor growth. Semin Cancer Biol 1992; 3:65-71.
35. Pasqualini R, Ruoslahti E. Organ targeting in vivo using phage display peptide libraries. Nature 1996; 380:364-6.

36. Fidler IJ. Review: biologic heterogeneity of cancer metastases. Breast Cancer Res Treat 1987; 9:17-26.
37. Fidler IJ. Critical factors in the biology of human cancer metastasis: twenty- eighth G.H.A. Clowes memorial award lecture. Cancer Res 1990; 50:6130-8.
38. Al-Mehdi AB, Tozawa K, Fisher AB, Shientag L, Lee A, Muschel RJ. Intravascular origin of metastasis from the proliferation of endothelium - attached tumor cells: a new model of metastasis. Nat Med 2000; 6: 100-2.
39. Chambers AF, MacDonald IC, Schmidt EE, Morris VL, Groom AC. Clinical targets for anti-metastasis therapy [In Process Citation]. Adv Cancer Res 2000; 79:91-121.
40. Yang M, Jiang P, Sun FX, et al. A fluorescent orthotopic bone metastasis model of human prostate cancer. Cancer Res 1999; 59:781-6.
41. Chambers AF, MacDonald IC, Schmidt EE, et al. Steps in tumor metastasis: new concepts from intravital videomicroscopy. Cancer Metastasis Rev 1995; 14:279-301.
42. MacDonald IC, Schmidt EE, Morris VL, Groom AC, Chambers AF. In vivo videomicroscopy of experimental hematogenous metastasis: cancer cell arrest, extravasation and migration. In: Wessels DRSaD, ed. Motion Analysis of Living Cells. New York: John Wiley and Sons, 1998:263-287.
43. Radinsky R, Fidler IJ. Regulation of tumor cell growth at organ-specific metastases. In Vivo 1992; 6:325-31.
44. Fidler IJ. Critical determinants of melanoma metastasis. J Investig Dermatol Symp Proc 1996; 1:203-8.
45. Day ML, Zhao X, Vallorosi CJ, et al. E-cadherin mediates aggregation-dependent survival of prostate and mammary epithelial cells through the retinoblastoma cell cycle control pathway. J Biol Chem 1999; 274:9656-64.
46. Guilford P. E-cadherin downregulation in cancer: fuel on the fire? Mol Med Today 1999; 5:172-7.
47. Giancotti FG, Ruoslahti E. Integrin signaling. Science 1999; 285:1028-32.
48. Arbiser JL, Moses MA, Fernandez CA, et al. Oncogenic H-ras stimulates tumor angiogenesis by two distinct pathways. Proc Natl Acad Sci U S A 1997; 94:861-6.
49. Holash J, Maisonpierre PC, Compton D, et al. Vessel cooption, regression, and growth in tumors mediated by angiopoietins and VEGF. Science 1999; 284:1994-8.
50. Maniotis AJ, Folberg R, Hess A, et al. Vascular channel formation by human melanoma cells in vivo and in vitro: vasculogenic mimicry [see comments]. Am J Pathol 1999; 155:739-52.
51. Turpeenniemi-Hujanen T, Thorgeirsson UP, Hart IR, Grant SS, Liotta LA. Expression of collagenase IV (basement membrane collagenase) activity in murine tumor cell hybrids that differ in metastatic potential. J Natl Cancer Inst 1985; 75:99-103.
52. Sidebottom E, Clark SR. Cell fusion segregates progressive growth from metastasis. Br J Cancer 1983; 47:399-405.
53. Ramshaw IA, Carlsen S, Wang HC, Badenoch-Jones P. The use of cell fusion to analyse factors involved in tumour cell metastasis. Int J Cancer 1983; 32:471-8.
54. Layton MG, Franks LM. Selective suppression of metastasis but not tumorigenicity of a mouse lung carcinoma by cell hybridization. Int J Cancer 1986; 37:723-30.
55. Sager R. Genetic suppression of tumor formation. Adv Cancer Res 1985; 44:43-68.
56. Harris H. The analysis of malignancy by cell fusion: the position in 1988. Cancer Res 1988; 48:3302-6.
57. Ichikawa T, Ichikawa Y, Isaacs JT. Genetic factors and suppression of metastatic ability of prostatic cancer. Cancer Res 1991; 51:3788-92.

58. Fournier RE, Ruddle FH. Microcell-mediated transfer of murine chromosomes into mouse, Chinese hamster, and human somatic cells. Proc Natl Acad Sci U S A 1977; 74:319-23.
59. Ege T, Ringertz NR, Hamberg H, Sidebottom E. Preparation of microcells. Methods Cell Biol 1977; 15:339-57.
60. Oshimura M. Lessons learned from studies on tumor suppression by microcell-mediated chromosome transfer. Environ Health Perspect 1991; 93:57-8.
61. Saxon PJ, Srivatsan ES, Leipzig GV, Sameshima JH, Stanbridge EJ. Selective transfer of individual human chromosomes to recipient cells. Mol Cell Biol 1985; 5:140-6.
62. Rinker-Schaeffer CW, Partin AW, Isaacs WB, Coffey DS, Isaacs JT. Molecular and cellular changes associated with the acquisition of metastatic ability by prostatic cancer cells. Prostate 1994; 25:249-65.
63. Chekmareva MA, Hollowell CM, Smith RC, Davis EM, LeBeau MM, Rinker-Schaeffer CW. Localization of prostate cancer metastasis-suppressor activity on human chromosome 17. Prostate 1997; 33:271-80.
64. Luu HH, Zagaja GP, Dubauskas Z, et al. Identification of a novel metastasis-suppressor region on human chromosome 12. Cancer Res 1998; 58:3561-5.
65. Chang ZF. Regulatory mechanisms of replication growth limits in cellular senescence. J Formos Med Assoc 1997; 96:784-91.
66. Vojta PJ, Futreal PA, Annab LA, Kato H, Pereira-Smith OM, Barrett JC. Evidence for two senescence loci on human chromosome 1. Genes Chromosomes Cancer 1996; 16:55-63.
67. Nakabayashi K, Ogata T, Fujii M, et al. Decrease in amplified telomeric sequences and induction of senescence markers by introduction of human chromosome 7 or its segments in SUSM-1. Exp Cell Res 1997; 235:345-53.
68. Matsuda T, Sasaki M, Kato H, et al. Human chromosome 7 carries a putative tumor suppressor gene(s) involved in choriocarcinoma. Oncogene 1997; 15:2773-81.
69. Murakami YS, Brothman AR, Leach RJ, White RL. Suppression of malignant phenotype in a human prostate cancer cell line by fragments of normal chromosomal region 17q. Cancer Res 1995; 55:3389-94.
70. Berube NG, Speevak MD, Chevrette M. Suppression of tumorigenicity of human prostate cancer cells by introduction of human chromosome del(12)(q13). Cancer Res 1994; 54:3077-81.
71. Pienta KJ, Coffey DS. Nuclear-cytoskeletal interactions: evidence for physical connections between the nucleus and cell periphery and their alteration by transformation. J Cell Biochem 1992; 49:357-65.
72. Miele ME, Robertson G, Lee JH, et al. Metastasis suppressed, but tumorigenicity and local invasiveness unaffected, in the human melanoma cell line MelJuSo after introduction of human chromosomes 1 or 6. Mol Carcinog 1996; 15:284-99.
73. You J, Miele ME, Dong C, Welch DR. Suppression of human melanoma metastasis by introduction of chromosome 6 may be partially due to inhibition of motility, but not to inhibition of invasion. Biochem Biophys Res Commun 1995; 208:476-84.
74. Miele ME, De La Rosa A, Lee JH, et al. Suppression of human melanoma metastasis following introduction of chromosome 6 is independent of NME1 (Nm23). Clin Exp Metastasis 1997; 15:259-65.
75. Phillips KK, Welch DR, Miele ME, Lee JH, Wei LL, Weissman BE. Suppression of MDA-MB-435 breast carcinoma cell metastasis following the introduction of human chromosome 11. Cancer Res 1996; 56:1222-7.

76. Nihei N, Ohta S, Kuramochi H, et al. Metastasis suppressor gene(s) for rat prostate cancer on the long arm of human chromosome 7. Genes Chromosomes Cancer 1999; 24:1-8.
77. Nihei N, Ichikawa T, Kawana Y, et al. Mapping of metastasis suppressor gene(s) for rat prostate cancer on the short arm of human chromosome 8 by irradiated microcell-mediated chromosome transfer. Genes Chromosomes Cancer 1996; 17:260-8.
78. Kuramochi H, Ichikawa T, Nihei N, et al. Suppression of invasive ability of highly metastatic rat prostate cancer by introduction of human chromosome 8. Prostate 1997; 31:14-20.
79. Nihei N, Ichikawa T, Kawana Y, et al. Localization of metastasis suppressor gene(s) for rat prostatic cancer to the long arm of human chromosome 10. Genes Chromosomes Cancer 1995; 14:112-9.
80. Rinker-Schaeffer CW, Hawkins AL, Ru N, et al. Differential suppression of mammary and prostate cancer metastasis by human chromosomes 17 and 11. Cancer Res 1994; 54:6249-56.
81. Ichikawa T, Ichikawa Y, Dong J, et al. Localization of metastasis suppressor gene(s) for prostatic cancer to the short arm of human chromosome 11. Cancer Res 1992; 52:3486-90.
82. Mashimo T, Watabe M, Cuthbert AP, et al. Human chromosome 16 suppresses metastasis but not tumorigenesis in rat prostatic tumor cells. Cancer Res 1998; 58:4572-6.
83. Backer JM, Mendola CE, Kovesdi I, et al. Chromosomal localization and nucleoside diphosphate kinase activity of human metastasis-suppressor genes NM23-1 and NM23-2. Oncogene 1993; 8:497-502.
84. Kantor JD, McCormick B, Steeg PS, Zetter BR. Inhibition of cell motility after nm23 transfection of human and murine tumor cells. Cancer Res 1993; 53:1971-3.
85. Leone A, Flatow U, King CR, et al. Reduced tumor incidence, metastatic potential, and cytokine responsiveness of nm23-transfected melanoma cells. Cell 1991; 65:25-35.
86. Parhar RS, Shi Y, Zou M, Farid NR, Ernst P, al-Sedairy ST. Effects of cytokine-mediated modulation of nm23 expression on the invasion and metastatic behavior of B16F10 melanoma cells. Int J Cancer 1995; 60:204-10.
87. de la Rosa A, Williams RL, Steeg PS. Nm23/nucleoside diphosphate kinase: toward a structural and biochemical understanding of its biological functions. Bioessays 1995; 17:53-62.
88. Baba H, Urano T, Okada K, et al. Two isotypes of murine nm23/nucleoside diphosphate kinase, nm23-M1 and nm23-M2, are involved in metastatic suppression of a murine melanoma line. Cancer Res 1995; 55:1977-81.
89. Lee JH, Welch DR. Identification of highly expressed genes in metastasis-suppressed chromosome 6/human malignant melanoma hybrid cells using subtractive hybridization and differential display. Int J Cancer 1997; 71:1035-44.
90. Freije JMP, MacDonald NJ, Steeg PS. Differential gene expression in tumor metastasis: Nm23. In: Gunthert U, Shlag PM, Birchmeier W, eds. Attempts to understand metastasis formation II: regulatory factors. Berlin: Springer-Verlag, 1996:215-232.
91. Freije JM, MacDonald NJ, Steeg PS. Nm23 and tumour metastasis: basic and translational advances. Biochem Soc Symp 1998; 63:261-71.

92. Leone A, Flatow U, VanHoutte K, Steeg PS. Transfection of human nm23-H1 into the human MDA-MB-435 breast carcinoma cell line: effects on tumor metastatic potential, colonization and enzymatic activity. Oncogene 1993; 8:2325-33.
93. Bhujwalla ZM, Aboagye EO, Gillies RJ, Chacko VP, Mendola CE, Backer JM. Nm23-transfected MDA-MB-435 human breast carcinoma cells form tumors with altered phospholipid metabolism and pH: a 31P nuclear magnetic resonance study in vivo and in vitro. Magn Reson Med 1999; 41:897-903.
94. Russell RL, Geisinger KR, Mehta RR, White WL, Shelton B, Kute TE. nm23--relationship to the metastatic potential of breast carcinoma cell lines, primary human xenografts, and lymph node negative breast carcinoma patients. Cancer 1997; 79:1158-65.
95. Fukuda M, Ishii A, Yasutomo Y, et al. Decreased expression of nucleoside diphosphate kinase alpha isoform, an nm23-H2 gene homolog, is associated with metastatic potential of rat mammary-adenocarcinoma cells. Int J Cancer 1996; 65:531-7.
96. Lim S, Lee HY, Lee H. Inhibition of colonization and cell-matrix adhesion after nm23-H1 transfection of human prostate carcinoma cells. Cancer Lett 1998; 133:143-9.
97. Hsu S, Huang F, Wang L, Banerjee S, Winawer S, Friedman E. The role of nm23 in transforming growth factor beta 1-mediated adherence and growth arrest. Cell Growth Differ 1994; 5:909-17.
98. Miyazaki H, Fukuda M, Ishijima Y, et al. Overexpression of nm23-H2/NDP kinase B in a human oral squamous cell carcinoma cell line results in reduced metastasis, differentiated phenotype in the metastatic site, and growth factor-independent proliferative activity in culture. Clin Cancer Res 1999; 5:4301-7.
99. Otsuki K, Alcalde RE, Matsumura T, Kimura N. Immunohistochemical analysis of nucleoside diphosphate kinases in oral squamous cell carcinomas. Oncology 1997; 54:63-8.
100. Kawana Y, Komiya A, Ueda T, et al. Location of KAI1 on the short arm of human chromosome 11 and frequency of allelic loss in advanced human prostate cancer. Prostate 1997; 32:205-13.
101. Dong JT, Lamb PW, Rinker-Schaeffer CW, et al. KAI1, a metastasis suppressor gene for prostate cancer on human chromosome 11p11.2 [see comments]. Science 1995; 268:884-6.
102. Gao AC, Lou W, Dong JT, Isaacs JT. CD44 is a metastasis suppressor gene for prostatic cancer located on human chromosome 11p13. Cancer Res 1997; 57:846-9.
103. Ueda T, Ichikawa T, Tamaru J, et al. Expression of the KAI1 protein in benign prostatic hyperplasia and prostate cancer. Am J Pathol 1996; 149:1435-40.
104. Rubinstein E, Le Naour F, Lagaudriere-Gesbert C, Billard M, Conjeaud H, Boucheix C. CD9, CD63, CD81, and CD82 are components of a surface tetraspan network connected to HLA-DR and VLA integrins. Eur J Immunol 1996; 26:2657-65.
105. Berditchevski F, Odintsova E. Characterization of integrin-tetraspanin adhesion complexes: role of tetraspanins in integrin signaling. J Cell Biol 1999; 146:477-92.
106. Maecker HT, Todd SC, Levy S. The tetraspanin superfamily: molecular facilitators. Faseb J 1997; 11:428-42.
107. Phillips KK, White AE, Hicks DJ, et al. Correlation between reduction of metastasis in the MDA-MB-435 model system and increased expression of the Kai-1 protein. Mol Carcinog 1998; 21:111-20.

108. Wei LL, Yang X, Phillips K, Weissman BE, Welch DR. Proc. Am. Assoc. Cancer Res. 1996; 37:76.
109. Takaoka A, Hinoda Y, Sato S, et al. Reduced invasive and metastatic potentials of KAI1-transfected melanoma cells. Jpn J Cancer Res 1998; 89:397-404.
110. Takaoka A, Hinoda Y, Satoh S, et al. Suppression of invasive properties of colon cancer cells by a metastasis suppressor KAI1 gene. Oncogene 1998; 16:1443-53.
111. Lombardi DP, Geradts J, Foley JF, Chiao C, Lamb PW, Barrett JC. Loss of KAI1 expression in the progression of colorectal cancer [In Process Citation]. Cancer Res 1999; 59:5724-31.
112. West A, Vojta PJ, Welch DR, Weissman BE. Chromosome localization and genomic structure of the KiSS-1 metastasis suppressor gene (KISS1). Genomics 1998; 54:145-8.
113. Lee JH, Miele ME, Hicks DJ, et al. KiSS-1, a novel human malignant melanoma metastasis-suppressor gene [see comments] [published erratum appears in J Natl Cancer Inst 1997 Oct 15;89(20):1549]. J Natl Cancer Inst 1996; 88:1731-7.
114. Lee JH, Welch DR. Suppression of metastasis in human breast carcinoma MDA-MB-435 cells after transfection with the metastasis suppressor gene, KiSS-1. Cancer Res 1997; 57:2384-7.
115. Seraj MJ, Samant RS, Verderame MF, Welch DR. Functional evidence for a novel human breast carcinoma metastasis suppressor, BRMS1, encoded at chromosome 11q13. Cancer Res 2000; 60:2764-9.
116. White RA, Hughes RT, Adkison LR, Bruns G, Zon LI. The gene encoding protein kinase SEK1 maps to mouse chromosome 11 and human chromosome 17. Genomics 1996; 34:430-2.
117. Yoshida BA, Dubauskas Z, Chekmareva MA, Christiano TR, Stadler WM, Rinker-Schaeffer CW. Mitogen-activated protein kinase kinase 4/stress-activated protein/Erk kinase 1 (MKK4/SEK1), a prostate cancer metastasis suppressor gene encoded by human chromosome 17 [In Process Citation]. Cancer Res 1999; 59:5483-7.
118. Whitmarsh AJ, Davis RJ. Transcription factor AP-1 regulation by mitogen-activated protein kinase signal transduction pathways. J Mol Med 1996; 74:589-607.
119. Ip YT, Davis RJ. Signal transduction by the c-Jun N-terminal kinase (JNK)--from inflammation to development. Curr Opin Cell Biol 1998; 10:205-19.
120. Gunthert U, Stauder R, Mayer B, Terpe HJ, Finke L, Friedrichs K. Are CD44 variant isoforms involved in human tumour progression? Cancer Surv 1995; 24:19-42.
121. Lou W, Krill D, Dhir R, et al. Methylation of the CD44 metastasis suppressor gene in human prostate cancer. Cancer Res 1999; 59:2329-31.
122. Noordzij MA, van Steenbrugge GJ, Schroder FH, Van der Kwast TH. Decreased expression of CD44 in metastatic prostate cancer. Int J Cancer 1999; 84:478-83.
123. Nagabhushan M, Pretlow TG, Guo YJ, Amini SB, Pretlow TP, Sy MS. Altered expression of CD44 in human prostate cancer during progression. Am J Clin Pathol 1996; 106:647-51.
124. De Marzo AM, Bradshaw C, Sauvageot J, Epstein JI, Miller GJ. CD44 and CD44v6 downregulation in clinical prostatic carcinoma: relation to Gleason grade and cytoarchitecture. Prostate 1998; 34:162-8.
125. Noordzij MA, van Steenbrugge GJ, Verkaik NS, Schroder FH, van der Kwast TH. The prognostic value of CD44 isoforms in prostate cancer patients treated by radical prostatectomy. Clin Cancer Res 1997; 3:805-15.

126. Kallakury BV, Sheehan CE, Ambros RA, et al. Correlation of p34cdc2 cyclin-dependent kinase overexpression, CD44s downregulation, and HER-2/neu oncogene amplification with recurrence in prostatic adenocarcinomas. J Clin Oncol 1998; 16:1302-9.
127. Sneath RJ, Mangham DC. The normal structure and function of CD44 and its role in neoplasia. Mol Pathol 1998; 51:191-200.
128. Schneider SS, Schick C, Fish KE, et al. A serine proteinase inhibitor locus at 18q21.3 contains a tandem duplication of the human squamous cell carcinoma antigen gene. Proc Natl Acad Sci U S A 1995; 92:3147-51.
129. Zou Z, Anisowicz A, Hendrix MJ, et al. Maspin, a serpin with tumor-suppressing activity in human mammary epithelial cells [see comments]. Science 1994; 263:526-9.
130. Seftor RE, Seftor EA, Sheng S, Pemberton PA, Sager R, Hendrix MJ. maspin suppresses the invasive phenotype of human breast carcinoma. Cancer Res 1998; 58:5681-5.
131. Umekita Y, Hiipakka RA, Liao S. Rat and human maspins: structures, metastatic suppressor activity and mutation in prostate cancer cells. Cancer Lett 1997; 113:87-93.
132. Pienta KJ, Partin AW, Coffey DS. Cancer as a disease of DNA organization and dynamic cell structure. Cancer Res 1989; 49:2525-32.
133. Guo X, Friess H, Graber HU, et al. KAI1 expression is up-regulated in early pancreatic cancer and decreased in the presence of metastases. Cancer Res 1996; 56:4876-80.
134. Guo XZ, Friess H, Di Mola FF, et al. KAI1, a new metastasis suppressor gene, is reduced in metastatic hepatocellular carcinoma. Hepatology 1998; 28:1481-8.
135. Yu Y, Yang JL, Markovic B, et al. Loss of KAI1 messenger RNA expression in both high-grade and invasive human bladder cancers. Clin Cancer Res 1997; 3:1045-9.
136. Adachi M, Taki T, Ieki Y, Huang CL, Higashiyama M, Miyake M. Correlation of KAI1/CD82 gene expression with good prognosis in patients with non-small cell lung cancer. Cancer Res 1996; 56:1751-5.
137. Uchida S, Shimada Y, Watanabe G, et al. Motility-related protein (MRP-1/CD9) and KAI1/CD82 expression inversely correlate with lymph node metastasis in oesophageal squamous cell carcinoma. Br J Cancer 1999; 79:1168-73.
138. Geradts J, Maynard R, Birrer MJ, et al. Frequent loss of KAI1 expression in squamous and lymphoid neoplasms. An immunohistochemical study of archival tissues. Am J Pathol 1999; 154:1665-71.
139. Chekmareva MA, Kadkhodaian MM, Hollowell CM, et al. Chromosome 17-mediated dormancy of AT6.1 prostate cancer micrometastases. Cancer Res 1998; 58:4963-9.
140. Holmgren L, O'Reilly MS, Folkman J. Dormancy of micrometastases: balanced proliferation and apoptosis in the presence of angiogenesis suppression [see comments]. Nat Med 1995; 1:149-53.
141. Welch DR, Chen P, Miele ME, et al. Microcell-mediated transfer of chromosome 6 into metastatic human C8161 melanoma cells suppresses metastasis but does not inhibit tumorigenicity. Oncogene 1994; 9:255-62.
142. Miele ME, Jewett MD, Goldberg SF, et al. A human melanoma metastasis-suppressor locus maps to 6q16.3-q23. Int J Cancer 2000; 86:524-8.
143. Welch DR, Goldberg SF. Molecular mechanisms controlling human melanoma progression and metastasis. Pathobiology 1997; 65:311-330.

144. Goldberg SF, Harms JF, Quon K, Welch DR. Metastasis-suppressed C8161 melanoma cells arrest in lung but fail to proliferate. Clinical and Experimental Metastasis 2000; in press.
145. Nelson PS, Ng WL, Schummer M, et al. An expressed-sequence-tag database of the human prostate: sequence analysis of 1168 cDNA clones. Genomics 1998; 47:12-25.
146. Hawkins V, Doll D, Bumgarner R, et al. PEDB: the Prostate Expression Database. Nucleic Acids Res 1999; 27:204-8.
147. Bissell MJ, Weaver VM, Lelievre SA, Wang F, Petersen OW, Schmeichel KL. Tissue structure, nuclear organization, and gene expression in normal and malignant breast. Cancer Res 1999; 59:1757-1763s; discussion 1763s-1764s.
148. Boudreau N, Bissell MJ. Extracellular matrix signaling: integration of form and function in normal and malignant cells. Curr Opin Cell Biol 1998; 10:640-6.
149. Hayward SW, Grossfeld GD, Tlsty TD, Cunha GR. Genetic and epigenetic influences in prostatic carcinogenesis (review). Int J Oncol 1998; 13:35-47.
150. Baskin LS, Hayward SW, DiSandro MS, Li YW, Cunha GR. Epithelial-mesenchymal interactions in the bladder. Implications for bladder augmentation [In Process Citation]. Adv Exp Med Biol 1999; 462:49-61.
151. Olumi AF, Grossfeld GD, Hayward SW, Carroll PR, Tlsty TD, Cunha GR. Carcinoma-associated fibroblasts direct tumor progression of initiated human prostatic epithelium. Cancer Res 1999; 59:5002-11.
152. Chung LW, Davies R. Prostate epithelial differentiation is dictated by its surrounding stroma. Mol Biol Rep 1996; 23:13-9.
153. Chung LW. The role of stromal-epithelial interaction in normal and malignant growth. Cancer Surv 1995; 23:33-42.
154. Chung LW. Implications of stromal-epithelial interaction in human prostate cancer growth, progression and differentiation. Semin Cancer Biol 1993; 4:183-92.
155. Welch DR. Technical considerations for studying cancer metastasis in vivo. Clin Exp Metastasis 1997; 15:272-306.
156. Day ML, Foster RG, Day KC, et al. Cell anchorage regulates apoptosis through the retinoblastoma tumor suppressor/E2F pathway. J Biol Chem 1997; 272:8125-8.
157. Kendrew J. The thread of life; an introduction to molecular biology. Cambridge, MA: Harvard University Press, 1996.
158. Schrodinger E. What is life? : the physical aspect of the living cell ; with, Mind and matter : & autobiographical sketches. Cambridge paperbacks. Science. New York: Cambridge University Press, 1992.
159. Coffey DS. Self-organization, complexity and chaos: the new biology for medicine [see comments]. Nat Med 1998; 4:882-5.
160. Schwab ED, Pienta KJ. Explaining aberrations of cell structure and cell signaling in cancer using complex adaptive systems. Advances in Molecular and Cell Biology. Vol. 24, 1997:207-247.

Chapter 2

THE ROLES OF MAP KINASES IN CONTROLLING CANCER METASTASIS

Alessandro Alessandrini

Medical Services, Massachusetts General Hospital, Department of Medicine, Harvard Medical School

INTRODUCTION

MEK (Map/Erk Kinase) family members are key components in an intracellular signaling pathway called the ERK (Extracellular-signal Regulated Kinases)/MAP kinase pathway, implicated in the transition of cells from Go to G1 in the cell cycle (1-3). Members of this kinase cascade are highly conserved between species from yeast to mammals (3). In addition, proteins with sequence similarity to ERK/MAP kinases and MEK participate in other cellular signaling pathways; for instance those responding to stresses such as osmotic shock and UV-induced DNA (4). The MAP/Erk kinase cascade is activated following stimulation of a wide variety of cell types with growth factors, hormones, or mitogens. The binding of these various ligands to the appropriate cell surface receptor results in receptor activation, which in turn leads to GTP binding of Ras complexed to members of the Raf family of serine/threonine kinases. Raf family members then activate the kinases MEK1 and MEK2 by phosphorylating them on serines 218 and 222 for MEK1, or 222 and 226 in the case of MEK2. MEKs, as dual specificity kinases, subsequently activate their downstream targets, ERK-1 and ERK-2, by phosphorylating them on threonine and tyrosine. ERKs then phosphorylate both cytoplasmic substrates and nuclear transcription factors, which, thus modified, contribute to the early response of the cell after stimulation. Although there are seven known members of the MEK family of kinases at present, only two of them, MEK1 and MEK2, have been shown to play a role in the ERK/MAP kinase pathway (4-10).

"Stress response protein kinases" (reviewed in (11, 12)), are only minimally activated by growth factors but are markedly activated by genotoxic stress, osmolar stress, and inflammatory cytokines (TNF and IL-1). The stress response MAP kinases, however, are also activated by agonists with heterotrimeric G protein-coupled receptors, including Ang II, ET-1, and a-adrenergic agents, which play important roles in hypertension, hypertrophy, and diabetic nephropathy. In addition, they are activated by cell stretch and shear

D.R. Welch (ed.), Cancer Metastasis – Related Genes, 35–50.

stress. One family, which has 54 and 48 kDa isoforms encoded by at least three genes, has been designated either stress activated protein kinases (SAPKs), since they are activated by cellular stress, or c-Jun N-terminal kinases (JNKs), based on the ability of the kinases to phosphorylate the amino terminus of c-Jun (11, 13, 14). The other stress kinase family includes p38a, the mammalian homolog of HOG-1, a yeast kinase involved in the response to osmolar stress, and three related kinases, p38ß, p38g, and p38d. Like ERK1/2, the SAPKs and p38 are proline directed and require phosphorylation on both tyrosine and threonine residues for activation (15). Unlike the TEY motif of the ERKs, the SAPKs contain a TPY motif and p38 a TGY motif within kinase subdomain VIII which, when phosphorylated, activates the kinases. Overall, there is 40-50% identity in the catalytic domains when comparing the ERKs, SAPKs, and p38 (16).

1. THE MAP KINASE PATHWAYS

Growth factors, such as platelet derived growth factor (PDGF), epidermal derived growth factor (EGF), fibroblast growth factor (FGF), insulin, insulin-like growth factor-1 (IGF-1), colony stimulating factor-1 (CSF-1), nerve growth factor (NGF), etc., as well as cellular stresses, have been shown to activate signal transduction pathways, such as the MAP kinase pathways. These growth factors may play an important role in cell growth and differentiation. This chapter will specifically deal with the role that MAP kinases play in metastasis.

1.1 The MEK/ERK Pathway (Figure 1)

The MAP/ERK kinase cascade is activated following stimulation of a wide variety of cell types with growth factors, hormones, or mitogens. The binding of these various ligands to the appropriate cell surface receptor results in receptor activation, which in turn leads to GTP binding of Ras complexed to members of the Raf family of serine/threonine kinases. Translocation of Raf-1 to the plasma membrane by Ras is necessary for the activation of Raf-1 (17, 18). However, co-incubation of purified GTP-loaded Ras with c-Raf-1 fails to fully activate the latter, suggesting that a membrane-localized co-factor may be necessary.

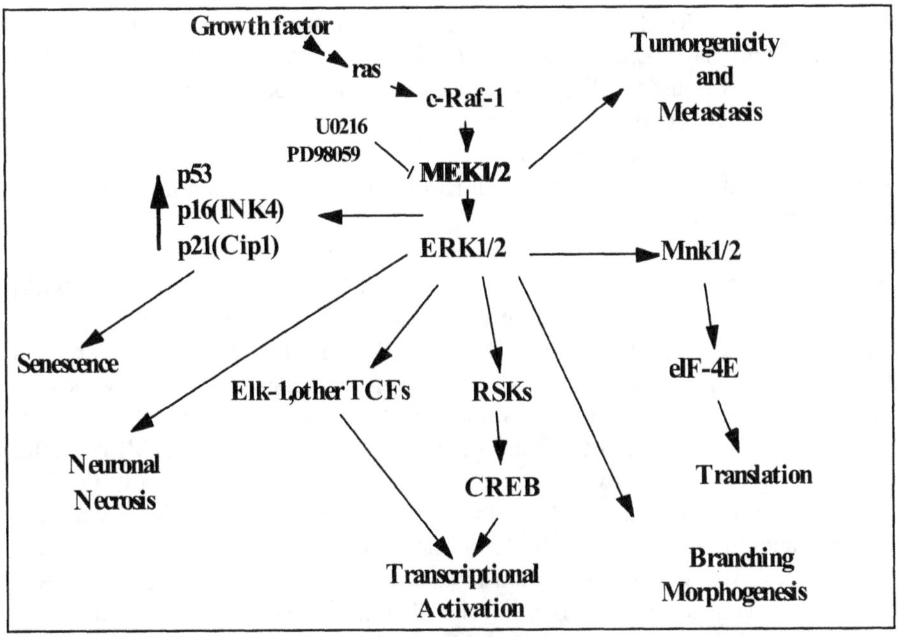

Figure 1. The various roles of the MEK/ERK Pathway.

The most likely candidates are the 14-3-3 proteins (19, 20). 14-3-3 proteins interact promiscuously with many different proteins, and the ß isoform of 14-3-3 associates with the N-terminal regulatory domain of c-Raf-1(21). Each 14-3-3 molecule can bind two c-Raf-1 molecules (reviewed in (22)). The 14-3-3 protein appears to allow c-Raf-1 molecules to interact more effectively when they are brought to the cell membrane by GTP-bound Ras. Oligomerization of c-Raf-1 is critical for activation of c-Raf-1 kinase activity (22). Raf family members activate the kinases MEK1 and MEK2 by phosphorylating them on serines 218 and 222 for MEK1, or 222 and 226 in the case of MEK2. MEKs, as dual specificity kinases, subsequently activate their downstream targets, ERK-1 and ERK-2, by phosphorylating them on threonine and tyrosine. MEK1 and MEK2 are approximately 90% similar and 80% identical; the differences in these proteins being mainly in the amino-terminal region outside the kinase domain, and in a proline-rich region between conserved kinase domains IX and X (5, 9, 10). These differences between MEK1 and MEK2 may contribute to differences in interactions and specificity.

The respective contributions of MEK1 versus MEK2 in the ERK/MAP kinase pathway are currently not well defined. Recently, it has been shown that MEK2 is more highly expressed during mouse embryogenesis than MEK1, suggesting that the former may play a key role in development (23). However, relative MEK specificities towards these two known substrates in *in vivo* systems have not yet been rigorously examined. Differential activation of MEK1 and

MEK2 by various Raf family members in HELA cells has been described by Wu et al., who find that MEK1 is activated *in vitro* by A-Raf, as well as Raf-1 and B-Raf; this is in contrast to MEK2, which is only phosphorylated by Raf-1 and B-Raf (24). Jelinek et al. reported that immobilized Ras-Raf-1 and Ras-B-Raf complexes bind MEK1 but not MEK2 (25). However, later studies indicated that both MEK1 and MEK2 interacted with Ras-bound Raf-1 through the proline-rich region in the kinase domain (26).

Another point that remains unclear is whether various external stimuli preferentially lead to activation of MEK1 versus MEK2 in cell lines and in tissue, or whether their functions are largely redundant. Stimulation of NIH3T3, Rat1, and PC12 cells with TPA, EGF, NGF, FGF, and PDGF failed to show a differential response between MEK1 and MEK2, suggesting that in some cell types, the functions of these two family members may be redundant. However, Downey et al. have reported that MEK2 is more active than MEK1 after stimulation of neutrophils with chemotactic peptides and is sensitive to the PI-3 kinase inhibitor, wortmannin (27). Interestingly, this inhibitor along with LY 294002, another PI-3 kinase inhibitor, have also been shown to inhibit the activation of proto-oncogene protein kinase B (PKB), also known as Akt/RAC. This family consists of Ser/Thr kinases that have been also been shown to respond to various growth factors. In addition, this pathway has been shown to be an anti-apoptotic pathway by phosphorylating Bad, resulting in the activation of Bcl2.

As mentioned previously, the activation of MEK1 and MEK2, leads to the phosphorylation and activation of the ERK/MAP kinases. ERKs then phosphorylate both cytoplasmic substrates, such as other kinases, and nuclear transcription factors, on specific amino acid sequences – (P/L)X(S/T)P. When modified, these proteins contribute to the early response of the cell after stimulation.

ERKs activate the p90 ribosomal S6 kinases, RSK1, RSK2 and RSK3, so named because their first identified substrate was the ribosomal S6 protein, although it has been shown that it is not an *in vivo* substrate. RSKs translocate into the nucleus and can also phosphorylate transcription factors (28, 29). RSK2 has been shown to play an important role in immediate early gene induction and mitogenesis by phosphorylating the transcription factor, CREB or cAMP response element binding protein at Ser133, resulting in the increase expression of c-*fos* in response to some growth factors (30).

Two additional ERK substrates have been identified and named MAP kinase-interacting serine/threonine kinase 1 and 2 (Mnk 1 and Mnk2) (31, 32). Mnk1 phosphorylates eukaryotic initiation factor-4E (eIF-4E) at Ser 209. eIF-4E is a translation initiation factor which binds the 7-methyl-guanosine cap on all eukaryotic mRNAs. This protein plays an important role in the regulation of translation in mammalian cells (33).

ERK proteins translocate to the nucleus upon activation in response to growth factors (34, 35). ERKs phosphorylate and activate a critical family of transcription factors, the ternary complex factors or TCFs (3, 36, 37). Elk-1 is a ternary complex factor that when phosphorylated, forms a complex with SRF (serum response factor). This complex binds to the promoter of a number of genes including c-*fos* that contain the serum response element or SRE (38). These activated transcription factors play critical roles in the induction of immediate early genes and in the mitogenic response.

Expression of the D-type cyclins, the regulatory (activating) subunits for the cyclin-dependent kinase 4 and 6 (cdk4 and cdk6) catalytic subunits, increases in response to growth factor stimulation (reviewed in (39)). Expression of dominant negative mutants of MEK-1 or ERK-1, or expression of the MAP kinase phosphatase, MKP-1, which dephosphorylates and inactivates the ERKs, inhibited growth factor-dependent expression of cyclin D1 (40-42). Activation of the MEK/ERK pathway also correlates with increased expression of cyclin E (the regulatory subunit of cdk2 which also promotes S phase progression) and decreased expression of the cyclin-dependent kinase inhibitor (CKI) $p27^{Kip1}$ (43). These data suggest the Raf/ERK cascade may interact at multiple sites to promote cell cycle progression. Recently, it has also been shown that constitutively active MEK1 and oncogenic cRaf1 result in the upregulation of the tumor suppressor genes, p53, p16(INK4) and p21(Cip1), in normal human fibroblasts, resulting in cell cycle arrest and senescence (44, 45).

1.2 The Stress-response Protein Kinases – The SAPKs/JNKs and p38 MAP Kinases (Figure 2)

The signaling cascades resulting in SAPK/JNK and p38 activation have direct parallels with the ERK cascade, but the cascades are relatively insulated from one another since the MEKs that activate one MAPK are much less effective at activating the others (Figure 12). Although there may be some activation of ERK1 or p38 by SEK1 (also known as MKK4, one of the MEKs upstream of the SAPKs) in cells overexpressing SEK1, at more physiologic levels of expression, there appears to be minimal cross-talk (46, 47). Additional evidence for the specificity of SEK1/MKK4 is the observation that a dominant negative construct of SEK1 blocks SAPK activation but does not alter agonist-induced ERK activation (46). Similarly, MKK7 is a specific activator of the SAPKs, and MKK3 and MKK6 are specific activators of p38.

Upstream of SEK1/MKK4 is MEKK1 (48). Although originally named because of its ability to activate MEK1, MEKK1 is a much more specific activator of SEK1/MKK4 than it is of MEK1 (49). MEKK1 is not the only activator of SEK1/MKK4 and the SAPK cascade. Multiple protein kinases from different families are capable of activating SEK1/MKK4 and thus function as MEKKs (11, 14). This diversity of upstream activators likely reflects the incredible diversity of stimuli that converge at activation of the SAPKs. MEKKs in the p38 cascade have not been clearly identified.

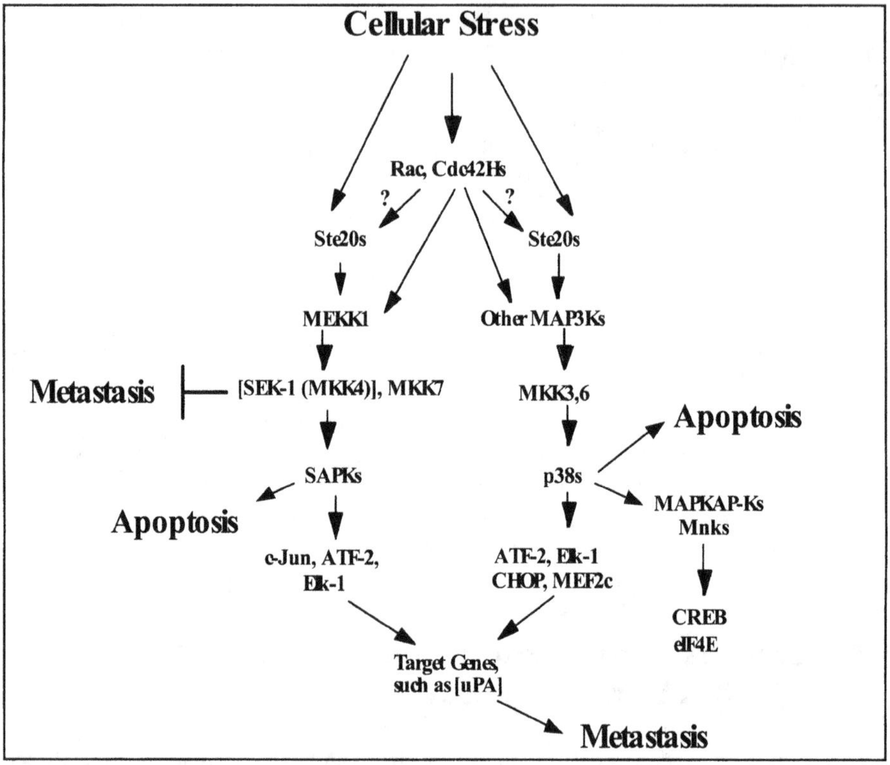

Figure 2. The SAPK and p38 pathways.

The signaling components upstream of the MEKK level are also likely to be quite different, depending on the stimulus (e.g. osmolar stress vs inflammatory cytokines). However, for many stimuli, one family of kinases highly conserved throughout evolution is likely to play a critical role. These kinases are from the Sterile20 (Ste20) family (so named because yeast with mutations in the STE20 gene do not mate normally in response to pheromone). Several Ste20s are capable of activating the SAPK cascade (50, 51). One of these kinases, Germinal center (GC) kinase, is a member of a growing subfamily of Ste20-like kinases, many of which activate the SAPKs. Although the physiologic role of most of these kinases is not known since activators have not been identified, GCK clearly plays a role in SAPK activation by TNFa and possibly other inflammatory cytokines (51). The other subfamily of Ste20-like kinases is the p21 activated kinase (PAK1) family. PAK1 is regulated by small G proteins of the Rho superfamily, Rac1 and Cdc42Hs.

Like the ERKs, following phosphorylation and activation, the SAPKs and p38 translocate to the nucleus where they phosphorylate and activate several transcription factors which play important roles in the response of the kidney to vasoactive peptides and cellular stresses such as ischemia (Figure 12 and reviewed in (52)). One of these transcription factors is c-Jun. The SAPKs phosphorylate c-Jun on Ser 63 and Ser 73, two residues within the transcriptional activation domain, and this enhances transactivating activity of c-Jun (53). The SAPKs also phosphorylate the transcription factor ATF-2 (activating transcription factor-2). Like c-Jun, ATF-2 contains an N-terminal transcriptional activation domain, and phosphorylation by the SAPKs (or p38) at Thr 69 and 71 within this domain enhances the transcriptional activating activity of ATF-2(54). ATF-2 can form homodimers, or heterodimers with other members of its family, ATF-3 and CREB, or with c-Jun or NF-kB, suggesting it may play a role in the activation of transcription from many promoters. For example, a c-Jun/ATF-2 dimer appears to control induction of the c-*jun* gene in response to cellular stresses, and it is likely that the SAPKs transduce this signal by phosphorylating both transcription factors.

p38 also plays a role in stress-induced c-Jun regulation since it phosphorylates and activates another transcription factor, MEF2C, which binds to AT-rich sequences in the promoter of the c-*jun* gene and enhances the transcription of c-*jun*. Once c-Jun protein has been synthesized, the SAPKs activate its transcriptional activating activity (55).

Like the ERKs, the SAPKs and p38 phosphorylate the ternary complex factor, Elk-1, within the C-terminal activation domain which enhances ternary complex formation, DNA binding, and transcriptional activating activity of Elk-1 (56). Since all three MAP kinases activate Elk-1, and since ternary complex formation at the SRE of the c-*fos* promoter controls, at least in part, c-*fos* induction, it is not surprising that c-*fos* is induced in response to a very wide variety of stimuli.

Although the SAPKs and p38s share many substrates, their substrate specificity is not identical since c-Jun is a SAPK (but not p38) target and the transcription factors SAP-1 (57) and CHOP (58), a transcription factor involved in genotoxic stress-induced G1 arrest, are p38 (but not SAPK) targets. p38 also phosphorylates and activates MAPKAP kinase (mitogen-activated protein kinase activated protein kinase)-2 and -3. Like the ERKs, which control activation of CREB via activation of the protein kinase, RSK2, the p38 pathway may control activation of CREB via MAPKAP kinase-2 (59). MAPKAP kinase-2 phosphorylates CREB at Ser133, increasing transcriptional activating activity of CREB (59). CREB is phosphorylated on Ser133 following cellular stress. Since the p38 inhibitor, SB203580, markedly inhibits CREB phosphorylation and activation by cellular stress(59), it is likely that this is mediated via activation of p38 and its target, MAPKAP kinase-2.

p38 has other substrates with roles in signal transduction and the response to stress, and identification of physiological roles of p38 has been greatly aided by the availability of what appears to be relatively specific inhibitors, SB203580 and related compounds. Studies employing this agent have identified an important role for p38 in the aggregation of platelets in response to certain agonists since pretreatment of platelets with SB203580 inhibits aggregation (60,

61). This has been postulated to be related to phosphorylation by p38 of a critical residue (Ser 505) of cytosolic phospholipase A_2 ($cPLA_2$), a residue that is also phosphorylated by the ERKs (61). Phosphorylation of this residue increases the activity of $cPLA_2$, releasing arachidonic acid, the precursor of thromboxane A2, a potent vasoconstrictor and pro-aggregatory eicosanoid.

2. THE MAP KINASES AS PROMOTERS OR SUPPRESSORS OF METASTASIS

As described above, mammalian cells express at least three families of MAP kinases – ERKs, SAPKs, and p38 MAP kinases. The molecular basis for cancer metastasis has not been fully elucidated. Previous studies have demonstrated that introduction of oncogenic forms of the ras oncogene confer both tumorigenic and metastatic potentials upon NIH3T3 cells. These results implicated events downstream in the regulation of metastasis. The current study was to identify key downstream components contributing to tumorigenicity and metastasis.

2.1 The MEK/ERK Pathway and Metastasis

Constitutively active variants of MEK1, a downstream component of the ras/MAPK pathway, have previously been shown to lead to cellular transformation (62-65). Previously, we showed that growth on soft agar did not correlate with ERK activity (62-65). Webb et al. used ras-transfectants to show that tumorigenicity occurred through both Raf-dependent and independent pathways. They observed that metastatic potential correlated with variants wthat were able to activate ERK1 activity (66). After 7-9 weeks following injected into mice, some of the ras-transfectants (that were originally nonmetastatic) formed lung colonies. This implies that some selection may have occurred and the authors suggest that increased expression of the Met receptor tyrosine kinase is responsible, though only a modest increase in ERK1/2 activity was observed in these metastatic variants. HGF, the ligand of the Met receptor, can induce both the MEK1 and PI3K pathways, leaving the possibility that the pathway responsible for acquisition of metastasis in these cells is MEK1 and/or PI3K dependent. In fact, we observed that in MEK1 transformants, metastatic potential appeared independent of ERK1/2 activity (67). While we are in agreement with Webb et al. concerning tumorigenicity and how the process may be ERK1/2-independent, we have contrasting results regarding metastatic potential and ERK1/2 activity. This discrepancy may be due to the fact that we specifically analyzed MEK1 transformants and not ras-transformants. It is possible that MEK1 induced transformation may occur through a pathway that is distinct from that of ras-transformation and may not necessarily require ERK1/2 activity to maintain transformation.

The mechanism by which MEK1 transformation confers metastasis is not known. One possibility may lie with MMP2, MMP9, and cathepsin L secretion.

We have shown that MEK1 transformants show an increase in MMP2 and MMP9 activities (67), two matrix metalloproteinases that have been implicated in metastasis (68-73). We also observed an increase in the expression of cathepsin L (67), which also has been correlated with metastatic potential in ras-transformed cells (74, 75). One way that cathepsin L may lead to an increase in metastatic potential is by allowing the transformed cells to evade the immune response by cleavage of the third component of complement (76). Another possibility may be the contribution of cyclooxygenase-2, COX-2, an enzyme that has been implicated in metastasis (77). We have recently observed an increase in COX-2 levels in MEK1-transformants (unpublished data, Sapirstein, Welch, and Alessandrini).

Analysis of 82 primary and metastatic prostate tumour specimens has revealed an association of increase ERK1/2 activities with advanced tumour grade and stage (78). The androgen-independent, fast growing human prostate cancer cell line, DU145, which serves as a model for advanced prostatic carcinoma, also exhibits constitutively active ERK2 activity. In addition, the androgen-insensitive, metastatic tumours exhibit low levels of E-cadherin, a Ca^{2+}-dependent cell adhesion molecule (79-81). Reintroduction of E-cadherin into prostate cancer cells that exhibit metastasis resulted in the attenuation of metastatic potential, in part by down regulating MMP2 activity (82). The MEK1/ERK pathway appears to play a role in the regulation of E-cadherin expression. Ras-transformation of MDCK renal epithelial cells, resulted in a change to a more fibroblastic phenotype, with a reduction in cell-cell adhesion due to a decrease in E-cadherin levels. Treatment of these cells with the MEK1 inhibitor, PD98059, reversed the cells to an epithelial phenotype, with reexpression of E-cadherin and increased cell-cell adhesion (83). From this it was concluded that activation of the MEK1/ERK pathway can regulate E-cadherin levels.

Ksr (kinase suppressor of Ras) was originally identified as a regulator of the Ras/MEK1/ERK pathway. Denouel et al. have shown that Ksr to interact with components of the MAP kinase pathway (84). Using the yeast two-hybrid system, they showed that Ksr-1 did not bind to Ras, B-Raf or Raf-1, but interacted strongly with both MEK-1 and MEK-2. This was further confirmed by co-immunoprecipitation experiments. The ability of Ksr to block MEK in an inactive form resulted in reduced Ras-induced transformation. Whether Ksr is able to inhibit MEK1-induced metastasis is yet to be determined, but the possibility is intriguing.

2.2 The SAPKs and p38 MAP Kinase and Metastasis

While the SAPKs and P38 MAP kinases have been implicated in the induction of apoptosis, recently it has been shown that p38a MAP kinase may play a role in the transcriptional activation of the urokinase plasminogen activator (uPA) and the urokinase-specific cell surface receptor (uPAR) in breast cancer cells (85). uPA is a serine protease that when bound to its receptor, uPAR, results in the cleavage and activation of MMPs, and has been implicated in the invasion process (86-88). The p38 MAP Kinase pathway has also been shown to play a role in the induction of several other proteases. p38 MAPK can mediate MMP-1 expression in fibroblasts in response to okadaic acid and EMMPRIN (extracellular matrix metalloproteinase inducer) treatments (89, 90). Addition of the p38 MAPK inhibitor, SB203580, to squamous carcinoma cells of head and neck, resulted in the inhibition of PMA-induced MMP9 expression (91).

Recently, Janulis et al. have suggested that a switch in protease expression during cellular invasion occurs through the regulation of SAPK, allowing for the invading cells to evade protease inhibitors of one class by switching to the expression of another class of proteases, in particular, a switch between uPA and cathepsin L (92).

But while the possible involvement of the SAPKs and p38 MAP kinase in the induction of various proteases involved in cellular invasion, it has been suggested that MKK4/SEK1, the upstream activator of SAPKs, may function as a metastasis suppressor in certain cell types (93, 94). How can this be explained? It is possible that the suppression of metastasis by MKK4 is cell type specific, as the authors suggest. It is also possible that while MKK4 acts as a suppressor, MKK7, the other upstream activator of SAPKs, may play a greater role in cellular invasion.

3. CONCLUDING REMARKS

The MAP kinases have been implicated in cell growth and differentiation. Dysregulation of these pathways results in the increased incidence of tumorgenicity and metastasis. The mechanisms by which these pathways lead to cancer have yet to be elucidated. Recently, Taguchi et al. have shown that blocking the interaction of the receptor for advanced glycation end products (RAGE) and its ligand, amphoterin by various strategies, such as the administration of the soluble, extracellular region of RAGE, resulted in the reduction of tumour volume and metastases in mice (95). Interestingly, the inhibition of RAGE-amphoterin interaction resulted in decreased ERK1/2, SAPK, and p38 MAP kinase activities, and reduced expression of MMP-2 and MMP-9 (95).

REFERENCES

1. Blenis J. Signal transduction via the MAP kinases: Proceed at your own RSK. Proc. Natl. Acad. Sci. 1993; 90: 5889-5892.
2. Davis R. The mitogen-activated protein kinase signal transduction pathway. J. Biol. Chem. 1993; 268: 14553-14556.
3. Seger R, Krebs EG. The MAPK signaling cascade. FASEB J. 1995; 9: 726-735.
4. Davis RJ. MAPKs: new JNK expands the group. TIBS. 1994; 19.

5. Brott BK, Alessandrini A, Largaespada DA et al. MEK2 is a kinase realted MEK1 and is differentially expressed in murine tissues. Cell Growth and Diff. 1993; 4: 921-929.
6. Crews CM, Alessandrini A, Erikson RL. The primary structure of MEK, a protein kinase that phosphorylates the *ERK* gene product. Science. 1992; 258: 478-480.
7. Seger R, Ahn NG, Posada J et al. Purification and characterization of mitogen-activated protein kinase activator(s) from epidermal growth factor-stimulated A431 cells. J. Biol. Chem. 1992; 267: 14373-14381.
8. Wu J, Dent P, Jelinek T, Wolfman A, Weber MJ, Sturgill TW. Inhibition of the EGF-activated MAP kinase signaling pathway by adenosine 3',5' - monophosphate. Science. 1993; 262: 1065-1069.
9. Wu J, Harrison JK, Dent P, Lynch KR, Weber MJ, Sturgill TW. Identification and characterization of a new mammalian mitogen-activated protein kinase kinase, MKK2. Mol. Cell. Biol. 1993; 13: 4539-4548.
10. Zheng C-F, Guan K-L. Cloning and characterization of two distinct human extracellular signal-regulated kinase activator kinases, MEK1 and MEK2. J. Biol. Chem. 1993; 268: 11435-11439.
11. Kyriakis JM, Avruch J. Protein kinase cascades activated by stress and inflammatory cytokines. BioEssays. 1996; 18: 567-577.
12. Force T, Pombo CM, Avruch JA, Bonventre JV, Kyriakis JM. Stress-activated protein kinases in cardiovascular disease. Circ. Res. 1996; 78: 947-953.
13. Kypta RM, Goldberg Y, Ulug ET, Courtneidge SA. Association between the PDGF receptor and members of the *src* family of tyrosine kinases. Cell. 1990; 62: 481-492.
14. Kyriakis JM, Avruch J. Sounding the alarm: protein kinase cascades activated by stress and inflammation. J. Biol. Chem. 1996; 271: 24313-24316.
15. Kyriakis JM, Avruch J. pp54 Microtubule-associated Protein 2 Kinase. J. Biol. Chem. 1990; 265: 17355-17363.
16. Kyriakis JM, Banerjee P, Nikolakaki E et al. The stress-activated protein kinase subfamily of c-Jun kinases. Nature. 1994; 369: 156-160.
17. Leevers SJ, Paterson HF, Marshall CJ. Requirement for Ras in Raf activation is overcome by targeting Raf to the plasma membrane. Nature. 1994; 369: 411-414.
18. Stokoe D, Macdonald SG, Cadwallader K, Symons M, Hancock JF. Activation of Raf as a result of recruitment to the plasma membrane. Science. 1994; 264: 1463-1467.
19. Liu D, Bienkowska J, Petosa C, Collier RJ, Fu H, Liddington R. Crystal structure of the zeta isoform of the 14-3-3 protein. Nature. 1995; 376: 191-194.
20. Xiao B, Smerdon SJ, Jones DH et al. Structure of a 14-3-3 protein and implications for coordination of multiple signaling pathways. Nature. 1995; 376: 188-191.
21. Li S, Janosch P, Tanji M et al. Regulation of Raf-1 kinase activity by the 14-3-3 family of proteins. EMBO J. 1995; 14: 685-696.
22. Marshall CJ. Raf gets it together. Nature. 1996; 383: 127-128.
23. Alessandrini A, Brott BK, Erikson RL. Differential expression of MEK1 and MEK2 during mouse development. Cell Growth Differ. 1997; 8: 505-511.
24. Wu X, Noh SJ, Zhou G, Dixon JE, Guan K-L. Selective actiation of MEK1 but not MEK2 by A-Raf from epidermal growth factor-stimulated Hela cells. J. Biol. Chem. 1996; 271: 3265-3271.

25. Jelinek T, Catling AD, Reuter CW, Moodie SA, Wolfman A, Weber MJ. RAS and RAF-1 form a signaling complex with MEK-1 but not MEK-2. Mol. Cell. Biol. 1994; 14: 8212-8218.
26. Catling AD. A proline-rich sequence unique to MEK1 and MEK2... Mol. Cell. Biol. 1995; 15: 5214-5225.
27. Downey GP, Butler JR, Brumell J et al. Chemotactic peptide-induced activation of MEK-2, the predominant isoform in human neutrophils. J. Biol. Chem. 1996; 271: 21005-21011.
28. Sturgill TW, Ray LB, Erikson E, Maller JL. Insulin-stimulated MAP-2 kinase phosphorylates and activates ribosomal protein S6 kinase II. Nature. 1988; 334: 715-718.
29. Zhao Y, Bjorbaek C, Weremowicz S, Morton C, Moller DE. RSK3 encodes a novel $pp90^{rsk}$ isoform with a unique N-terminal sequence: growth factor stimulated kinase function and nuclear translocation. Mol. Cell. Biol. 1995; 15: 4353-4363.
30. Xing J, Ginty DD, Greenberg ME. Coupling of the RAS-MAPK pathways to gene activation by RSK2, a growth factor-regulated CREB kinase. Science. 1996; 273: 959-960.
31. Fukunaga R, Hunter T. Mnk1, a new MAP kinase-activated protein kinase, isolated by a novel expression screening method for identifying protein kinase substrates. EMBO J. 1997; 16: 1221-1233.
32. Waskiewicz AJ, Flynn A, Proud CG, Cooper JA. Mitogen-activated protein kinases activate the serine/threonine kinases Mnkl and Mnk2. EMBO J. 1997; 16: 1909-1920.
33. Flynn A, Proud CG. The role of eIF4 in cell proliferation. Cancer Surv. 1996; 27: 293-310.
34. Lenormand P, Sardet C, Pages G, L'Allemain G, Brunet A, Pouyssegur J. Growth factors induce nuclear translocation of MAP kinases (p42 and p44) but not of their activator MAP kinase kinase (p45) in fibroblasts. J. Cell Biol. 1993; 122: 1079-1088.
35. Gonzalez F, Seth A, Raden D, Bowman D, Fay F, Davis R. Serum-induced translocation of mitogen-activated protein kinase to the cell surface ruffling membrane and the nucleus. J. Cell Biol. 1993; 122: 1089-1101.
36. Treisman R. Regulation of transcription by MAP kinase cascades. Curr. Opin. Cell. Biol. 1996; 8: 205-215.
37. Treisman R. Journey to the surface of the cell: Fos regulation and the SRE. EMBO J. 1995; 14: 4905-4913.
38. Gille H, Sharrocks AD, Shaw PE. Phosphorylation of transcription factor $p62^{TCF}$ by MAP kinase stimulates ternary complex formation at c-*fos* promoter. Nature. 1992; 358: 414-416.
39. Grana X, Reddy EP. Cell cycle control in mammalian cells. Oncogene. 1995; 11: 211-219.
40. Lavoie JN, L'Allemain G, Brunet A, Muller R, Pouyssegur J. Cyclin D1 expression is regulated positively by the p42/44 MAPK and negatively by the p38/HOG MAPK pathway. J. Biol. Chem. 1996; 271: 20608-20616.
41. Woods D, Parry D, Cherwinski H, Bosch E, Lees E, McMahon M. Raf-induced proliferation or cell cycle arrest is determined by the level of Raf activity with arrest mediated by p21Cip1. Mol Cell Biol. 1997; 17: 5598-5611.

42. Sewing A, Wiseman B, Lloyd AC, Land H. High-intensity Raf signal causes cell cycle arrest mediated by p21Cip1. Mol Cell Biol. 1997; 17: 5588-5597.
43. Alessandrini A, Chiaur DS, Pagano M. Regulation of the cyclin-dependent kinase inhibitor p27 by degradation and phosphorylation. Leukemia. 1997; 11: 342-345.
44. Zhu J, Woods D, McMahon M, Bishop JM. Senescence of human fibroblasts induced by oncogenic Raf [In Process Citation]. Genes Dev. 1998; 12: 2997-3007.
45. Lin AW, Barradas M, Stone JC, van AL, Serrano M, Lowe SW. Premature senescence involving p53 and p16 is activated in response to constitutive MEK/MAPK mitogenic signaling [In Process Citation]. Genes Dev. 1998; 12: 3008-3019.
46. Sánchez I, Hughes RT, Mayer BJ et al. Role of SAPK/ERK kinase-1 in the stress-activated pathway regulating transcription factor c-jun. Nature. 1994; 372: 794-797.
47. Dérijard B, Hibi M, Wu IH et al. JNK1: A protein kinase stimulated by UV light and Ha-Ras that binds and phosphorylates the c-Jun activation domain. Cell. 1994; 76: 1025-1037.
48. Lange-Carter CA, Johnson GL. Ras-Dependent Growth Factor Regulation of MEK kinase in PC12 Cells. Science. 1994; 265: 1458-1461.
49. Yan M, Dai T, Deak JC et al. Activation of stress-activated protein kinase by MEKK1 phosphorylation of its activator SEK1. Nature. 1994; 372: 798-800.
50. Zhang S, Han J, Sells MA et al. Rho family GTPases regulate p38 MAP kinase through the downstream mediator Pak1. J. Biol. Chem. 1995; 270: 23934-23936.
51. Pombo C, Kehrl J, Sanchez I et al. Activation of the SAPK pathway by the human STE20 homolog germinal centre kinase. Nature. 1995; 377: 750-754.
52. Force T, Bonventre JV. Growth factors and mitogen-activated protein kinases. Hypertension. 1998; 31 (part 2): 152-161.
53. Pulverer B, Kyriakis JM, Avruch J, Nikolakaki E, Woodgett JR. Phophorylation of c-*jun* mediated by MAP kinases. Nature. 1991; 353: 670-674.
54. Morooka H, Bonventre JV, Pombo CM, Kyriakis JM, Force T. Ischemia and reperfusion enhance ATF-2 and c-Jun binding to cAMP response elements and to an AP-1 binding site from the c-jun promoter. J. Biol. Chem. 1995; 270: 30084-30092.
55. Cohen P. The search for physiological substrates of mitogen- and stress-activated protein kinases in mammalian cells. Trends Cell Biol. 1997; 7: 353-361.
56. Whitmarsh AJ, Yang S-H, Su M-S, Sharrocks AD, Davis RJ. Role of p38 and JNK mitogen-activated protein kinases in activation of ternary complex factors. Mol. Cell. Biol. 1997; 17: 2360-2371.
57. Whitmarsh AJ, Yang SH, Su MS, Sharrocks AD, Davis RJ. Role of p38 and JNK mitogen-activated protein kinases in the activation of ternary complex factors. Mol Cell Biol. 1997; 17: 2360-2371.
58. Wang XZ, Ron D. Stress-induced phosphorylation and activation of the transcription factor CHOP (GADD153) by p38 MAP Kinase. Science. 1996; 272: 1347-1349.
59. Tan Y, Rouse J, Zhang A, Cariati S, Cohen P, Comb M. FGF and stress regulate CREB and ATF-1 via a pathway involving p38 MAP kinase and MAPKAP kinase-2. Embo J. 1996; 15: 4629-4642.
60. Saklatvala J, Rawlinson L, Waller RJ et al. Role for p38 mitogen-activated protein kinase in platelet aggregation caused by collagen or a thromboxane analogue. J. Biol.Chem. 1996; 271: 6586-6589.

61. Kramer RM, Roberts EF, Um SL et al. p38 mitogen-activated protein kinase phosphorylates cytosolic phospholipase A2 in thrombin-stimulated platelets. J. Biol. Chem. 1996; 271: 27723-27729.
62. Alessandrini A, Greulich H, Huang W, Erikson RL. Mekl phosphorylation site mutants activate Raf-1 in NIH 3T3 cells. J Biol Chem. 1996; 271: 31612-31618.
63. Brunet A, Pages G, Pouyssegur J. Constitutively active mutants of MAP kinase kinase (MEK1) induce growth factor-relaxation and oncogenicity when expressed in fibroblasts. Oncogene. 1994; 9: 3379-3387.
64. Cowley S, Paterson H, Kemp P, Marshall CJ. Activation of MAP kinase kinase is necessary and sufficient for PC12 differentiation and for transformation of NIH3T3 cells. Cell. 1994; 77: 841-852.
65. Mansour SJ, Matten WT, Hermann AS et al. Transformation of mammalian cells by constitutively active MAP kinase kinase. Science. 1994; 265: 966-970.
66. Webb CP, Van AL, Wigler MH, Woude GF. Signaling pathways in Ras-mediated tumorigenicity and metastasis. Proc Natl Acad Sci U S A. 1998; 95: 8773-8778.
67. Welch DR, Sakamaki T, Pioquinto R et al. Transfection of constitutively active mitogen-activated protein/extracellular signal-regulated kinase kinase confers tumorigenic and metastatic potentials to NIH3T3 cells. Cancer Res. 2000; 60: 1552-1556.
68. Yoshizaki T, Sato H, Furukawa M, Pagano JS. The expression of matrix metalloproteinase 9 is enhanced by Epstein- Barr virus latent membrane protein 1. Proc Natl Acad Sci U S A. 1998; 95: 3621-3626.
69. Tanaka K, Iwamoto Y, Ito Y et al. Cyclic AMP-regulated synthesis of the tissue inhibitors of metalloproteinases suppresses the invasive potential of the human fibrosarcoma cell line HT1080. Cancer Res. 1995; 55: 2927-2935.
70. Repassy G, Forster HC, Juhasz A, Adany R, Tamassy A, Timar J. Expression of invasion markers CD44v6/v3, NM23 and MMP2 in laryngeal and hypopharyngeal carcinoma. Pathol Oncol Res. 1998; 4: 14-21.
71. Saad Z, Bramwell VH, Wilson SM, O'Malley FP, Jeacock J, Chambers AF. Expression of genes that contribute to proliferative and metastatic ability in breast cancer resected during various menstrual phases [published erratum appears in Lancet 1998 Aug 1; 352(9125): 408] [see comments]. Lancet. 1998; 351: 1170-1173.
72. Kawamata H, Uchida D, Hamano H et al. Active-MMP2 in cancer cell nests of oral cancer patients: correlation with lymph node metastasis. Int J Oncol. 1998; 13: 699-704.
73. Kawamata H, Nakashiro K, Uchida D, Harada K, Yoshida H, Sato M. Possible contribution of active MMP2 to lymph-node metastasis and secreted cathepsin L to bone invasion of newly established human oral- squamous-cancer cell lines. Int J Cancer. 1997; 70: 120-127.
74. Silberman S, Janulis M, Schultz RM. Characterization of downstream Ras signals that induce alternative protease-dependent invasive phenotypes [published erratum appears in J Biol Chem 1997 Apr 25; 272(17): 11670]. J Biol Chem. 1997; 272: 5927-5935.
75. Casson AG, Wilson SM, McCart JA et al. ras mutation and expression of the ras-regulated genes osteopontin and cathepsin L in human esophageal cancer. Int J Cancer. 1997; 72: 739-745.

76. Frade R, Rodrigues LF, Huang S, Xie K, Guillaume N, Bar EM. Procathepsin-L, a proteinase that cleaves human C3 (the third component of complement), confers high tumorigenic and metastatic properties to human melanoma cells. Cancer Res. 1998; 58: 2733-2735.
77. Yokota J. Tumor progression and metastasis. Carcinogenesis. 2000; 21: 497-503.
78. Gioeli D, Mandell JW, Petroni GR, Frierson HJ, Weber MJ. Activation of mitogen-activated protein kinase associated with prostate cancer progression. Cancer Res. 1999; 59: 279-284.
79. Bussemakers MJ, van MR, Giroldi LA et al. Decreased expression of E-cadherin in the progression of rat prostatic cancer. Cancer Res. 1992; 52: 2916-2922.
80. Cheng L, Nagabhushan M, Pretlow TP, Amini SB, Pretlow TG. Expression of E-cadherin in primary and metastatic prostate cancer. Am J Pathol. 1996; 148: 1375-1380.
81. Umbas R, Schalken JA, Aalders TW et al. Expression of the cellular adhesion molecule E-cadherin is reduced or absent in high-grade prostate cancer. Cancer Res. 1992; 52: 5104-5109.
82. Luo J, Lubaroff DM, Hendrix MJ. Suppression of prostate cancer invasive potential and matrix metalloproteinase activity by E-cadherin transfection. Cancer Res. 1999; 59: 3552-3556.
83. Lu Q, Paredes M, Zhang J, Kosik KS. Basal extracellular signal-regulated kinase activity modulates cell- cell and cell-matrix interactions. Mol Cell Biol. 1998; 18: 3257-3265.
84. Denouel GA, Douville EM, Warne PH et al. Murine Ksr interacts with MEK and inhibits Ras-induced transformation. Curr Biol. 1998; 8: 46-55.
85. Huang S, New L, Pan Z, Han J, Nemerow GR. Urokinase plasminogen activator/urokinase-specific surface receptor expression and matrix invasion by breast cancer cells requires constitutive p38alpha mitogen-activated protein kinase activity. J Biol Chem. 2000; 275: 12266-12272.
86. Aznavoorian S, Murphy AN, Stetler SW, Liotta LA. Molecular aspects of tumor cell invasion and metastasis. Cancer. 1993; 71: 1368-1383.
87. Wang Y. The role and regulation of urokinase-type plasminogen activator receptor gene expression in cancer invasion and metastasis. Med Res Rev. 2001; 21: 146-170.
88. Stetler SW, Aznavoorian S, Liotta LA. Tumor cell interactions with the extracellular matrix during invasion and metastasis. Annu Rev Cell Biol. 1993; 9: 541-573.
89. Westermarck J, Holmstrom T, Ahonen M, Eriksson JE, Kahari VM. Enhancement of fibroblast collagenase-1 (MMP-1) gene expression by tumor promoter okadaic acid is mediated by stress-activated protein kinases Jun N-terminal kinase and p38. Matrix Biol. 1998; 17: 547-557.
90. Lim M, Martinez T, Jablons D et al. Tumor-derived EMMPRIN (extracellular matrix metalloproteinase inducer) stimulates collagenase transcription through MAPK p38. Febs Lett. 1998; 441: 88-92.
91. Simon C, Goepfert H, Boyd D. Inhibition of the p38 mitogen-activated protein kinase by SB 203580 blocks PMA-induced Mr 92,000 type IV collagenase secretion and in vitro invasion. Cancer Res. 1998; 58: 1135-1139.
92. Janulis M, Silberman S, Ambegaokar A, Gutkind JS, Schultz RM. Role of mitogen-activated protein kinases and c-Jun/AP-1 trans-activating activity in the regulation

of protease mRNAs and the malignant phenotype in NIH 3T3 fibroblasts. J Biol Chem. 1999; 274: 801-813.
93. Teng DH, Perry WL, Hogan JK et al. Human mitogen-activated protein kinase kinase 4 as a candidate tumor suppressor. Cancer Res. 1997; 57: 4177-4182.
94. Yoshida BA, Dubauskas Z, Chekmareva MA, Christiano TR, Stadler WM, Rinker SC. Mitogen-activated protein kinase kinase 4/stress-activated protein/Erk kinase 1 (MKK4/SEK1), a prostate cancer metastasis suppressor gene encoded by human chromosome 17. Cancer Res. 1999; 59: 5483-5487.
95. Taguchi A, Blood DC, del TG et al. Blockade of RAGE-amphoterin signalling suppresses tumour growth and metastases. Nature. 2000; 405: 354-360.

Chapter 3

TUMOR METASTASIS-ASSOCIATED HUMAN MTA1 GENE: ROLE IN EPITHELIAL CANCER CELL PROLIFERATION AND REGULATION

Garth L. Nicolson,[1] Akihiro Nawa,[2] Yasushi Toh,[3] Shigeki Taniguchi,[4] Katsuhiko Nishimori[2]

*[1]The Institute for Molecular Medicine, Huntington Beach, California 92649 USA; [2]Department of Obstetrics and Gynecology, Nagoya University School of Medicine, Nagoya 466, Japan; [3]Department of Gastroenterorogic Surgery, National Kyushu Cancer Center, 811-1395 Japan; [4]Laboratory of Molecular Biology, Graduate School of Agricultural Science, Tohoku University, Sendai 981-8555, Japan; *Correspondence to: Prof. Garth L. Nicolson, The Institute for Molecular Medicine, 15162 Triton Lane, Huntington Beach, California 92649, USA; Tel: +1-714-903-2900; Fax +1-714-379-2082. E-mail: gnicolson@immed.org*

Key words: metastasis, nuclear regulatory protein; gene expression; antisense oligonucleotides; cell proliferation; nucleosome remodeling histone deacetylase complex; gene structure; breast cancer, gastrointestinal cancer

INTRODUCTION

Several genes have been identified as metastasis-associated genes (1, 2). It has been reported that at least the following genes are associated with progression or metastasis of carcinoma cells: *mst1, nm23, WDNM1, WDNM2, pGM21, stromelysin-3, KAI-1, BRMS1, KiSS1* and *MKK4* genes (3-10). Although for the most part direct evidence for the roles of these specific genes and their encoded products in particular steps of the metastatic process is not available, these genes are generally over-expressed or under-expressed in metastatic cells compared to their nonmetastatic counterparts.

Previously we cloned a novel candidate metastasis-associated gene, *mta1* (11, 12), which was isolated by differential cDNA library screening using the 13762NF rat mammary adenocarcinoma metastatic system (13). We found that *mta1* mRNA was differentially expressed in highly metastatic rat mammary

D.R. Welch (ed.), Cancer Metastasis – Related Genes, 51–63.

```
  1 MAANMYRVGDYVYFENSSSNPYLIRRIEELNKTANGNVEAKVVCFYRRRD  50  HuMTA1
    ||||||||||||||||||||||||||||||||||||||||||||||||||
  1 MAANMYRVGDYVYFENSSSNPYLIRRIEELNKTANGNVEAKVVCFYRRRD  50  Ratmta1

 51 ISSTLIALADKHATLSVCYKAGPGADNGEEGEIEEEMENPEMVDLPEKLK 100
    |||.|||||||||||||||:||||||.|||||:|||:|||||||||||||
 51 ISSSLIALADKHATLSVCYRAGPGADTGEEGEVEEEVENPEMVDLPEKLK 100

101 HQLRHRELFLSRQIESLPATHIRGKCSVTLLNETESLKSYLEREDFFFYS 150
    ||||||||||||||||||||||||||||||||||||||||||||||||||
101 HQLRHRELFLSRQIESLPATHIRGKCSVTLLNETESLKSYLEREDFFFYS 150

151 LVYDPQQKTLLADKGEIRVGNRYQADITDLLKEGEEDGRDQSRLETQVWE 200
    |||||||||||||||||||||||||||||||||:||||||||:|||.|||
151 LVYDPQQKTLLADKGEIRVGNRYQADITDLLKDGEEDGRDQSKLETKVWE 200

201 AENPLTDKQIDQFLVVARSVGTFARALDCSSSVRQPSLHMSAAAASRDIT 250
    |||||.||||||||||||||||||||||||||||||||||||||||||||
201 AENPLVDKQIDQFLVVARSVGTFARALDCSSSVRQPSLHMSAAAASRDIT 250

    *      *      *      *
251 LFHAMDTLHKNIYDISKAISALVPQGGPVLCRDEMEEWSASEANLFEEAL 300
    ||||||||||||||||||||||||||||||||||||||||||||||||||
251 LFHAMDTLHKNIYDISKAISALVPQGGPVLCRDEMEEWSASEANLFEEAL 300

301 EKYGKDFTDIQQDFLPWKSLTSIIEYYYMWKTTDRYVQQKRLKAAEAESK 350
    ||||||||||||||||||||||||||||||||||||||||||||||||||
301 EKYGKDFTDIQQDFLPWKSLTSIIEYYYMWKTTDRYVQQKRLKAAEAESK 350

                                               +  +
351 LEQVYIPNYNKPNPNQISVNNVKAGVVNGTGAPGQSPGAGRACESCYTTQ 400
    ||||||||||||||||||||.|||:||||||.||||||||||||||||||
351 LEQVYIPNYNKPNPNQISVNSVKASVVNGTGTPGQSPGAGRACESCYTTQ 400

                 +  +
401 SYQWYSNGPPNMQCRLCASCWTYWKKYGGLKMPTRLDGERPGPNRSNMSP 450
    ||||||||||||||||||||||||||||||||||||||||||||||.|||
401 SYQWYSNGPPNMQCRLCASCWTYWKKYGGLKMPTRLDGERPGPNRNNMSP 450

451 NGLPARSSGSPKFAMKTRQAFYLHTTKLTRIARRLCREILRPWHAARHPY 500
    ||:|||||||||||||||||||||||||||||||||||||||||||||||
451 NGIPARSSGSPKFAMKTRQAFYLHTTKLTRIARRLCREILRPWHAARHPY 500

501 LPINSAAIKAECTARLPEASQSPLVLKQAVRKPLEAVLRYLETHPRPPKP 550
    :||||||||||||||||||||||||||||.|||||||||||||||||||||
501 MPINSAAIKAECTARLPEASQSPLVLKQVVRKPLEAVLRYLETHPRPPKP 550

551 DPVKSVSSVLSSLTPAKVAPVINNGSPTILGKRSYEQHNGVDGNMKKRLL 600
    ||||| ||||||||||| ||||||||||||||||||||||||
551 DPVKSSSSVLSSLTPAKSAPVINNGSPTILGKRSYEQHNGVD........ 592

601 MPSRGLANHGQTRHMGPSRNLLLNGKSYPTKVRLIRGGSLPPVKRRRMNW 650
        ||||||||||||||||||||||||||||||||||||||||||||||
593 ....GLANHGQTRHMGPSRNLLLNGKSYPTKVRLIRGGSLPPVKRRRMNW 638

651 IDAPDDVFYMATEETRKIRKLLSSSETKRAARRPYKPIALRQSQALPPRP 700
    ||||||||||||||||||||||||||||||||||||||||||||||| ||
639 IDAPDDVFYMATEETRKIRKLLSSSETKRAARRPYKPIALRQSQALPLRP 688

701 PPPAPVNDEPIVIED* 716
    |||||||||||||||
689 PPPAPVNDEPIVIED* 704
```

Figure 1. Alignment of the predicted amino acid sequences of the human MTA1 and rat mta1 proteins. The identical amino acid residues (96%) between human MTA1 and rat mta1 proteins are indicated by (|), well-conserved replacements by (:) and less conserved by (.) (11, 12). The underlined polypeptide sequences 251-273 are characteristic of a lucine zipper motif. The underlined and italic polypeptide sequences 393-417 are characteristic of a GATA-type zinc finger motif. Five SPXX motifs are also present and conserved in both human MTA1 and rat mta1 proteins. Three and two putative nuclear localization sequences (shown in underlined and italic figures) are in human MTA1 and rat mta1, respectively. The C-terminal proline rich region found previously starts at amino acid residue 696 of the human MTA1 protein (from ref. 14, with permission).

adenocarcinoma cell lines (11-13); however, the function of the *mta1* gene product was unknown. We have now cloned the human homologue *MTA1* gene, characterized this gene and investigated the putative function of its encoded product (14, 15).

Recently, two groups reported that nucleosome remodeling histone deacetylase complex (NuRD complex), which is involved in chromatin remodeling, contains MTA1 protein or a MTA1-related protein (MTA2) (16, 17). Thus a possible function for the MTA1 protein has been reported; however, the exact role of the MTA1 protein in tumor progression and metastasis must still be determined. Here we will discuss the structure and possible function of the *MTA1* gene and its encoded product

SEQUENCE ANALYSIS OF MTA1 GENE

The nucleotide sequence of the human *MTA1* gene (accession number U35113) was 88% identical to the rat *mta1* sequence. The human *MTA1* gene encoded a putative protein of 715 amino acid residues with a predicted molecular weight of ~82 kDa. The amino acid sequences of the rat and human proteins were 96% identical and 98% similar (Figure 1) (11, 12). Similar to the rat Mta1 protein, the human MTA1 protein contained a proline-rich stretch (LPPRPPPPAP) at the carboxy-terminal end of the molecule at residues 696-705. This sequence completely matched the consensus sequence for the src homology 3 domain-binding site, XPXXPPPFXP (18) or XpFPpXP (19) (where X stands for nonconserved residues, P for proline, p for residues that tend to proline, and F for hydrophobic residues) (Figure 2) (15).

Recently our attention has turned toward possible DNA binding and nuclear transcription factors. In this analysis of the human MTA1 protein, we also found a putative zinc finger DNA binding motif Cys-X2-Cys-X17-Cys-X2-Cys (20) beginning at residues 393, and a leucine zipper motif (21) beginning at residue 251. These sequences were also conserved in the rat Mta1 protein (Figure 1). The human MTA1 protein was rich in SPXX motifs, and these are known to occur frequently in gene regulatory and DNA-binding proteins (22). The human

MTA1 *Figure 2. Comparison of the Src homology 3 domain-binding site in rat*

SH3-Binding Motifs in Some Common Proteins

Gene	*Sequence No.*	*SH3-Binding Motifs*											
Consensis 1		X	P	X	X	P	P	P	Φ	X	P		
Consensis 2		-	-	-	X	P	Φ	P	P	X	P		
Rat Mta1	684-695	L	P	L	R	P	P	P	P	A	P	V	N
Human MTA1	787-797	L	P	L	R	P	P	P	P	A	P	V	N
Murine 3BP1	267-277		P	T	M	P	P	P	L	P	P	V	P
Murine 3BP2	2-12		P	A	Y	P	P	P	P	V	P	V	P
Murine SOS1	1146-1156		E	V	P	V	P	P	P	V	P	P	P
Murine SOS1	1174-1184		H	L	D	S	P	P	A	I	P	P	R
Murine SOS1	1285-1295		H	S	I	A	G	P	P	V	P	P	R
Murine Formin	873-883		P	P	T	P	P	P	L	P	P	P	L
Rat m4 mAchR	277-286		P	A	L	P	P	P	V	R	P	V	P
Human PI3K p85	91-101		P	P	R	P	L	P	P	A	P	G	S
Human PI3K p85	303-313		P	A	P	A	L	P	V	K	P	P	K
Human GAP CDC42	250-260		A	P	K	P	M	P	P	R	P	P	L
Human HK2	93-103		G	V	R	P	L	P	P	L	P	D	P
Human proacrosin	333-343		P	P	R	P	L	P	P	R	P	P	A
Human dynamin	785-795		A	P	A	V	P	P	A	R	P	G	S
Human dynamin	811-821		G	A	P	P	V	P	S	R	P	G	A
v-Fgr	12-22		R	P	R	P	L	P	P	L	P	P	T

Mta1, human MTA1 and various other proteins (from ref. 15 with permission).

protein contained five SPXX sequences (Figure 1), corresponding to frequencies of 7.09 x 10^{-3} which is ~2.5 times the average protein frequency (2.89 x 10^{-3}). Furthermore, the MTA1 protein encoded three putative nuclear localization sequences (using the PSORT prediction software) (Figure 1) (11, 12, 14). We have also found a SANT domain in MTA1, and this type of domain was recently reported to be similar to the DNA-binding domain of myb-related proteins. A SANT domain has been identified in SWT3, a yeast component of the SWI/SNF complex (24), along with ADA2, a component of the histone deacetylase complex (25), N-CoR, a nuclear hormone co-repressor (26), and TFIIIB subunit B, a basal pol III transcription factor in yeast (27). The SANT domain has also been referred to as the MFY domain since it has many aromatic amino acid residues (16).

Finally, there are two highly acidic regions located in the 200 MTA1 amino-terminal residues. These highly negatively charged regions are characteristic of the acidic activation domains of many transcription factors (28).

To assess the extent of evolutionary conservation of the *MTA1* gene we analyzed genomic DNA of several species by Southern blot analysis. Strong genomic signals were detected in monkey and yeast, moderate signals in human, rat, mouse, dog, cow and rabbit, and weak signals were detected in chicken (11). Thus the *MTA1* gene was conserved in all species examined.

EXPRESSION OF MTA1 GENE IN HUMAN CELL LINES

Previously we found that the rat *mta1* gene was over-expressed in highly metastatic rat mammary carcinoma cells compared to poorly metastatic or nonmetastatic rat mammary cells (11-13). To determine the expression of the human *MTA1* gene in non-tumorigenic and tumorigenic cells, we examined 14 cell lines of human origin. Transcripts for the *MTA1* gene were found in virtually all cell lines analyzed (15). Interestingly, human breast cancer MDA-MB-231 cells of high metastatic potential strongly expressed the *MTA1* gene, whereas MDA-MB-435 cells of poor metastatic potential (29) expressed the *MTA1* gene at very low levels (11, 12). The expression level of the *MTA1* gene in a normal breast epithelial cell line (Hs578Bst) with slow growth rate was from one-third to one-half that seen in breast adenocarcinoma cells and atypical breast cells (HBL-100) with a rapid growth rate (15). The relative expression (normalized with respect to GADPH expression) of the *MTA1* gene in various human cell lines from highest to lowest was as follows: MDA-MB-231, HeLa > SKOV-3, ZR-75-1, HBL-100, A2058 > OVCA-433, OVCA-432, Ovcar-3, HT-29, KM-12C, Hs578Bst > MBA-MD-435, OVCA-429 (15). Thus the *MTA1* gene was expressed at various levels among different cell lines. Although the expression of the *MTA1* gene in animal and human cell lines generally followed metastatic potential, there were some exceptions to this that may reflect differences in the cells or alterations during cell culture (15).

Expression of MTA1 Gene in Human Tumors

The expression of the *MTA1* gene was examined in tumor biopsy specimens obtained from various epithelial cancers. We initially focused our attention on breast cancers because preliminary results suggested over-expression of the *MTA1* gene in malignant breast carcinomas compared to surrounding normal tissues. The majority of 20 invasive breast carcinomas over-expressed *MTA1* gene (tumor/normal ratio >2) compared to surrounding normal tissue. Similarly, *MTA1* gene was over-expressed in malignant gastric and esophageal carcinomas (30, 31). In 14/36 colorectal carcinomas and 13/34 gastric carcinomas the *MTA1* gene was over-expressed (tumor/normal ratio >2). Tumors that over-expressed *MTA1* RNA showed significantly higher rates of invasion and lymph node metastasis and tended to have higher rates of vascular involvement.

Localization of MTA1 Protein in Breast Cancer Cells

Using indirect immunofluorescence we found that the MTA1 protein accumulated in the nucleus of breast cancer cells (14). This nuclear immunoreactivity was present in many large, intense foci that were not detected near the nuclear membrane. However, using the anti-MTA1 reagent the nucleolus region was negative for fluorescence (14).

ANTISENSE MTA1 GENE AND CELL GROWTH

To directly demonstrate a role for the *MTA1* gene in breast cancer cells we used the technique of antisense inhibition of *MTA1* gene expression. Although this procedure is not without its limitations (32), we employed phosphorothioate oligonucleotides (PONs) as antisense oligodeoxynucleotides with prolonged lifetime (33, 34). Human breast cancer MDA-MB-231 and MDA-MB-435 cells were treated with PONs for 4 hours, after which their proliferation was monitored for several days (35). Antisense PONs markedly inhibited the cell growth of MDA-MB-231 cells to 22% of mock-treated cells and to 28% of sense PONs treated cells at 72 hours, respectively (Figure 3A). In contrast, antisense PONs did not affect the growth of the nonmetastatic MDA-MB-435 cells (Figure 3B).

Effect of MTA1 Antisense Oligonucleotides on MTA1 Protein Levels

To confirm that the inhibition of cell proliferation by antisense PONs was involved in the suppression of the MTA1 protein, the MTA1 protein levels of MDA-MB-231 cells treated with antisense PONs and sense control were examined by Western blot analysis using anti-MTA1 protein polyclonal antibodies. The anti-MTA1 protein recognized a 83 kDa protein band that was identified as identical in migration to the MTA1 protein. This band disappeared after preincubation of anti-MTA1 protein with the immunogen oligopeptides (14, 15). To verify that the antisense effects were due to inhibition of target gene expression we quantitated MTA1 protein expression in MDA-MB-231 cells treated with antisense PONs or sense control PONs. Antisense PONs inhibition of the *MTA1* gene resulted in an approximately 70% reduction of MTA1 protein levels as detected by Western blot analysis in the antisense treated cells after two days, whereas in cells treated with sense control PONs there was no effect on the MTA1 protein levels (Figure 4).

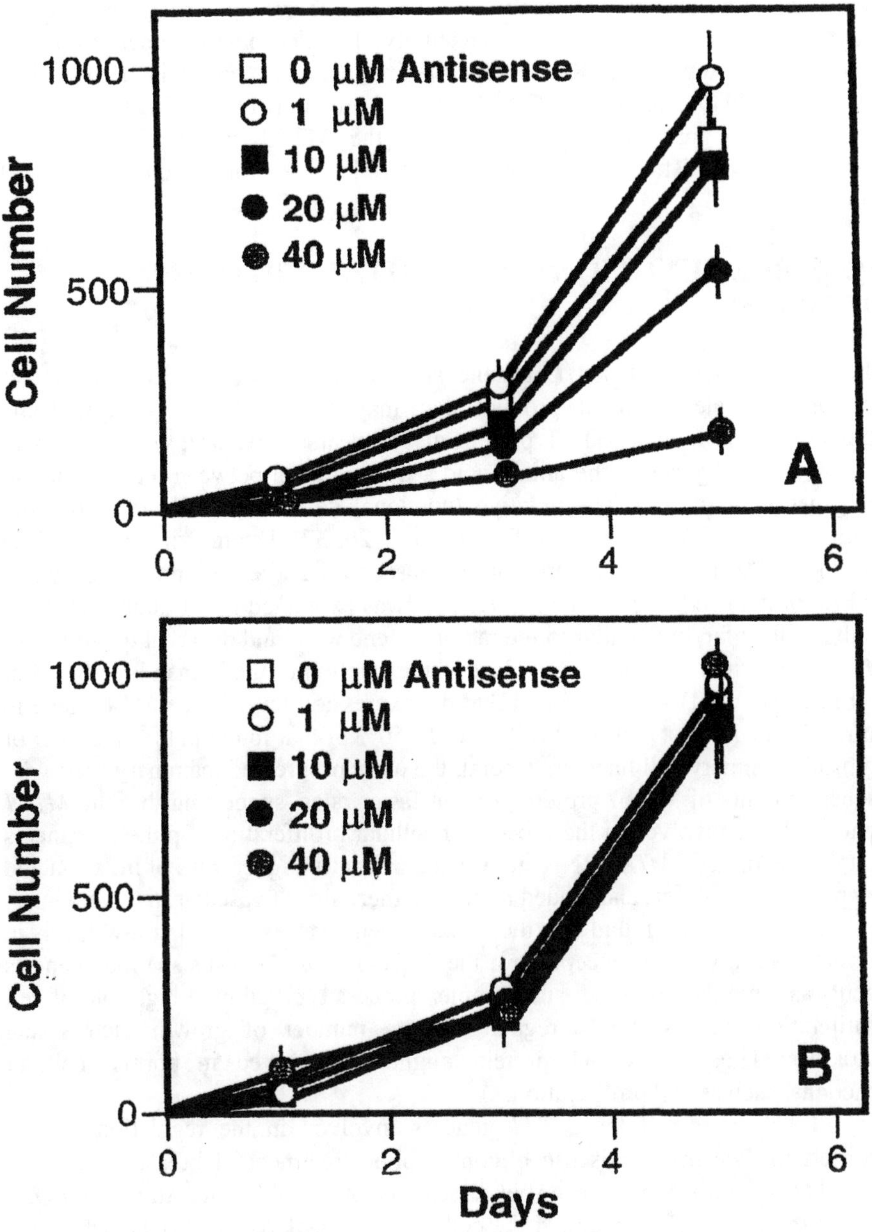

Figure 3. Effect of MTA1 antisense PONs on the growth of MDA-MB-231 (A) or MDA-MB-435 (B) cells. Time-course of cell growth inhibition using MTA1 antisense PONs were evaluated as described in Materials and Methods. Only the MDA-MB-231 cells were growth inhibited by the MTA1 antisense PONs (from ref. 15, with permission).

Since the MTA1 protein was localized in the nucleus, we sought to determine if it interacted with nuclear proteins. Interestingly, a nucleosome remodeling histone deacetylase complex (NuRD complex) involved in chromatin remodeling, contains the MTA1 protein or a MTA1-related protein (MTA2) (16, 17). Using a double-labeling procedure we found that the MTA1 protein is physically associated with histone deacetylase 1 (HDAC1) in a protein complex (NuRD complex) (36).

POSSIBLE FUNCTION OF THE MTA1 PROTEIN IN METASTASIS

The metastasis-associated *MTA1* gene is a novel, highly conserved gene that encodes a nuclear protein product that may be involved in chromosome alterations. The human MTA1 protein also appears to be well conserved with only a 4% divergence at the amino acid sequence level between the human and rat genes (11, 12, 15). The putative functional domains like the SH3-binding motif (18, 19), GATA-type zinc finger motif (20, 37), leucine zipper motif (21) and the SPXX motifs (22) were highly conserved between the predicted human and rat protein sequences. The *MTA1* gene was expressed in all tumor cell lines analyzed thus far, but similar to the rat *mta1* gene we found different quantities of *MTA1* transcripts in various cells. With the exception of the human breast cancer cell line MDA-MB-435, we found that the expression level of the *MTA1* gene in untransformed breast epithelial cells was 28-50% of that found in breast cancer or atypical mammary cell lines. In general, the more progressed mammary cells with higher amounts of MTA1 protein grow at faster rates, suggesting that the *MTA1* gene might be involved in the process of cellular proliferation. Epithelial cancers that over-expressed *MTA1* RNA showed significantly higher rates of invasion and lymph node metastasis and tended to have higher rates of vascular involvement.

We previously found that the rat *mta1* gene was expressed at low levels in normal tissues, with the exception of the testis (11). In the testis spermatogenesis occurs as a highly controlled and complex process typified by a high rate of cell proliferation that is tightly regulated by a number of growth factors and cytokines. Thus the MTA1 protein might be involved in normal cellular functions, such as cell proliferation.

To determine if the *MTA1* gene is involved in the regulation of cell proliferation we used antisense oligonucleotide treatment of breast cancer cell lines that show different levels of expression of the *MTA1* gene. Antisense PONs against the *MTA1* gene inhibited the cell growth of MDA-MB-231 breast cancer cells. Specific inhibition of gene expression by the use of antisense PONs has been extensively (37-39) but these procedures are not without their problems (32-34). We found that after transfection of antisense PONs MTA1 protein decreased to 20-30% of that found in sense-treated cells within 2 days, but we failed to find significant changes in the amounts of MTA1 protein with the sense sequence PONs, indicating that the growth inhibition by antisense PONs was a sequence-

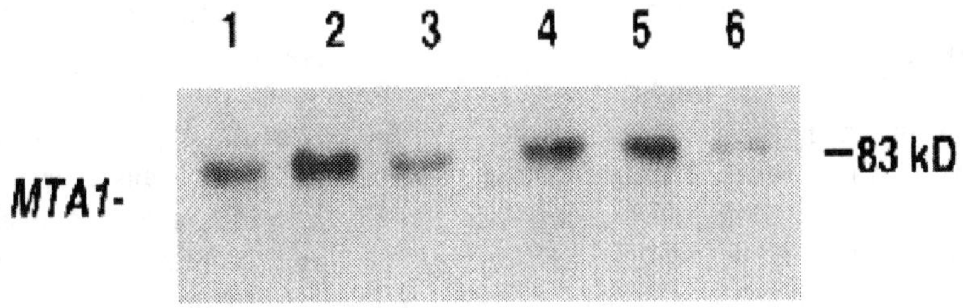

Figure 4. Effect of MTA1 antisense PONs on the expression of the MTA1 gene in MDA-MB-231 cells. Cells were cultured for 24 (lanes 1-3) or 48 (lanes 4-6) hours after transfected MTA1 antisense PONs as described in Materials and Methods. The cells were harvested, lysed, proteins separated by SDS-PAGE, and the gel was immunoblotted with anti-human MTA1 antibody as described in Materials and Methods. Ten μg of cell protein extract was loaded for each sample. Lanes 1 and 4, no PONs; Lanes 2 and 5, MTA1 sense PON; Lanes 3 and 6, MTA1 antisense PON. In this figure, the level of MTA1 protein in cells treated with MTA1 antisense decreased to 30% of the level without PON or using MTA1 sense PON at 48 hours (from ref. 15 with permission).

specific effect. Also, we could not demonstrate an effect of antisense PONs on the growth inhibition of cells with a low level of MTA1 protein expression.

The MTA1 protein may be used in a complex to remodel chromosomes. When the *MTA1* gene was transfected and expressed in 293Tcells, the MTA1 protein localized within the nuclear matrix. A nucleosome remodeling histone deacetylase complex (NuRD complex) involved in chromatin remodeling contains the MTA1 protein or a MTA1-related protein (MTA2) (16, 17). Trichostatin A, a potent specific inhibitor of histone deacetylase (HDAC), causes G1/G2 arrest in fibroblasts (40). Moreover, HDAC1 which is a component of the NuRD complex has been shown to interact with Rb to repress transcription (41-43). Acetylation of the C-terminal of p53 modified its ability to bind to DNA (44); therefore, the MTA1 protein might interact with the histone deacetylase and act as a co-activator of this complex. In support of this MTA1 has a unique protein primary structure that suggests that it might function in signal transduction and DNA-binding. The MTA1 protein is the first mammalian protein found that contains the motif Cys-X2-Cys-X17-Cys-X2-Cys, which is a zinc-finger domain that also appears in GATA transcription factors. The same zinc-finger domain configuration has been found in *GLN3*, *areA* and *nit-2*, major regulatory factors for nitrogen metabolism in *Saccharomyces cerevisiae, Aspergillus* and *Neurospora,* respectively (37, 45, 46). The *nit-2* protein recognizes an identical core sequence of TATCTA, and a recent study has also shown that the GLN3 protein binds the nitrogen upstream activation sequence of

GLN1, the gene that encodes glutamine synthetase (46). Thus it is plausible that the MTA1 protein binds to a specific sequence of DNA and regulates gene expression.

The MTA1 protein might interact with specific genes involved in cellular regulation. In preliminary experiments we have found that the expression of the *MTA1* gene was increased four-times in *c-erbB2/neu* stable transfectants of MDA-MB-435 cells compared to untransfected cells. The MTA1 protein is also associated with histone deacetylase 1 (HDAC1) in a NuRD protein complex (30). Determination of the role of the MTA1 protein in nuclear protein complexes will be necessary to confirm our notion that the *MTA1* gene is involved in gene and cell growth regulation and the progression of epithelial cancers.

SUMMARY

Using homology to the rat *mta1* gene we cloned the human *MTA1* gene. We found a close similarity between the human *MTA1* and rat *mta1* genes and their predicted amino acid sequences. Both genes encode novel proteins that contain a proline rich region (SH3 binding motif), a putative zinc finger motif, a leucine zipper motif and 5 copies of the SPXX motif often found in gene regulatory proteins. Using Southern blot analysis the *MTA1* gene was found to be highly conserved among all species examined, and using Northern blot analysis *MTA1* transcripts were found in virtually all cell lines of human origin that were analyzed, including melanoma and breast, cervix and ovarian carcinoma cells and normal breast epithelial cells. However, the expression level of the *MTA1* gene in rapidly growing human cell lines and normal breast epithelial cells was approximately 50% of that found in metastatic cell lines and invasive and metastatic tumors. Tumors that over-expressed *MTA1* RNA showed significantly higher rates of invasion and lymph node metastasis and tended to have higher rates of vascular involvement.

Experimental inhibition of MTA1 protein expression using antisense phosphorothioate oligonucleotides resulted in growth inhibition of human MDA-MB-231 breast cancer cells with relatively high expression of the *MTA1* gene. Furthermore, the MTA1 protein was localized in the nuclei of cells transfected using a mammalian expression vector containing the full-length *MTA1* gene. In vitro studies indicate that the MTA1 protein binds to a nucleosome remodeling histone deacetylase (NuRD) complex. The results suggest that the MTA1 protein may function as a nuclear regulatory factor in cellular signaling and chromosome alterations that are important in the progression and growth of cancer cells.

ACKNOWLEDGMENTS

The authors thank Drs. K. Yokoyama and K. Uchida for helpful suggestions on designing the antisense oligonucleotides. We also thank Dr. R. C. Bast, Jr. (Division of Medicine, The University of Texas M. D. Anderson Cancer Center) for providing us with ovarian cancer cell lines. We acknowledge the excellent technical assistance of Science Tanaka Co., LTD. This work was supported by NCI Grant R01-CA63045 (to G.L.N).

REFERENCES

1. Moustafa A., Nicolson G.L. Breast cancer metastasis-associated genes: prognostic significance and therapeutic implications. Oncol Res 1998; 9: 505-525.
2. Welch D.R., Steeg P.S., Rinker-Schaeffer C.W. Genetic regulation of human breast carcinoma metastasis. Biology of Breast Cancer Metastasis 2000; 2: 408-416.
3. Steeg P.S., Bevilacqua G., Kopper L., Thorgeirsson U.P., Talmadge J.E., Liotta L.A., Sobel M.E. Evidence for a novel gene associated with low tumor metastatic potential. J Natl Cancer Inst 1998; 80: 200-204.
4. Ebralidze A., Tulchinsky E., Grigorian M., Afanasyeva A., Senin V., Revazova E., Lukanidin E. Isolation and characterization of a gene specially expressed in different metastatic cells and whose deduced gene product has a high degree of homology to Ca^{2+} binding protein. Genes Dev 1989; 3: 1086-1093.
5. Dear T.N., Ramshaw I.A., Kefford R.F. Differential expression of a novel gene, *WDNM1*, in nonmetastatic rat mammary adenocarcinoma cells. Cancer Res 1988; 48: 5203-5209.
6. Dear T.N., McDonald D.A., Kefford R.F. Transcriptional down-regulation of a rat gene, *WDNM2*, in metastatic DMBA-8 cells. Cancer Res 1989; 49: 5323-5328.
7. Phillips S.M., Bendall A.J., Ramshaw I.A. Isolation of gene associated with high metastatic potential in rat mammary adenocarcinoma. J Natl Cancer Inst 1990; 82: 199-203.
8. Bisset P., Bellocq J.P., Wolf C., Stoll I., Hutin P., Limacher J.M., Podlhajcer O.L., Chenerd M.P., Rio M.C., Chambon P. A novel metalloproteinase gene specifically expressed in stromal cells of breast carcinomas. Nature 1990; 348: 699-704.
9. Dong J.T., Lamb P.W., Rinker-Schaeffer C.W., Vukanovic J., Ichikawa T., Isaacs J.T., Barrett J.C. KAI1, a metastasis suppressor gene for prostate cancer of human chromosome 11p11.2. Science 1995; 268: 884-886.
10. Seraj M.J., Samant R.S., Verderame M.F., Welch D.R. Functional evidence for a novel human breast carcinoma metastasis suppressor, BrMS1, encoded at chromosome 11q13. Cancer Res 2000; 60: 2764-2769.
11. Toh Y., Pencil S.D., Nicolson G.L. A novel candidate metastasis-associated gene, *mta1*, differentially expressed in highly metastatic mammary adenocarcinoma cell lines. J Biol Chem 1994; 269: 22958-22963.
12. Toh Y., Pencil S.D., Nicolson G.L. Analysis of the complete sequence of the novel gene *mta1* differentially expressed in highly metastatic mammary adenocarcinoma and breast cancer cell lines and clones. Gene 1995; 159: 99-104.
13. Pencil S.D., Toh Y., Nicolson G.L. Candidate metastasis-associated genes of the rat 13762NF mammary adenocarcinoma. Breast Cancer Res Treat 1993; 25: 165-174.
14. Nawa A., Nishimori K., Lin P., Maki Y., Moue K., Sawada H., Toh Y., Funitaka K., Nicolson G.L. Tumor metastasis-associated human *MTA1* gene: its deduced protein sequence, localization and association with breast cancer cell proliferation using antisense phosphorothioate oligonucleotides. J Cell Biochem 2000; 79: 202-212.

15. Nawa A., Sawada H., Toh Y., Nishimori K., Nicolson G.L. Tumor metastasis-associated human *MTA1* gene: effects of antisense oligonucleotides on cell growth. Intern J Med Biol Environ 2000; 28(1): 33-39.
16. Xue Y., Wong J., Moreno G.T., Young M.K., Cote J., Wang W. NURD, a novel complex with both ATP-dependent chromatin-remodeling and histone deacetylase activities. Mol Cell 1989; 2: 851-861.
17. Zhang Y., LeRoy G., Seelig H.P., Lane W.S., Reinberg D. The dermatomyositis-specific autoantigen Mi2 is a component of a complex containing histone deacetylase and nucleosome remodeling activities. Cell 1998; 95: 279-289.
18. Ren R., Mager B.J., Cicchetti P., Baltimore D. Identification of a ten-amino acid proline-rich SH3 binding site. Science 1993; 259: 1157-1161.
19. Yu H., Chen J.K., Feng S., Dalgarno D.C., Brauer A.W., Schreiber S.L. Structual basis for the binding of proline-rich peptides to SH3 domains. Cell 1994; 76: 933-945.
20. Martin D.I.K., Orkin S.H. Transcriptional activation and DNA binding by the erythroid factor GF-1/NF-E1/Eryf. Genes Dev 1990; 4: 1886-1898.
21. Vinson C.R., Sigler P.B., McKnight S.L. Scissors-Grip model for DNA recognition by a family of leucine zipper proteins. Science 1989; 246: 911-916.
22. Suzuki M. SPXX, a frequent sequence motif in gene regulatory proteins. J Mol Biol 1989; 207: 61-84.
23. Aasland, R. The SANT domain: putative DNA-binding domain in the SWI-SNF and ADA complexes, the transcriptional co-repressor N-CoR and TFIIIB. Trends Biochem Sci 1996; 21: 87-88.
24. Peterson C. L., Tamkun J. W. The SWI-SNF complex: a chromatin remodeling machine? Trends Biochem Sci 1995; 20: 143-146.
25. Horiuchi J., Silverman N., Marcus G. A., Guarente L. ADA3, a putative transcriptional adaptor, consists of two separable domains and interacts with ADA2 and GCN5 in a trimeric complex. Mol Cell Biol 1995; 15: 1203-1209.
26. Horlein A. J., Naar A. M., Heinzel T. et al. Ligand-independent repression by the thyroid hormone receptor mediated by a nuclearreceptor co-repressor. Nature 1995; 377: 397-404.
27. Kassavetis G. A., Nguyen S. T., Kobayashi R., Kumar A., Geiduschek E. P., Pisano M. Cloning, expression, and function of TFC5, the gene encoding the B" component of the Saccharomyces cervisiae RNA polymerase III transcription factor TFIIIB. Proc Natl Acad Sci USA 1995; 92: 9786-9790.
28. Ptashne, M. How eukaryotic transcriptional activators work. Nature 1988; 335: 683-689.
29. Price J.E., Polyzos A., Zhang R.D., Daniels L.M. Tumorigenicity and metastasis of human breast carcinoma cell lines in nude mice. Cancer Res 1990; 50: 717-721.
30. Toh Y., Oki E., Oda S., Tokunaga E., Ohno S., Maehara Y., Nicolson G.L., Sugimachi K. Overexpression of *MTA1* gene in colorectal and gastrointestinal carcinomas: correlation with invasion and metastasis. Intern J Cancer 1997; 74: 459-463.
31. Toh Y., Kuwano H., Mori M., Nicolson G.L., Sugimachi K. Overexpression of metastasis-associated *MTA1* mRNA in invasive oesophageal carcinomas. Brit J Cancer 1999; 79: 1723-172626.
32. Stein C.A. Does antisense exist? Nature Med 1995; 1: 1119-1121.

33. Bennett C.F., Chiang M.Y., Chan H., Shoemaker J.E.E., Mirabelli C.K. Cationic lipids enhance cellular uptake and activity of phosphorothioate antisense oligonucleotides. Mol Pharmacol 1992; 41: 1023-1033.
34. Yaswen P., Stampfer M.R., Ghosh K., Cohen J.S. Effect of sequence of thioated oligonucleotides on cultured human mammary epithelial cells. Antisense Res Dev 1993; 3: 67-77.
35. Cavanaugh P.G., Nicolson G.L. Purification and some properties of a lung-derived growth factor that differentially stimulates the growth of tumor cells metastatic to the lung. Cancer Res 1989; 49: 3928-3933.
36. Toh Y., Kuninaka S., Endo H., Ohshiro T., Ikeda Y., Nakashima H., Baba H., Kohnoe S., Okamura T., Nicolson G.L., Sugimachi K. Molecular analysis of a candidate metastasis-associated gene *MTA1*: Interaction with histone deacetylase 1. J Exp Clin Cancer Res 2000; 19: 105-111.
37. Kudla B., Caddick M.X., Langdon T., Martinez-Rossi N.M., Bennet C.F., Sibley S., Davies R.W., Arst Jr. H.N. The regulatory gene *areA* mediating nitrogen metabolite repression in *Aspergillus nidulans*. Mutations affecting specificity of gene activation alter a loop residue of a putative zinc finger. EMBO J 1990; 9: 1355-1364.
38. Takeuchi K., Sato N., Kasahara H., Funayama N., Nagafuchi A., Yonemura S., Tukita S., Tukita S. Perturbation of Cell Adhesion and Microvilli Formation by Antisense Oligonucleotides to ERM Family Members. J Cell Biol 1994; 125: 1371-1384.
39. Thompson M.E., Jensen R.A., Obermiller P.S., Page D.L., Holt J.T. Decreased expression of BRCA1 accerates growth and is often present during sporadic breast cancer progression. Nature Genet 1994; 9: 444-450.
40. Yoshida M., Beppu T. Reversible arrest of proliferation of rat 3Y1 fibroblasts in both the G1 and G2 phases by trichostatin A. Exp Cell Res 1998; 177: 122-131.
41. Brehm A., Miska E.A., MaCane D.J., Reid J.L., Bannister A.J., Kouzarides T. Retinoblastoma protein recruits histone deacetylase to repress transcription. Nature 1998; 391: 597-601.
42. Luo R.X., Postigo A.A., Dean D.C. Rb interacts with histone deacetylase to repress transcription. Cell 1998; 92: 463-473.
43. Magnaghi-Jaulin L., Groisman R., Naguibneva I., Robin P., Lorain S., Le Villain J.P., Troalen F., Trouche D., Harel-Bellan A. Retinoblastoma protein represses Transcription by recruiting a histone deacetylase. Nature 1998; 391: 601-605.
44. Gu W., Roeder R.G. Activation of p53 sequence-specific DNA binding by acetylation of the p53 C-terminal domain. Cell 1997; 90: 595-606.
45. Fu Y.H., Marzluf G.A. *nit-2*, the major positive-acting nitrogen regulatory gene of Neurospora crassa, encodes a sequence-specific DNA-binding protein. Proc Natl Acad Sci USA 1990; 87: 5331-5335.
46. Minehart P.L., Magasanik B. Sequence and expression of GLN3, a positive nitrogen regulatory gene of Saccharomyces cerevisiae encoding a protein with a putative zinc finger DNA-binding domain. Mol Cell Biol 1991; 11: 62 16-6228.

Chapter 4

COOPERATIVE INTEGRIN INTERACTIONS IN THE REGULATION OF TUMOR METASTASIS

Peter C. Brooks

New York University School of Medicine, Kaplan Cancer Center, Rusk Building RM. 806. Departments of Radiation Oncology and Cell Biology 400 East 34th Street, New York, NY 10016, Tel: (212) 263-3021, Fax: (212) 263-3018, Email: peter.brooks@med.nyu.edu

1. INTRODUCTION

The spread of tumor cells to distant sites is one of the leading causes of death associated with human cancer. Thus, an in-depth understanding of the mechanisms by which regulatory molecules facilitate tumor dissemination is of paramount importance. Over the last 25 years an explosion of information has occurred concerning the identification of molecular regulators of metastasis. Many of the molecules thought to contribute to tumor cell metastasis can be loosely grouped into 5 distinct categories including, growth factors and their receptors, transcription factors, enzymes, cell adhesion molecules and extracellular matrix components (Table 1). While these families of molecules play critical roles in normal physiological processes, aberrant expression and altered function likely contributes to invasive cellular behavior. While collectively, these families of molecules may contribute to tumor progression, it is important to point out that none of these families function in isolation, but rather work cooperatively to facilitate tumor cell invasion and metastasis. Thus, studying tumor metastasis within the context of molecular cooperativity may provide unique insight into novel approaches for the treatment of malignant tumors.

Tumor cell metastasis is not a single static event, but is rather a continuum of interconnected cellular processes that are linked both temporally and spatially (1-3). In order for tumor cells to establish viable secondary foci, they must successfully complete all the steps within the metastatic cascade (Figure 1). In general, the metastatic cascade is initiated by accumulation of genetic alterations, which result in the modifications of cell-cell adhesion, allowing dissociation of tumor cells from the primary tumor mass. Next, alterations in the ratio of proteases to protease inhibitors occur, resulting in interstitial matrix remodeling and local tumor invasion. During hematogenous metastasis, the tumor cells gain access to the host circulation by a complex process of intravasation which

D.R. Welch (ed.), Cancer Metastasis – Related Genes, 65–87.

involves penetration of the vasculature. Once in the circulation, the tumor cells must survive the physical shear forces and evade the host immune surveillance. Ultimately tumor cells arrest at distinct microvascular beds by a combination of mechanical trapping and specific adhesive interactions. Following arrest, the tumor cells undergo the process of extravasation leaving the microvasculature and invade the local interstitial matrix. Finally, in order for the metastatic deposits to survive, grow and expand beyond a minimal size, angiogenesis is induced to provide new blood vessels for the exchange of nutrients and removal of wastes.

Given the breadth and diversity of the potential molecular regulators that contribute to the metastatic cascade, we will limit our discussion to that of cell adhesion molecules and their associated regulatory proteins. Thus, the primary focus of this chapter will be to examine tumor cell metastasis in terms of the emerging concept of molecular cooperativity involving adhesion receptors and associated molecules.

2. CELL ADHESION MOLECULES AND METASTASIS

In order for tumor cells to invade and metastasize they must have the capacity to sense and in turn respond to both their cellular and acellular microenvironments. A major group of molecules that facilitate communication between tumor cells and the extracellular environment are cell adhesion molecules (4, 5). Cell adhesion receptors play critical roles either directly or indirectly in many if not all the steps of metastasis. A variety of cell adhesion molecules have been described with an array of structural features and distinct functions. In general, cell adhesion molecules can be organized into at least 4 broad categories including Selectins, Cadherins, Ig Super Family, and Integrins (4-7). Historically, cell adhesion molecules were thought to function primarily as "molecular glue" mediating interactions between cells and the extracellular environment. However, with the emergence of more sophisticated molecular techniques, this limited view of cell adhesion molecules has disappeared. In fact, many families of cell adhesion molecules such as the integrins are now widely regarded as multi-functional proteins capable of regulating a variety of cellular processes such as proliferation, cell cycle and signal transduction. While the molecular mechanisms by which these adhesive events regulate tumor progression are not completely understood, a number of potential mechanisms have been proposed. While a brief introduction and summary will be provided for all 4 classes of cell adhesion molecules, we have chosen to focus our discussion to that of the integrins and their associated regulatory molecules in tumor metastasis.

3. SELECTINS

Multiple changes in the expression of cell surface molecules have been observed on metastatic tumor cells. Interestingly, studies have suggested that changes in lectin and carbohydrate expression may correlate with malignant transformation (8-10). For example, investigators have identified a variety of tumor associated carbohydrate antigens including molecules that contain the tetrasaccharide Sialyl Lewis X (8-10). Many epithelial tumors as well as some melanomas express elevated levels of sialyl lewis X and exhibit enhanced tumor growth and metastasis *in vivo* (8-12). Moreover, patients whose lung and colon carcinomas expressed high levels of sialyl lewis X tended to have poor clinical prognosis and enhanced metastasis (8-12). Taken together these studies suggest that certain carbohydrate antigens such as Sialyl Lewis X may play a role in tumor metastasis.

To this end, Selectins are a family of calcium dependent transmembrane glycoproteins that can specifically bind to many carbohydrate moieties including sialyl lewis X (13). Selectins have a conserved structural motif, which includes an N-terminal calcium dependent lectin-like domain, an EGF-like domain, multiple complement regulatory-like regions, a transmembrane domain and a short cytoplasmic tail (13). The most well characterized members of this family include E, P and L selectins (13). Selectins mediate dynamic but weak cell-cell interactions and leukocyte rolling (14). Given these physiological functions, in conjunction with the fact that many metastatic tumor cells expressed sialyl lewis X antigen, it is possible selectins may play a role in mediating tumor cell interactions with the microvasculature and thereby contribute to the metastatic cascade. In fact, specific function blocking monoclonal antibodies (Mabs) directed to selectins as well as specific peptide mimics block tumor cell metastasis *in vivo* (10, 15). In addition, selectins may play a role in signal transduction (16). In fact, leukocyte adhesion to cytokine activated endothelium resulted in fluxes in intracellular calcium and activation of MAP kinase pathways (16). Activation of MAP kinase signaling cascades have been shown to regulate cellular proliferation, migration and survival. Importantly, all these cellular processes play critical roles in tumor metastasis.

4. CADHERINS

A second family of cell adhesion molecules that mediate cell-cell interactions are cadherins (17, 18). This large family of adhesion receptors shares a similar domain type- structure (17, 18). Recent studies have revealed that cadherins contain a large extracellular region that is composed of at least 5 ectodomains (17, 18). In addition, these receptors have a transmembrane domain and a cytoplasmic tail, which is known to bind to a family of proteins, termed catenins (17, 18). The functional activity of cadherins is in part dependent on association with α, β and γ catenins (17, 18). Catenins can mediate protein connections to the actin cytoskeleton and facilitate signal transduction. Moreover, recent crystallographic data has suggested that higher order structural dimers of cadherins may form within the same cell to provide functional stability for mediating cell-cell interactions (17, 18). The functional cadherin-catenin complexes have been strongly implicated in tumor invasion and metastasis. For example, immunohistochemical analysis of epithelial carcinomas suggested an inverse correlation between the expression of E-cadherins and tumor invasiveness and metastatic capacity (19-21). Following these early observations, many experimental findings began to support a link between the functional loss of E-cadherin and the gain of metastatic phenotype. In fact, treatment of MDCK cells with an anti-E-cadherin antibody induced an invasive and migratory response (22, 23). Moreover, transfection of E-cadherin negative cells with E-cadherin cDNA resulted in inhibition of motility and invasion (22, 23).

Interestingly, while the loss of E-cadherin is thought to play a role in the transition from a benign to a more invasive and metastatic phenotype, the expression of another cadherin appears to have the opposite effect (24). In studies by Hazan and others, N-cadherin expression appeared upregulated in invasive breast carcinoma cell lines that lacked E-cadherin (24). Moreover, transfection of N-cadherin in breast tumor cell lines that expressed E-cadherin, resulted in enhanced invasion and metastasis *in vivo* (24). The enhanced capacity to metastasize was thought to be due in part to the upregulation of the matrix metalloproteinase MMP-9 following FGF stimulation (24). Thus, expression of N-cadherin may modulate the susceptibility of tumor cells to growth factor induced MMP production. Taken together, cadherins may regulate tumor cell metastasis by a number of mechanisms including altering tight cell-cell interactions and sensitizing tumor cells to growth factor induced upregulation of proteolytic enzymes.

5. Ig SUPER FAMILY

In addition to selectins and cadherins, a third family of adhesion molecules that mediate cell-cell interactions is the Immunoglobulin Super Family (IgSF) (4, 25). Members of this family have an extracellular domain composed on a number of immunoglobulin homology units stabilized by disulfide bonds (4, 25). Some members of this family such as N-CAM have a transmembrane domain and a cytoplasmic tail, while others are anchored to the cell surface through lipid interactions. Some of the well-characterized members include, ICAM-1, VCAM, NCAM, CEA, DCC, and MUC18 (4-6, 25, 26). Many of these molecules have been shown to function in immune cell recognition, leukocyte trafficking and neural development (12, 25, 26). However, as with many of the other families of cell adhesion molecules, some members of the IgSF may play functional roles in tumor invasion and metastasis.

A variety of studies have implicated members of the IgSF in the regulation of tumor metastasis. For example, immunohistochemical studies have indicated that ICAM-1 expression levels are elevated in metastatic hepatocarcinomas as compared to more benign tumors (27). Moreover, other studies have also indicated a positive correlation between expression of CEA in colon carcinoma progression and up regulation of MUC18 in malignant melanomas (4, 27-29). Interestingly, loss of DCC expression was correlated with the development of colorectal carcinomas (30). Other evidence to support a role for IgSF adhesion molecules in tumor invasion and metastasis comes from antibody and peptide inhibition experiments where specific antagonists of IgSF molecules inhibited tumor invasion and metastasis *in vivo* (31). Finally, recent studies from ICAM-1 knockout mice have demonstrated a resistance to T-cell lymphoma metastasis (32). Taken together, these findings strongly suggest that IgSF family members play an important role in the regulation of tumor invasion and metastasis.

6. INTEGRINS

Integrins are a large family of heterodimers that are composed of noncovalently associated α and β chains. Currently, there are at least 14 α chain and 9 β chains that can combine to provide a wide variety of integrin dimers, each with their own unique cellular and adhesive specificity (33, 34). Integrins are transmembrane cell surface receptors that are expressed on all cells of the body. Members of the integrin family can be organized according to their chain composition as well as their tissue distribution. For example, the β2 subfamily of integrins are expressed primarily on cells of hemopoetic origin, and can mediate heterophillic interactions with a variety of ligands including members of the IgSF (33, 34). In contrast, β1 containing integrins, are more widely distributed and can mediate both cell-cell and cell-matrix interactions (33, 34). While the expression of integrins such as α6β4 are more restricted to epithelial cell types, the αIIbβ3 integrin has been suggested to be primarily expressed on platelets (33-36). However, recent studies have indicated that it can also be expressed on some tumor cells such as invasive melanomas and prostate carcinomas (35, 36).

7. GENERAL INTEGRIN STRUCTURE

The α chains of integrin can be organized into a heavy and light chain that are disulfide linked. The N-terminal region of the α chains contain several genetic repeats, some of which share homology with the calcium binding EF-hand like repeats (33, 34). In addition, an insertion or I-like domain of about 200 amino acids can be found in the extracellular domain of many leukocyte integrins (37). The α chains have a hydrophobic transmembrane domain and a short cytoplasmic tail, which has been shown to mediate interactions with cytoskeletal proteins (37). The β chains contain critical cysteine residues which contribute to its highly folded and disulfide bonded conformation. The β chains also have a transmembrane domain and a short cytoplasmic tail, which can also mediate interactions with cytosolic proteins. Importantly, the structural integrity and conformation of this complex heterodimer is critical for ligand binding and specificity. In fact, cation binding at the EF- hand like motifs may be associated with changes in conformation that modify the affinity and specificity of integrin binding (37). These specific conformationally mediated changes in affinity and specificity allow an increase in the diversity and utility of these integrin receptors for binding to the vast array of ligands. Taken together, the structural conformation in conjunction with the variety of α and β chain combinations provide for a highly diverse set of adhesion receptors that could regulate a number of distinct processes during tumor invasion and metastasis.

8. INTEGRIN – ECM INTERACTIONS AND MATRIX ASSEMBLY

As mentioned earlier, integrins can mediate both cell-cell and cell-ECM interactions. While a major function for these transmembrane adhesion receptors includes facilitating cell adhesion to the extracellular environment, integrin receptors are capable of mediating a variety of cellular functions in addition to simple cell adhesion. Some of the well characterized functions include regulations of cell migration, invasion and matrix assembly. For example, integrin αvβ3 can facilitate cellular migration on the ECM protein vitronectin (37, 38). Interestingly, elevated expression of αvβ3 on melanoma cells has been correlated with the vertical growth phase of melanoma and enhanced invasion and metastasis (39). Moreover, function-blocking Mab directed to αvβ3 can inhibit cellular migration and invasion both *in vitro* and *in vivo* (37-40). A variety of other integrin receptors such as the fibronectin receptor α5β1, the collagen receptor α2β1 and the laminin receptor α6β1 have all been shown to help facilitate cellular migration on their respective ligands (33, 34). Importantly, function-blocking antibodies directed to these integrins were shown to block tumor cell adhesion and migration *in vitro* and tumor metastasis *in vivo* (33, 34, 41-43).

While some of the amino acid sequences that are recognized by integrins are known, many others are still undefined. Integrin-ECM interactions depend on the proper orientation and conformation of the ligands. For example, the tripeptide sequence RGD is best recognized by integrin receptors such as αvβ3 when it is exposed in loop-type structures (44). In contrast, many of the RGD sites within the triple helix of collagen are not exposed in the appropriate orientation and are thus cryptic. However, following denaturation or proteolysis, the three dimensional structure of the collagen is altered allowing exposure of the RGD sites in a manner which can be recognized (45, 46). Recently, a variety of other cryptic sites within ECM molecules have been defined that require structural modification in order to be recognized by integrins. Some important examples include cryptic RGD sites in vitronectin, collagen, fibronectin, osteopontin, and tenascin (46-50). Specific exposure of these cryptic sites and subsequent integrin interactions may play a role in regulating invasive cellular processes during tumor growth and metastasis.

Finally, studies have also implicated integrin-ligand interactions in the regulation of proper ECM assembly. Assembly of matrix components into complex three dimensional ECM structures may play an important role in regulating tumor growth and metastasis (51-54). For example, proper assembly of fibronectin provides cells with critical attachment sites necessary for initiating signaling cascades that control migration, proliferation and survival. Interestingly, integrin $\alpha5\beta1$ has been suggested to play an important role in fibronectin matrix assembly, since antibodies directed to this integrin disrupt fibrilogenesis (53, 54). Other $\beta1$ containing integrins have also been suggested to play a significant role in proper assembly of collagen type-IV and laminin into the basement membrane (54). A disorganized basement membrane may expose unique cryptic sites that are recognized by integrin receptors leading to enhanced migration and invasion. Thus, integrin receptors may modulate invasive cell behavior by regulating both ECM assembly and recognition of novel cryptic sites within the ECM.

9. INTEGRIN SIGNALING AND REGULATION OF PROLIFERATION

In addition to mediating physical interactions between tumor cells and the acellular environment, integrin ligation can lead to the initiation of distinct signaling cascades which can ultimately regulate proliferation, gene expression and cell survival. Since integrin receptors do not possess kinase or phosphatase activity, two enzymatic activities that are thought to contribute to signaling, a great deal of effort has been directed at understanding the molecular mechanisms by which integrins facilitate signal transduction. While integrin mediated signaling is not completely understood, some general concepts are beginning to emerge. Besides having the ability to specifically bind to ECM proteins via their extracellular domains, integrins can also interact with a variety of cytosolic proteins via their cytoplasmic tails (55, 56). Thus, integrin are capable of transferring information from outside the cell through the plasma membrane to the inside of the cells. Information transfer from inside the cell to outside the cell can also be mediated by integrins (55, 56).

It would be beyond the scope of this chapter to discuss in detail integrin signaling, thus we will only briefly discuss the basic molecular machinery thought to be involved. Ligation of integrin receptors on the cell surface can lead to receptor clustering within the plane of the membrane (55, 56). These physical clustering events help bring cytoplasmic tails of the integrins into close proximity with assessory and adapter proteins. For example, studies have indicated that the cytoplasmic domains of integrins can bind to a number of cytosolic proteins including talin, vinculin, and paxilin (55, 56). Importantly, these interactions can be mediated by cytoplasmic tails of both the α and β subunits. In fact, some aspects of signaling specificity may arise from the different combination of accessory and adapter proteins that are bound to either the α and/or β tail of a particular integrin (55, 56). Interestingly, these cytosolic proteins also have binding sites for other adapter proteins and intracellular kinase such as Focal Adhesion Kinase (FAK) and Src (55, 56). Activation of specific kinases as well as phosphatases can lead to modifications of these proteins resulting in formation of signaling complexes (55, 56). These signaling complexes ultimately lead to the activation of distinct signaling pathways such as the Ras/ERK pathway (55, 56). Studies have indicated that integrin mediated ligation events that lead to activation of Shc, cooperate with growth factor induced signaling to promote transcription of Fos, which can regulate cell cycle progression and proliferation (55, 56). In contrast, integrin mediated interactions with ECM ligands that do not activate Shc, leads to growth arrest and differentiation (55, 56). Thus, integrin interactions with ECM components may specifically influence tumor cell proliferation both positively and negatively.

10. INTEGRINS AND THE REGULATION OF GENE EXPRESSION

Integrin ligation has also been shown to regulate gene expression and cell survival. To this end, studies have shown that ligation of fibronectin via interactions with α5β1 can lead to increased expression of bcl-2, a suppressor of apoptosis (57). Moreover, ligation of integrin αvβ3 was shown to be associated with down regulation of bax leading to a high bcl-2 to bax ratio (58, 59). Importantly, a high bcl-2 to bax ratio is thought to promote cell survival by inhibiting apoptosis. These findings are of critical importance to tumor growth and metastasis since the capacity of a tumor to grow and disseminate may depend in part on the balance between cell proliferation and cell death (58, 59).

An additional example of the critical role that integrin signaling and gene expression may have on tumor invasion and metastasis involves the coordinated expression of proteolytic enzymes. An elegant example of this concept involves integrin-mediated regulation of the expression of a family of matrix degrading enzymes termed Matrix Metalloproteinases or MMPs (60). Studies have indicated α5β1 ligation of the 120kd cell binding domain of fibronectin leads to up regulation of a number of MMPs including MMP-1, MMP-9 and Stromelysin (61). In contrast, α4β1 mediated ligation of the CS-1 region of fibronectin resulted in decreased expression of MMPs (61). In still other studies, ligation of integrin αvβ3 was shown to upregulate the expression of MMP-2, while α6β1 integrin mediated ligation of laminin upregulated MMP-9 and the serine protease urokinase plasminogen activator (uPA) (62, 63). Finally, expression of the collagen receptor α1β1 was shown to upregulate MMP-7 and MMP-9 (64). Importantly, elevated expression of MMPs have been correlated with increased tumor cell invasion and metastasis. Moreover, specific antagonists that inhibit either the expression or function of MMPs block tumor growth and metastasis *in vivo* (65-67). While enhanced expression of these proteolytic enzyme are clearly important during pathological processes, the activation state of the proteases is also crucial. Interestingly, recent studies have indicated that integrin such as αvβ3 may potentiate the activation of MMPs such as MMP-2 and MT1-MMP (68). Thus, integrin mediated regulation of MMP expression and/or activity may play critical roles in tumor dissemination.

11. INTEGRINS AND METASTASIS

Cellular interactions with the extracellular matrix (ECM) has been shown to regulate nearly all events thought to contribute to tumor invasion and metastasis including, gene expression, cellular proliferation, adhesion, migration and invasion (33, 34, 69). Given the diverse set of integrins expressed in tissues, coupled with the numerous roles integrins play in regulating cellular behavior, it is likely that integrins may play a major role in regulating tumor cell metastasis. In fact, a variety of studies have implicated a number of distinct integrin receptors in regulating tumor cell invasion and metastasis. One of the best characterized examples is that of integrin $\alpha v\beta 3$. Studies have demonstrated an enhanced expression of $\alpha v\beta 3$ integrin in vertically invasive and metastatic human melanomas as compared to benign non-invasive tumors (39). Moreover, transfection of $\alpha v\beta 3$ negative tumor cells with $\alpha v\beta 3$ resulted in increased tumor invasion and metastasis *in vivo* (70). Finally, peptide and antibody antagonists directed to $\alpha v\beta 3$ have been shown to inhibit metastasis in animal models (71). A similar vitronectin binding integrin $\alpha v\beta 5$ has also been shown to play a role in tumor metastasis. Interestingly, simple expression of $\alpha v\beta 5$ was not sufficient to promote tumor cell migration and metastasis *in vivo*, but required growth factor stimulation and subsequent activation (72). A third vitronectin binding integrin thought to contribute to tumor metastasis is the platelet receptor $\alpha IIb\beta 3$. While this integrin is primarily expressed on platelets it has also been shown to be expressed on some invasive tumor cells including some melanomas and prostate carcinomas (35, 36). Trikha and colleagues recently showed that the highly invasive and metastatic prostate carcinoma cells DU-145 expressed elevated levels of $\alpha IIb\beta 3$ as compared to the non-metastatic PC-3 prostate carcinoma cells (35). Moreover, function blocking antibodies directed to $\alpha IIb\beta 3$ inhibited experimental metastasis of DU-145 cells *in vivo* (35).

In addition to vitronectin binding integrins, a variety of $\beta 1$ integrins have been implicated in tumor metastasis. For example, transfection of $\alpha 2$ integrin subunit into rhabdomyosarcoma cells to allow functional expression of $\alpha 2\beta 1$, resulted in enhanced tumor invasion and metastasis (73). In contrast, transfection of the $\alpha 4$ integrin subunit into B16 melanoma cells to allow functional expression of $\alpha 4\beta 1$, resulted in reduced metastasis (74). Thus, function expression of particular $\beta 1$ integrins can regulate metastasis both positively and negatively. Additional examples of $\beta 1$ containing integrins that may contribute to tumor metastasis include $\alpha 6\beta 1$ and $\alpha 3\beta 1$. Vogelmann and others showed that function blocking antibodies directed to either $\alpha 6$ or $\beta 1$ integrins could significantly inhibit PaTu8988 prostate carcinoma cell metastasis (42). This inhibition was thought to be primarily due to disruption of the ability of integrin α6β1 to interact with the basal lamina (42). Finally, a variety of studies have implicated α3β1 in tumor metastasis. In fact, studies have indicated that function blocking antibodies to α3 integrin can block tumor invasion and metastasis *in vivo* (75). The mechanisms by which α3β1 regulates metastasis may be complex, since studies have indicated that α3β1 can form complexes with members of the tetraspanin family of cell surface molecules which may in turn regulate signal transduction, proliferation and migration (76). Taken together, it is clear that a number of distinct integrins play important regulatory roles during the metastatic cascade. Interestingly, the activation state, function and distribution of integrins can be influenced by a number of assessory molecules. Thus, the molecular cooperation between integrins and other regulatory proteins may have a significant impact on the metastatic cascade.

12. COOPERATIVE INTERACTIONS BETWEEN INTEGRINS AND ACCESSORY MOLECULES

As one can see from the brief discussion of the general adhesive and signaling functions of integrins, there are many steps along the metastatic cascade in which integrins may be critical regulators of cellular invasion. While integrins are thought to play a role in tumor growth and metastasis, they do not exist or function in isolation. In fact, a variety of molecules may work in cooperation with integrins to regulate their expression, function and distribution during pathological processes. Some critical examples of these integrin regulatory molecules include, membrane lipids, growth factors and their receptors, proteolytic enzymes, ECM components, tetraspanin molecules and other cell adhesion molecules (Figure 2). Thus, by studying integrin associated molecules and understanding cooperative molecular interaction between integrins and other molecules, one may gain unique insight into the role of integrins in tumor metastasis. The remainder of this chapter will focus on some interesting and potentially important cooperative interactions between integrins and other regulatory molecules, and how these unique molecular associations may regulate tumor invasion and metastasis.

13. INTEGRIN COOPERATION WITH GROWTH FACTORS AND GROWTH FACTOR RECEPTORS

Growth factor mediated signaling cascades have long been known to regulate a variety of cellular events thought to potentiate tumor invasion and metastasis including modulation of adhesion, migration, proliferation and survival. Interestingly, recent evidence has suggested that integrins may function cooperatively with growth factors and growth factor receptors in the regulation of many of these cellular processes (55, 56). In fact, it has been suggested that certain signaling cascades activated by growth factors can be coordinately regulated by integrin receptors. For example, integrin mediated signaling through the Fyn-Shc pathway can activate Ras signaling, while growth factor induced signaling also activates Ras pathways (56, 77). Moreover, studies have indicated that EGF can stimulate an αvβ5-dependent migratory response in carcinoma cells while having little if any effect on αvβ5-dependent adhesion (78). In other studies, Insulin-like growth factor receptor mediated signaling resulted in αvβ5-dependent tumor cell metastasis *in vivo* by a mechanism involving enhanced motility (72).

Integrin function can be regulated by growth factor stimulation. For example Trusolino and others showed that hepatocyte growth factor stimulation of HTU-5 thyroid cells caused activation of integrin αvβ3 mediated adhesion (79). Intriguing new evidence has also suggested that integrins may directly or indirectly associate with growth factors and growth factor receptors and thereby modulate cellular behavior. For example, Schneller and others used immunoprecipitaion assays to demonstrate a physical association between αvβ3 and both insulin and PDGF receptors (80). While the molecular mechanisms that mediate these unique associations are not completely understood, complex formation appeared to require prior stimulation with growth factors (80). These unique integrin-growth factor receptor complexes were associated with enhanced mitogenicity and motility *in vitro* (80). Thus, growth factor-integrin complex formation may regulate cellular events know to facilitate tumor invasion and metastasis.

Additional examples of cooperative interactions between growth factor receptors and integrins include associations between VEGFR-2 (KDR/Flk-1) and αvβ3 integrin. Soldi and colleagues showed that integrin αvβ3 and phosphorylated VEGFR-2 co-immunoprecipitated following VEGF stimulation of human endothelial cells, while other integrin receptors such as α2β1 and αvβ5 were not detected (81). Antibodies directed to the β3 subunit significantly reduced tyrosine phosphorylation of PI3 kinase, a down stream effector molecule involved in integrin mediated signaling (81). These findings suggest that integrin αvβ3 may function cooperatively with VEGF to regulate activation of VEGFR-2 in these cells. Finally, interesting studies by Munger and others demonstrated a unique association between integrin αvβ6 and TGFβ1 LAP (82). These studies demonstrated that binding of integrin αvβ6 to latent TGFβ1 LAP promoted conversion of latent TGFβ1 to its active form (82). Active TGFβ1 could then bind its cognate TGFβ1 receptors to facilitate signaling. The interaction between the LAP peptide region and αvβ6 appears to be dependent on recognition of an RGD site within the C-terminus of the peptide (82). Other RGD binding integrins have also been shown to exhibit some binding to the LAP-peptide such as αvβ1, αIIbβ3 and to some extent αvβ5 (82). However, little is known concerning the ability of these integrins to activate TGFβ-1. The αvβ6 mediated activation of latent TGFβ1 not only required binding to the LAP peptide but also required αvβ6 association with the actin cytoskeleton (82). Taken together, these examples suggest that integrins and growth factors can function cooperatively to regulate either the activation of integrins and/or the activation state of growth factors and growth factor receptors. In turn, regulation of the activity of these molecules may then potentiate signaling cascades that promote cellular proliferation and migration, critical processes involved in tumor invasion and metastasis.

14. COOPERATIVE INTERACTIONS BETWEEN INTEGRINS AND TETRASPAN MOLECULES

An expanding set of studies have provided evidence that integrin receptors can associate with several members of the Tetraspanin super family of transmembrane proteins (76). The tetraspanin super family, is a group of membrane proteins that are characterized by having 4 hydrophobic transmembrane domains (76). Studies have predicted that these membrane proteins are oriented in the lipid bilayer with small and large extracellular loop-type structures and a short cytoplasmic tail. A number of highly conserved polar amino acids, in conjunction with conserved cysteine residues are thought to contribute to the unique folding pattern which may allow for proper function and orientation within the cell membrane (76). Since their initial description in the early 1990's, the number of individual members have grown to at least 20 with more likely to be described in the coming years (76). This large family of cell surface proteins have a broad tissue distribution and are expressed in a wide variety of cells ranging from cells of hemopoetic lineage, to neuronal, stromal and tumor cells (76). Moreover, studies have suggested that this important group of membrane proteins may regulate a diverse set of cellular functions including adhesion, migration, proliferation and signal transduction. (76). As was discussed above, these same cellular functions can be regulated by integrins. Thus, integrins and tetraspanins may function cooperatively in the regulation of these cellular processes.

Interestingly, members of the tetraspanin family are known to form complexes with a variety of molecules including other tetraspanins, MHC molecules as well as integrins (76). In fact, the ability of these unique proteins to form molecular associations with other proteins, in conjunction with their capacity to regulate a number of cellular functions, have lead many investigators to view them as "molecular facilitators" (76). Early studies utilizing co-immunoprecipitation assays from cultured cells revealed a number of unique β1 integrin-tetraspanin complexes. For example, α3β1 integrin was shown to form specific complexes with a number of tetraspanins including, CD9, CD63, CD81, CD82, and CD151/PETA-3 (76, 83). The laminin receptor α6β1 forms a complex with, CD9, CD63, CD81, and CD82. Additional integrins that associate with tetraspanins include αIIbβ3, α4β1, and α4β7 (76, 83). While many β1 containing integrins have been shown to specifically associate with tetraspanins, others do not such as α1β1 and α2β1, implying that these unique integrin-tetraspanin complexes form specifically.

It is of interest to note that many of the integrins that form unique complexes with tetraspanins such as $\alpha IIb\beta 3$, $\alpha 3\beta 1$, $\alpha 4\beta 1$, and $\alpha 6\beta 1$ have also been suggested to play a role in tumor cell metastasis. Thus, tetraspanins may regulate metastasis by modulating integrin functions. Interestingly, when MDA-MB-231 breast cancer cells were allowed to bind to a mixture of immobilized anti-tetraspanin antibodies and collagen, the resulting ligation event significantly enhanced collagen induced phoshoylation of Focal Adhesion Kinase (FAK) (84). FAK is a known down stream effector molecule of integrin mediated signaling pathways and is known to regulate integrin functions such as adhesion and motility (85). Moreover, FAK has also been implicated in the regulation of cell survival and tumor metastasis (86). Further experimental evidence to support a role for tetraspinins in the regulation of integrin functions come from recent studies concerning cellular adhesion and motility. Anti-CD9 antibodies have been shown to stimulate B- cell adhesion mediated by $\alpha 4\beta 1$ (87). Interestingly, enhanced motility on fibronectin and laminin was shown to be regulated by integrin $\alpha 4\beta 1$ following transfection of CD9 into CD9-negative B-cell lines (88). In contrast, transfection of CD9 into certain non-lymphoid cells caused reduction of cell migration (89). Thus, the modulating effects of tetraspanins on integrins may be cell type specific. Finally, antibodies directed to CD81 promoted B- cell adhesion to fibronectin via integrin $\alpha 4\beta 1$ and has also been shown to exhibit anti-proliferative effects (89). Importantly, both $\alpha 4\beta 1$ and $\alpha 6\beta 1$ have been shown to play a role in tumor metastasis *in vivo*, since either modulating the expression or blocking the function of these integrins impacted tumor metastasis *in vivo* (42, 74).

An active role for tetraspanins in the regulation of tumor invasion and metastasis has been suggested as early as the mid 1980's (76). In fact, CD-9 expression in certain melanoma cells was inversely correlated with melanoma metastasis (90). Reduced expression of CD-9 was also correlated with a poor clinical prognosis in breast cancer (91). Moreover, studies indicated that transfection of CD-9 into metastatic prostate cancer cells suppressed tumor cell motility and metastasis, prompting some to consider CD-9 as a metastasis suppressor gene (91, 92). Other studies have also suggested that the tetraspanin CD82 may also play a critical role in metastasis since it was isolated as a metastasis suppressor gene from prostate cancer cells (92). Finally, in recent studies by Testa and colleagues, the functional expression of CD151/PETA-3 was shown to play a crucial role in the metastasis of the human epidermoid carcinoma Hep-3 *in vivo* (93). In these studies, the immunological technique of subtractive immunization was used to isolate antibodies directed to metastasis associated cell surface antigens. The antigen recognized by these antibodies turned out to be the tetraspanin CD151/PETA-3 (93). CD151/PETA-3 was shown to be in a complex with the integrin $\alpha 3\beta 1$ which has been shown previously to play a role in metastasis (93). Moreover, both Mabs directed to CD151/PETA-3 (Mabs 50-6 and IA5) potently inhibited tumor metastasis *in vivo* (93). Interestingly, these Mabs significantly inhibited tumor cell motility, but did not affect cellular adhesion or proliferation, suggesting that CD151/PETA-3 may modulate metastasis by regulating integrin dependent migratory events. In fact, transfection of HELA cells with CD151/PETA-3 caused enhanced motility, which could be blocked by anti-CD151/PETA-3 antibodies (93). Thus, it is likely given the accumulating evidence, that tetraspanin-mediated modulation of integrin function may play an important role in the metastatic cascade.

15. COOPERATIVE INTERACTIONS BETWEEN INTEGRINS, PROTEOLYTIC ENZYMES AND THEIR RECEPTORS

Proteolytic remodeling of the extracellular matrix (ECM) is known to be a critical event in facilitating tumor cell invasion and metastasis. Two major groups of proteolytic enzymes thought to contribute to matrix remodeling include the serine protease and matrix metalloproteinase (MMP) families (60, 94). Important members of the serine protease family thought to contribute to tumor invasion and metastasis include, urokinase plasminogen activator (uPA), and Plasmin (94). uPA, which is known to bind to a GPI-linked cell surface receptor (uPAR) facilitates conversion of plasminogen to the broad spectrum serine protease Plasmin. Plasmin in turn, can degrade a number of ECM components including fibrinogen, fibrin, and fibronectin (94). The MMP family is composed of a large group of metal dependent enzymes, which collectively can degrade nearly all the components of the ECM including a variety of distinct types of triple helical collagen (60). Important examples of the MMP family that are thought to contribute to tumor invasion and metastasis include interstitial collagenase (MMP-1), Stromelysin and the gelatinases MMP-2 and MMP-9 (60). Studies have demonstrated a significant upregulation of many of these proteolytic enzymes and/or their cognate receptors within highly invasive and metastatic tumors as compared to either normal tissues or benign non-invasive tumors (60, 94). Moreover, specific antagonist that block the functional activity of both serine proteases and MMPs have been shown to significantly inhibit tumor invasion and metastasis in multiple animal models (60, 94).

While significant insight has been made into the molecule mechanisms by which proteolytic activity contributes to tumor invasion and metastasis, it is still not completely understood. Proteolytic enzymes may function by degrading restrictive matrix barriers, thereby generating an extracellular environment that is conducive to tumor invasion and motility (60, 94). Matrix degrading enzymes may also function by facilitating proteolytic release of matrix sequestered growth factors that could in turn, bind tumor cells stimulating proliferation and migration. In addition, proteolytic remodeling of the three dimensional structure of ECM proteins may expose cryptic regulatory regions of ECM proteins that are normally hidden. Exposure of these unique cryptic regions may provide new binding sites for integrin receptors resulting in altered integrin ligation events thereby initiating signaling cascades required for tumor progression. Finally, proteolytic enzymes may also degrade matrix components and release soluble peptides that bind tumor cells thereby modulating adhesion, migration and proliferation. While it is likely, that a combination of all these mechanisms contribute to tumor metastasis, proteolytic activity must be specifically regulated and coordinated with adhesion and migration in order to promote metastasis.

Interestingly, recent studies have lead to the discovery of a novel family of membrane proteins called ADAMs (95). This novel family of proteins were characterized as containing a disintegrin-like domain and a metalloproteinase-like domain and therefore, may possess both adhesive and proteolytic functions within one protein (95). The discovery of these unique membrane proteins provided evidence that coordinated adhesive and proteolytic functions at the cell surface may play an important role in regulating cellular behavior. In this regard, recent studies have demonstrated functional associations between integrins and both proteases or protease receptors (96, 97). These unique molecular associations may function in a similar way as the ADAMs to coordinate and regulate adhesive and proteolytic activity at the cell surface. While the functional significance and molecular mechanisms by which integrin-protease complexes regulate cellular behavior are not completely understood, several possibilities exist. For example, integrin protease interactions may help localize proteolytic activity to close cell-substratum contact points to facilitate directed proteolysis. Interactions of proteases with integrins may also help potentiate activation of latent enzymes. Moreover, protease binding to integrins may disrupt integrin recognition of ECM ligands. Alternatively, protease binding may induce a conformational change in the integrin, which could activate integrin binding. Finally protease or protease receptor associations with integrins could also modulate signal transduction pathways. While many of these possibilities are speculative and have not been studied in detail, some evidence has been provided that a unique integrin-uPAR complex may modulate signaling cascades as well as regulating integrin associated adhesive and migratory events (98).

An important example of the link between proteolytic and adhesive systems comes from the observations that the uPA receptor (uPAR) can associate with a number of integrins. In fact, studies have suggested that uPAR may associate with integrins from the β1, β2 and β3 subfamilies (96). Several lines of experimental data supporting this unique association include co-immunoprecipitation assays, Resonance Energy Transfer (RET) experiments and direct binding assays (96). Moreover, recent studies have identified a putative region of the integrin receptor that may mediate the association between uPAR and β2 integrins (96). In fact, a synthetic peptide corresponding to a region just outside the I-like domain of the α chain of the integrin, disrupted integrin –uPAR associations (99). A number of elegant studies are now beginning to shed some new light on the potential biological significance of the uPAR integrin complexes. For example, Sitrin and others demonstrated that binding of αMβ2 integrin to uPAR stimulated adhesion of monocytes to fibrinogen but not to other β1 integrin dependent ligands such as fibronectin (100). In addition, studies by

Aguirre and colleagues showed that uPAR forms a complex with the fibronectin receptor α5β1 in Hep-3 cells (98). Formation of this complex was associated with enhanced Hep-3 cell adhesion to fibronectin substrates (98). These findings suggest that associations between uPAR and integrin receptors can modulate integrin dependent adhesive events that may be necessary for proper function and/or stabilization of active integrin conformations.

Additional studies have suggested that uPAR-integrin complexes may help facilitate signaling pathways that regulate migration and proliferation. For example, Wei and others recently demonstrated that membrane protein caveolin associates with β1 integrins and this association may regulate uPAR-β1 integrin mediated signal transduction (101). In fact, these investigators suggest that caveolin may function in conjunction with β1 integrin –uPAR complexes in assembling signaling kinases such as Src, which in turn may facilitate FAK phosphorylation (101). Aguirre and colleagues showed that uPAR forms a complex with the fibronectin receptor α5β1 in metastatic Hep-3 carcinoma cells which exhibited strong binding to fibronectin. However, non-metastatic variants of Hep-3 cells showed little if any uPAR- α5β1complex and demonstrated little capacity to bind to fibronectin (98). These investigators suggested that functional association between α5β1 and uPAR not only results in enhanced adhesion to fibronectin, but results in ERK activation (98). The activation of ERK appeared to be specific to uPAR- α5β1 integrin complex formation in these cells since treating cells with either an antibody or a peptide that disrupts uPAR- α5β1 interactions inhibited ERK activation in metastatic Hep-3 cells (98). Importantly, the uPAR- α5β1 dependent ERK activation was associated with a conversion of dormant Hep-3 cells to highly tumorigenic and metastatic phenotype *in* vivo (98). Thus, these findings suggest that uPAR- α5β1 complex formation may not only regulate integrin dependent adhesive processes but also facilitate tumor progression *in vivo*. In addition to integrin associations with serine protease receptors such as uPAR, recent studies suggest that integrins can also bind to proteases themselves, not simply their receptors. For example, Byzova and Plow demonstrated that integrin αvβ3 could specifically bind to the serine protease prothrombin (102). This interaction was shown to be dependent on an RGD site in prothrombin. However, the capacity of αvβ3 to bind to prothrombin appeared to depend on the activation state of αvβ3 (102).

Besides serine proteases, previous studies have suggested that certain MMPs may also associate with integrins. Studies have suggested that the matrix metalloproteinase MMP-2 can associate with integrin αvβ3 on the surface of both tumor cells and endothelial cells (97, 103). While the molecular mechanisms that facilitate this unique protease-integrin association are not completely understood, recent evidence suggests that it may depend in part on sequences present within the C-terminal hemopexin-like domain of MMP-2. (97, 103) In fact, mutant forms of MMP-2 that lack the hemopexin domain fail to bind to αvβ3 on the cell surface (97, 103). The interaction between MMP-2 and αvβ3 may provide a mechanism to localize proteolytic activity to the surface of invasive cells and thereby potentiate tumor invasion and metastasis. Importantly, αvβ3 bearing melanoma cells pretreated with MMP-2 resulted in enhanced degradation of collagen *in vitro* (97, 103). Moreover, these same cells have been shown to aggressively metastasize *in vivo* as compared to variant cell lines that lacked cell surface αvβ3 and MMP-2 (72, 103). Taken together, unique interactions between integrins and proteases and protease receptors may provide crucial functions during tumor progression.

16. CONCLUSIONS

It has been known for sometime that a major impediment to an effective treatment for many human tumors involves the spread of the malignant tumors from the primary site to distant sites. Thus, an in-depth understanding of the biochemical and molecular mechanisms regulating tumor metastasis is of particular importance in the development of effective therapies for malignant cancer. It is becoming increasingly clear that the array of molecular regulators of tumor metastasis do not necessarily function in isolation, but are rather interconnected in a continuum of biochemical and cellular events. Understanding these unique molecular interactions involved in metastasis may provide novel approaches for the design of effective new drugs to regulate malignant human diseases. In this regard, an alternative approach for directly targeting proteolytic activity during tumor dissemination, might be to prevent the localization of these enzymes to the cell surface by blocking unique protease / integrin interactions. However, the critical keys to establishing these novel types of therapies involves an in-depth understanding of the basic molecular mechanisms that facilitate these protein-protein interactions.

In this regard, we have discussed tumor metastasis in terms of molecular cooperation between integrins and a number of metastasis regulatory molecules, including growth factors and their receptors, proteolytic enzymes and their receptors, tetraspanins and extracellular matrix proteins. Understanding the functions of integrins in the context molecular cooperativity and how unique molecular associations between integrins and assessory proteins regulate cellular invasion, may shed new light on the fascinating and complex process of metastasis.

ACKNOWLEDGMENTS

The author would like to thank Kathryn Carner for her help in preparation of this manuscript. This work was supported in part by grants CA74132-01 and CA86140-01 to PCB.

REFERENCES

1. Liotta LA, Steeg PS, Stetler-Stevenson WG. Cancer metastasis and angiogenesis: an imbalance of positive and negative regulation. Cell. 1991; 64: 327-336.
2. Wyckoff JB, Jones JG, Condeelis JS, Segall JE. A critical step in metastasis: *in vivo* analysis of intravasation at the primary tumor. Cancer Res. 2000; 60: 2504-2511.
3. Kurschat P, Mauch C. Mechanisms of metastasis. Clinic Exp. Dermatol. 2000; 25: 482-489.
4. Albelda SM. Role of integrins and other cell adhesion molecules in tumor progression and metastasis. Lab Invest. 1993; 68: 4-14.
5. Johnson JP. Cell adhesion molecules in the development and progression of malignant melanoma. Cancer Met. Rev. 1999; 18: 345-357.
6. Johnson JP. Cell adhesion molecules of the immunoglobulin supergene family and their role in malignant transformation and progression to metastatic disease. Cancer Met. Rev. 1991; 10: 11-22.
7. Drillenburg P, Pals ST. Cell adhesion receptors in lymphoma dissemination. Blood. 2000; 95:1900-1910.
8. Stanford DR, Starkey JR, Magnuson JA. The role of tumor-cell surface carbohydrate in experimental metastasis. Int. J. Cancer. 1986; 37: 435-444.
9. Kishimoto T, Ishikura H, Kimura C, Takahashi T, Kato H, Yoshiki T. Phenotypes correlating to metastatic properties of pancreas adenocarcinoma *in vivo*: the importance of surface sialy lewis antigen. Int. J. Cancer. 1996; 69: 290-294.
10. Fukuda MN, Ohyama C, Lowitz K, Matsuo O, Pasqualini R, Ruoslahti E, Fukuda M. A peptide mimic of E-selectin ligand inhibits sialy lewis X-dependent lung colonization of tumor cells. Cancer Res. 2000; 60: 450-456.
11. Lotan R, Raz A. Endogenous lectins as mediators of tumor cell adhesion. J. Cell. Biochem. 1988; 37: 107-117.
12. Matsushita Y, Cleary KR, Ota DM, Hoff SD, Irimura T. Sialyl-dimeric lewis-X antigen expressed on mucin-like glycoproteins in colorectal cancer metastases. Lab Invest. 1990; 63: 780-791.
13. Kansas GS. Selectins and their ligands: current concepts and controversies. Blood. 1996; 88: 3259-3269.
14. Somers WS, Tang J, Shaw GD, Camphausen RT. Insights into the molecular basis of leukocyte tethering and rolling revealed by structures of P-and E-selectin bound to SLE^X and PSGL-1. Cell. 2000; 103: 467-479.
15. Kobayashi K, Matsumoto S, Morishima, T, Kawabe T, Okamoto T. Cimetidine inhibits cancer cell adhesion to endothelial cells and prevents metastasis by blocking E-selectin expression. Cancer Res. 2000; 60: 3978-3984.
16. Hu Y, Kiely J, Szente BE, Rosenzweig A, Gimbrone MA. E-selectin-dependent signaling via the mitogen activated protein kinase pathway in vascular endothelial cells. J. Immunol. 2000; 170: 2142-2148.
17. Nollet F, Kools P, van Roy F. Phylogenetic analysis of the cadherin superfamily allows identification of six major subfamilies besides several solitary members. J. Mol. Biol. 2000; 299: 551-572.
18. Leckband D, Sivasankar S. Mechanism of homophillic cadherin adhesion. Curr. Opin. Cell Biol. 2000; 12: 587-592.
19. Asgeirsson KS, Jonasson JG, Tryggvadottir L, Olafsdottir K, Sigurgeirsdottir JR, Ingvarsson S, Ogmundsdottir HM. Altered expression of E-cadherin in breast

cancer: patterns, mechanisms and clinical significance. Eur. J. Cancer. 2000; 36: 1098-1106.

20. Bukholm IK, Nesland JM, Borresen-Dale AL. Re-expression of E-cadherin, α-catenin and β-cateinin, but not γ-catenin, in metastatic tissue from breast cancer patients. J. Pathol. 2000; 190: 15-19.
21. Hodivala KJ, Watt FM. Evidence that cadherins play a role in the down regulation of integrin expression that occurs during keratinocyte terminal differentiation. J. Cell Biol. 1994; 124: 589-600.
22. Behrens J, Birchmeier W, Goodmann SL, Imhof BA. Dissociation of madin-darby canine kidney epithelial cells by the monoclonal antibody anti-arc-1: mechanistic aspects and identification of the antigen as a component related to Uvomorulin. J. Cell Biol. 1985; 101: 1307-1315.
23. Behrens J, Macreel MM, VanRoy FM, Birchmeier W. Dissection of tumor cell invasion: epithelial cells aquire invasive properties after loss of uvomorulin mediated cell-cell adhesion. J. Cell. Biol. 1989; 108: 2435-2447.
24. Hazan RB, Phillips GR, Qiao RF, Norton L, Aaronson SA. Exogenous expression of N-cadherin in breast cancer cells induces cell migration, invasion and metastasis. J. Cell Biol. 2000; 148: 779-790.
25. Williams AF, Burclay AN. The immunoglobulin superfamily-domains for cell surface recognition. Ann. Rev. Immunol. 1988; 6: 381-405.
26. Gonzalez R,Gonzalez F, Madrid F. Adhesion molecules in inflammatory diseases. Drugs. 1998; 56: 977-988.
27. Sun JJ, Zhou XD, Liu YK, Tang ZY, Fen JX, Zhou G, Xue Q, Chen J. Invasion and metastasis of liver cancer: expression of intercellular adhesion molecule-1. J. Cancer Res. Clin. Oncol. 1999; 125: 28-34.
28. Denton KJ, Stretch JR, Gatter KC, Harris AL. A study of adhesion molecules as markers of progression in malignant melanoma. J. Pathol. 1992; 167: 187-191.
29. Lehmann JM, Riethmuller G, Johnson JP. Muc18: a marker of tumor progression in human melanoma shows sequence similarity to neural cell adhesion molecules of the immunoglobulin superfamily. Proc. Natl. Acad. Sci USA. 1989; 86: 9891-9895.
30. Fearon JM, Hamilton SR, Preisinger AC, Thomas G, Kinzler KW, Vogelstein B. Identification of a chromosome 18q gene that is altered in colorectal cancers. Science. 1990; 247: 49-56.
31. Sun JJ, Zhou XD, Liu YK, Tang RX, Zhao Y, Uemura T. Inhibitory effects of synthetic β peptide on invasion and metastasis of liver cancer. J.Cancer Res. Clin. Oncol. 2000; 126: 595-600.
32. Aoudjit F, Potworowski EF, Springer TA, St-Pierre Y. Protection from lymphoma cell metastasis in ICAM-1 mutant mice: a posthoming event. J. Immunol. 1998; 161: 2333-2338.
33. Hynes RO. Integrins: versatility, modulation, and signaling in cell adhesion. Cell. 1992; 69: 11-25.
34. Holly SP, Larson MK, Parise LV. Multiple roles of integrins in cell motility. Exp. Cell. Res. 2000; 261: 69-74.
35. Trikha M, Raso E, Cai Y, Fazakas Z, Paku S, Porter, AT, Timar J, Honn KV. Role of αIIbβ3 integrin in prostate cancer metastasis. The Prostate. 1998; 35: 185-192.
36. Trikha M, Timar J, Lundy SK, Szekeres R, Cai Y, Honn HV. The high affinity αIIbβ3 integrin is involved in invasion of human melanoma cells. Cancer Res. 1997; 57: 2522-2528.

37. Cheresh DA, Mecham RP. Integrins: molecular and biological responses to the extracellular matrix. San Diego CA: Academic Press 1994.
38. Leavesley DL, Ferguson GD, Wayner EA, Cheresh DA. Requirement of the integrin β3 subunit for carcinoma cell spreading or migration in vitronectin and fibrinogen. J. Cell Biol. 1992; 117: 1101-11107.
39. Nip J, Brodt P. The role of the integrin vitronectin receptor αvβ3 in melanoma metastasis. Cancer Met. Rev. 1995; 14: 241-252.
40. Seftor REB, Seftor EA, Genhlsen KR, Stetler-Stevenson WG, Brown PD, Ruoslahit E, Hendrix MJC. Role of the αvβ3 integrin in human melanoma cell invasion. Proc. Natl. Acad. Sci. US. 1992; 89:1557-1561.
41. Nakahara H, Nomizu M, Akiyama SK, Yamada Y, Yeh Y, Chen WT. A mechanism for regulation of melanoma invasion. J. Biol. Chem. 1996; 271: 27221-27224.
42. Vogelmann, R, Kreuser ED, Adler G, Lutz MP. Integrin α6β1 role in metastatic behavior of human pancreatic carcinoma cells. Int. J. Cancer. 1999; 80: 791-795.
43. Ruoslahti R. Fibronectin and its integrin receptors in cancer. Adv. Cancer Res. 1999; 65: 1-20.
44. Pfaff M, Tangemann K, Muller B, Gurrath M, Muller G, Kessler H, Timpl R, Engel J. Selective recongnition of cyclic RGD peptides of NMR defined conformation by αIIbβ3, αvβ3 and α5β1 integrins. J. Biol. Chem. 1994; 269: 20233-20238.
45. Davis GE. Affinity of integrins for damaged extracellular matrix: αvβ3 binds to denatured collagen type-I through RGD sites. Biochem. Biophys. Res. Comm. 1992; 182: 1025-1031.
46. Montgomery AMP, Reisfeld RA, Cheresh DA. Integrin αvβ3 rescues melanoma cells from apoptosis on three-dimensional dermal collagen. Proc. Natl. Acad. Sci USA. 1994; 91: 8856-8860.
47. Seiffert D, Smith JW. The cell adhesion domain in plasma vitronectin is cryptic. J. Biol. Chem. 1997; 272: 13705-13710.
48. Clark RAF, Wikner NE, Doherty DE, Norris DA. Cryptic chemotactic activity of fibronectin for human monocytes resides in the 120-kda fibroblastic cell-binding fragment. J. Biol. Chem. 1988; 263: 12115-12123.
49. Smith LL, Cheung HK, Ling LE, Chen J, Sheppard D, Pytela R, Giachelli CM. Osteopontin N-terminal domain contains a cryptic adhesive sequence recognized by α9β1 integrin. J. Biol. Chem. 1996; 271: 28485-28491.
50. Denda S, Muller U, Crossin KL, Erickson HP, Reichardt LF. Utilization of a soluble integrin-alkaline phosphatase chimera to characterize integrin α8β1 receptor interactions with tenascin: murine α8β1 binds to a RGD site in tenascin-C fragment, but not native tenascin-C. Biochemistry. 1998; 37: 5464-5474.
51. Yurchenco PD, Orear JJ. Basal lamina assembly. Curr. Biol. 1994; 6: 674-681.
52. Sasaki T, Forsgerg E, Bloch W, Addicks K, Fassler R, Timpl R. Deficiency of β1 integrins in teratoma interferes with basement membrane assembly and laminin-1 expression. Exp. Cell Res. 1998; 238: 70-81.
53. Pankov R, Cukierman E, Katz B, Matsumoto K, Lin DC, Lin S, Hahn C, Yamada KM. Integrin dynamics and matrix assembly: tensin-dependent translocation of α5β1 integrins promotes early fibroblast fibrillogenesis. J. Cell Biol. 2000; 148: 1075-1090.
54. Hocking DC, Smith RK, McKeown-Longo PJ. A novel role for the integrin-binding III-10 module in fibronectin matrix assembly. J. Cell Biol. 1996; 133: 431-444.
55. Giancotti FG, Ruoslahti E. Integrin signaling. Science. 1999; 285: 1028-1032.

56. Ginacotti FG. Complexity and specificity of integrin signaling. Nat. Cell Biol. 2000; 2: 13-14.
57. Zhang Z, Vuori K, Reed JC, Ruoslahti E. The $\alpha5\beta1$ integrin supports survival of cells on fibronectin and up-regulates bcl-2 expression. Proc. Natl. Acad. Sci USA. 1995; 92: 6161-6165.
58. Stromblad S, Becker JC, Yebra M, Brooks PC, Cheresh DA. Suppression of p53 activity and $p21^{WAF1/CIP1}$ expression by vascular cell integrin $\alpha v\beta3$ during angiogenesis. J. Clin. Invest. 1996; 98: 1-8.
59. Petitclerc, E, Stromblad S, von Schalscha TL, Mitjanns F, Piulats J, Montgomery AMP, Cheresh DA, Brooks PC. Integrin $\alpha v\beta3$ promotes M21 melanoma growth in human skin by regulating tumor cell survival. Cancer Res. 1999; 59: 2724-2730.
60. Curran S, Murray GI. Matrix metalloproteinase; molecular aspects of their roles in tumor invasion and metastasis. Eur. J. Cancer. 2000; 36: 1621-1630.
61. Huhtala P, Humphries MJ, McCarthy JB, Tremble PM, Werb Z, Damsky CH. Cooperative signaling by $\alpha5\beta1$ and $\alpha4\beta1$ integrins regulate metalloproteinase gene expression in fibroblasts adhering to fibronectin. J. Cell Biol. 1995; 129: 867-879.
62. Khan KMF, Falcone DJ. Role of laminin in matrix induction of macrophage urokinase-type plasminogen activator and 92kd metalloproteinase expression. J. Biol. Chem. 1997; 272: 8270-8275.
63. Seftor REB, Seftor E, Stetler-Stevenson WG, Hendrix MJC. The 72kd type-IV collagenase is modulated via differential expression of $\alpha v\beta3$ and $\alpha5\beta1$ integrins during human melanoma cell invasion. Cancer. Res. 1993; 53: 3411-3415.
64. Pozzi A, Moberg PE, Miles LA, Wagner S, Soloway P, Gardner HA. Elevated matrix metalloproteinase and angiostatin levels in integrin $\alpha1$ knockout mice cause reduced tumor vascularization. Proc. Natl. Acad. Sci USA. 2000; 97: 2202-2207.
65. Stetler-Stevenson WG. Dynamics of matrix turnover during pathological remodeling of the extracellular matrix. Am. J. Pathol. 1996; 148: 1345-1350.
66. Tsuchiya Y, Sato H, Endo Y, Okada Y, Mai M, Sasaki T, Seiki M. Tissue inhibitor of metalloproteinase 1 is a negative regulator of the metastatic ability of a human gastric cancer cell line, KKLS in the chick embryo. Cancer Res. 1993; 52: 1397-1402.
67. Coussens LM, Tinkle CL, Hanahan D, Werb Z. MMP-9 supplied by bone marrow-derived cells contributes to skin carcinogenesis. Cell. 2000; 103: 481-490.
68. Hofmann UB, Westphal JR, Van Kraats AA, Rutter DJ, Muijen GNP. Expression of integrin $\alpha v\beta3$ correlates with activation of membrane type matrix metalloproteinase-1 (MT1MMP) and matrix metalloproteinase-2 (MMP-2) in human melanoma cells *in vitro* and *in vivo*. Int. J. Cancer. 2000; 87: 12-19.
69. Juliano RL, Haskill S. Signal transduction from the extracellular matrix. J. Cell Biol. 1993; 120: 577-585.
70. Hsu MY, Shih DT, Meier FE, Van Belle P, Hsu JY, Elder DE, Buck CA, Herlyn M. Adenoviral gene transfer of $\beta3$ integrin subunit induces conversion from radial to vertical growth phase in primary human melanoma. Am. J. Pathol. 1998; 253: 1435-1442.
71. Seftor REB. Role of the $\beta3$ integrin subunit in human primary melanoma progression. Am. J. Pathol. 1998; 153: 1347-1351.
72. Brooks PC, Klemeke RL, Schon S, Lewis JM, Schwartz MA, Cheresh DA. Insulin-like growth factor receptor cooperates with integrin $\alpha v\beta5$ to promote tumor cell dissemination *in vivo*. J. Clin. Invest. 1997; 99: 1390-1398.

73. Chan BMC, Matura N, Takada Y, Zetter BR, Hemler ME. *In vitro* and *in vivo* consequences of VLA-2 expression on rhabdomyosarcoma cells. Science. 1991; 251: 1600-1602.
74. Qian F, Vaux DL, Weissman IL. Expression of the integrin α4β1 on melanoma cells can inhibit the invasive stage of metastasis formation. Cell. 1994; 77: 335-347.
75. Morini M, Mottolese M, Ferrari N, Ghiorzo F, Buglioni S, Mortarini R, Noonan DM, Natali PG, Albini A. The α3β1 integrin is associated with mammary carcinoma cell metastasis, invasion and gelatinase B (MMP-9) activity. Int. J. Cancer. 2000; 87: 336-342.
76. Maecker HT, Todd SC, Levy S. The tetraspanin superfamily: molecular facilitators. FASB. 1997; 11: 428-442.
77. Wary KK, Mainiero F, Isakoff SJ, Marcantonio EE, Giancotti FG. The adaptor protein Shc couples a class of integrins to the control of cell cycle progression. Cell. 1996; 87: 733-743.
78. Klemeke RL, Yebra, M, Bayna EM, Cheresh DA. Receptor tyrosine kinase signaling required for integrin alpha v beta 5-directed cell motility but not adhesion on vitronectin. J. Cell Biol. 1994; 127: 859-866.
79. Trusolino L, Serini G, Cecchini G, Besati C, Ambesi-Impiombato FS, Marchisio PC, Filippi R. Growth factor-dependent activation of αvβ3 integrin in normal epithelial cell: implications for tumor invasion. J. Cell Biol. 1998; 142: 11450-1156.
80. Schneller M, Vuori K, Ruoslahti E. αvβ3 integrin associates with activated insulin and PDGF receptors and potentiates the biological activity of PDGF. EMBO.J. 1997; 16: 5600-5607.
81. Soldi R, Mitola S, Strasly M, Defilippi P, Tarone G, Bussolino F. Role of αvβ3 integrin in the activation of vascular endothelial growth factor receptor-2. EMBO. J. 1999; 18: 882-892.
82. Muger JS, Huang X, Kawakatsu H, Griffths MJD, Dalton SL, Wu J, Pittet JF, Kaminski N, Garat C, Matthay MA, Rifkin DB, Sheppard D. The integrin αvβ6 binds and activates latent TGFβ1: a mechanism for regulating pulmonary inflammation and fibrosis. Cell. 1999; 96: 319-328.
83. Berditchevski F, Zutter MM, Hemler ME. Characterization of novel complexes on the cell surface between integrins and proteins with 4 transmembrane domains (TM4 proteins). Mol. Cell. Biol. 1996; 7: 193-207.
84. Berditchevski F, Odintsova E. Characterization of integrin-tetraspanin adhesion complexes: role of tetraspanin in integrin signaling. J. Cell Biol. 1999. 146: 477-492.
85. Guan JL. Role of focal adhesion kinase in integrin signaling. Int. J. Biochem. Cell Biol. 1997; 29: 1085-1096.
86. Kornberg LJ. Focal adhesion kinase and its potential involvement in tumor invasion and metastasis. Head & Neck. 1998; 16: 745-752.
87. Masellis-Smith A, Shaw AR, CD9-regulated adhesion: anti-CD9 monoclonal antibody induced pre-B cell adhesion to bone marrow fibroblast through de novo recognition of fibronectin. J. Immunol. 1994; 152: 2768-2777.
88. Shaw AR, Domanska A, Mak A, Gilchrist A, Dobler K, Visser L, Poppema S, Fliegel L, Letarte M, Wilett BJ. Ectopic expression of human and feline CD9 in human B cell line confers beta 1 integrin-dependent motility on fibronectin and laminin substrates and enhances tyrosine phosphorylation. J. Biol. Chem. 1995; 267 24092-24099.

89. Ikeyama S, Koyama M, Yamaoko M, Sasada R, Mitake M. Suppression of cell invasion and metastasis by transfection with human motility-related protein-1(MRP-1/CD9) DNA. J. Exp. Med. 1993; 177: 1231-1237.
90. Behr S, Schriever F. Engaging CD9 or target of an anti-proliferative antibody on human B lymphocyte induces binding of B cells to interfollicular stroma of human tonsils via integrin α4β1 and fibronectin. J. Exp. Med. 1995; 182: 1191-1199.
91. Si Z, Hersey P. Expression of neuroglandular antigen and analogues in melanoma, CD9 expression appears inversely related to metastatic potential of melanoma. Int. J. Cancer. 1993; 54: 37-43.
92. Miyake M, Nakano K, Itoi SI, Koh T, Taki I. Motility-related proteins-I (MRP-I/CD9) reduction as a factor of poor prognosis in breast cancer. Cancer Res. 1996; 56: 1244-1249.
93. Dong JT, Lamb PW, Rinber-Schaeffer CW, Vekanovic J, Ichikowu T, Isaacs JT, Barrett JC. KAI1, a metastasis suppressor gene for prostate cancer on chromosome 11p11.2. Science. 1995; 268: 884-886.
94. Testa JE, Brooks PC, Lin JM, Quigley JP. Eukaryotic expression cloning with an anti-metastatic monoclonal antibody identifies a tetraspanin (PETA-3/CD151) as an effector of human tumor cell migration and metastasis. Cancer Res. 1999; 59: 3812-3820.
95. Mazar AP, Henkin J, Goldfarb RH. The urokinase plasminogen activator system in cancer: implications for tumor angiogenesis and metastasis. Angiogenesis. 1999; 3: 15-32.
96. Wolfsberg TG, Primakoff P, Myles DG, White JM. ADAM, a novel family of membrane proteins containing a disintegrin and metalloproteinase domain: multipotential functions in cell-cell and cell-matrix interactions. J. Cell Biol.1995; 131: 275-278.
97. Ossowski L, Aguirre-Ghiso JA. Urokinase receptor and integrin partnership: coordination of signaling for cell adhesion, migration and growth. Curr. Opin. Cell Biol. 2000; 12: 613-620.
98. Brooks PC, Stromblad S, Sanders LC, Stetler-Stevenson WG, Quigley JP, Cheresh DA. Localization of matrix metalloproteinase MMP-2 to the surface of invasive cells by interaction with integrin αvβ3. Cell. 1996; 85: 683-693.
99. Aguirre JA, Kovalski K, Ossowski L. Tumor dormancy induced by down regulation of urokinase receptor in human carcinoma involves integrin and MAPK signaling. J. Cell Biol. 1999; 147: 89-103.
100. Simon DI, Wei Y, Zhang L, Rao NK, Xu H, Chen Z, Liu Q, Rosenberg S, Chapman HA. Identification of a urokinase receptor-integrin interaction site. J. Biol. Chem. 2000; 275: 10228-10234.
101. Sitrin RG, Todd RF, Petty HR, Brock TG, Schollenberger SB, Albrecht E, Gyetko MR. The urokinase receptor (CD87) facilitates CD11b/CD18-mediated adhesion of human monocytes. J. Clin. Invest. 1996; 97: 1942-1951.
102. Wei Y, Yang X, Liu Q, Wilkins JA. Chapman HA. A role for caveolin and the urokinase receptor in integrin-mediated adhesion and signaling. J Cell Biol. 1999; 144: 1285-1294.
103. Byzova TV, Plow EF. Activation of αvβ3 on vascular cells controls recognition of prothrombin. J. Cell Biol. 1998; 143: 2081-2092.
104. Brooks PC, Silletti S, von Schalscha TL, Friedlander M, Cheresh, DA. Disruption of angiogenesis by PEX, a noncatalytic metalloproteinase fragment with integrin binding activity. Cell. 1998; 92: 391-400.

Chapter 5

BRAIN-METASTASIS ASSOCIATED GENES

Dario Marchetti

Department of Neurosurgery, The University of Texas-Houston Health Science Center, Houston, TX

1. INTRODUCTION

Brain metastases represent an intriguing and significant area of cancer research due to their increasing incidence and to the fact that, despite the relevance of brain metastases as a clinical problem, relatively few research efforts have been directed toward understanding their origins and biology, either at the basic or clinical science level. Since the brain microvasculature represents an exceedingly small fraction of the total microvasculature, much smaller than the relative proportion of tumors that metastasize to the brain, the process of brain metastasis is thought to be selective and nonrandom (1-5). Although metastasis formation in other organs may be tolerated or remain asymptomatic, once metastatic cells colonize the brain, tumor growth often results in a rapid decline in the quality of life and ensuing death. Problems in the difficulty of treating brain metastases include obstacles in directing drug delivery as well as problems like cerebral edema and/or conditions detrimental to patient survival (1).

The brain is a unique target for tumor invasion and metastasis formation (2-4). Besides being confined by the skull, it is highly sensitive to the slightest change in the local microenvironment. The brain is also surrounded by a formidable barrier, called blood-brain barrier (BBB). The BBB is a physiologic and anatomic structure that is defined by tight junctions between the brain endothelial cells, a relatively thick basement membrane, and an underlying layer of astrocytes that strictly regulate the flow of ions, nutrients, and cells into the brain. To metastasize to the brain, blood-borne malignant tumor cells must therefore attach to microvessel endothelial cells, respond to brain-derived survival and growth factors, invade the BBB, and relate with the normal cells present in the brain microenvironment (3). This implies that specific tumor-brain cell interactions, mediated by defined sets of properties and factors such as those involved in malignant adhesion, invasion, and growth, are important and critical. The involvement of particular molecular determinants and their roles in the development of brain metastases are the subject of this review.

D.R. Welch (ed.), Cancer Metastasis – Related Genes, 89–108.

2. MOLECULAR DETERMINANTS OF BRAIN METASTASES IN MELANOMA

A malignancy with one of the highest frequencies of brain metastasis formation and one of the most rapidly increasing cancers among young adults is malignant melanoma. Patients with disseminated malignant melanoma frequently develop metastatic lesions in the brain and spinal cord that can result in severe and debilitating neurological complications (1, 2). While approximately 13% of all cancer patients will present clinically with symptomatic complications related to brain metastases, almost 40% of melanoma patients will be treated for complications due to lesions to the central nervous system (CNS). At autopsy, 70-80% of malignant melanoma patients have CNS metastatic deposits (6, 7).

Melanomas undergo progressive changes during their malignant progression. Of the phenotypic changes that occur during their metastatic spreading, differences in the expression of receptors for paracrine growth factors and their production at the target organ sites are important (8-10). While advances have been made toward understanding the roles of various growth factors and receptors in the progression of malignant melanoma to highly aggressive brain-metastatic disease, underlying mechanisms remain largely unknown. We have demonstrated that a family of neurotrophic factors, the neurotrophins (NT), and their receptors (NTR), modulate the invasion of melanoma cells to the brain and the activity of an extracellular matrix (ECM) degradative enzyme, called heparanase (11-13). Heparanase is a determinant for the successful degradation of the BBB basement membrane and the tumor-surrounding ECM: its activity degrades heparan sulfate proteoglycans (HSPGs) (14, 15), which are important BM/ECM components (16), at sites within HS linear chains.

The significance of the NT/NTR axis in modulating the brain-metastatic properties exhibited by melanoma cells is thought to be critical in allowing malignant cells to survive and invade such an unusual organ compartment like the brain, where paracrine growth factors differ from most other organs and where elevated NT synthesis occurs.

We have formulated the hypothesis that brain metastases essentially represent a traumatic event related to brain-injury processes (17). It centers on experimental evidences from our and other laboratories demonstrating that NT/NTR are important in melanoma progression to the brain and that heparanase is critical in the metastatic process being a potential target for antimetastasis drugs. Following mechanical/chemical brain insults increased NT/NTR presence is imperative in neuronal regeneration (17-19). These changes can be paralled by brain invasive melanoma cells whose colonization within the brain microenvironment trigger NT production and their secretion by surrounding brain cells as a response to the invasion event. Similarly, melanoma cells overexpressing NTR can benefit from such a synergistic microenvironment in terms of survival, growth, and further invasion into the brain parenchyma. NTR present on the the cell-surface of melanoma cells thus play important roles in melanoma progression to the brain while NT-regulated heparanse can be critical to this proces.

Of equal importance, we have postulated two roles for NT-regulated heparanase in metastatic melanoma: one is that heparanase is relevant for the extravasation of melanoma cells into secondary organs by degrading HSPGs at HS intrachain sites, while the second pertains on heparanase ability to release HS-bound angiogenic factors at the metastatic site, with the brain being the ideal environment for its high NT levels.

3. ROLES OF NEUROTROPHINS, NEUROTROPHIN RECEPTORS IN BRAIN METASTATIC EVENTS

Because of their many effects on neuronal cells, neurotrophins (NT) are one of the best example of neurotrophic factors (18, 19). Neurotrophins can promote the survival, differentiation, and targeted tissue invasion of certain neuronal subpopulations (17). NT are a family of small, highly basic homodimers (~26.5 kDa, P.I. = 9.4) (19-23). In addition to nerve growth factor (NGF), the prototypic NT, mammalian NT family consist of three other members: brain-derived neurotrophic factor (BDNF), neurotrophin-3 (NT-3) and NT-4/5. All NT exhibit neurotrophic properties and are highly conserved in the amino acid sequences (19, 23, 24). Additionally, they are produced in the brain at relative high levels and, at least for NGF, they are under the regulation of astrocytic cytokines (22). The essential role of NT is illustrated by the fact that homozygous NT knockout mutations in mice have proven lethal during early post-natal development (20). These and previous observations have substantiated the so-called 'neurotrophic factor hypothesis'. This hypothesis was originally presented to account for the massive neuronal cell death in developing embryos and based on the premise that the survival of developing neurons is dependent on the supply of NT synthesized in limiting quantities from regions innervated by these neurons (20). It addressed two aspects of neuronal dependence on target tissues they innervate: one is neuronal survival, the other is growth and arborization of neuronal axons and dendrites. Additional recent evidences also suggest that neurons may be supplied with critical NT through both autocrine and paracrine mechanisms, the latter from cells along the axonal shaft or in the vicinity of neuronal cell bodies and dendritic networks (for review see 20). Applying these concepts to the large number of metastatic cells in the blood circulation or at the implantation sites can partially explain the survival and invasive capabilities of only certain neoplastic cells with specific properties to survive under growth-limiting conditions and in a specific microenvironment. This may be particularly true for melanoma cells that, as tumor of neuroectodermally-derived melanocytes, share a common embryologic origin with the most studied and NT-responsive neuronal populations.

NT biological effects are mediated through two unrelated classes of cell surface membrane receptors. NT receptors (NTR) include two affinity types, a high-affinity receptor ($K_D \sim 1 \times 10^{-11}$ M) and a low-affinity receptor class

($K_D \sim 2 \times 10^{-9}$ M) (6, 9). The gene for the human low-affinity NGF receptor, called $p75^{NTR}$, was cloned by Moses Chao's group (21) and encodes a 75 kDa cell surface glycoprotein (~ 90 kDa in melanoma due to augmented receptor glycosilation) that is capable of binding all NT members. Sequence analysis of $p75^{NTR}$, however, indicates that the molecule lacks a tyrosine kinase consensus sequence in the cytoplasmic domain (21). The search for high-affinity NTR with tyrosine kinase activity involved several independent observations that identified the Trk family as the high-affinity NTR (23-26). The Trk family consists of multiple molecules (mainly three that are called TrkA, TrkB, and TrkC respectively) with degrees of binding specificity for the different members of the NT family. They all encode cell surface molecules of 140-145 kDa that are known to also exist as truncated forms or contain inserts in their tyrosine kinase domains (23).

It is now apparent that $p75^{NTR}$ has alternative functions from Trk that depend on the cellular context in which it is expressed. It can possess a signaling function operating independently of Trk presence (24-27) and involving ceramide (26), a metabolic product of gangliosides (28). In addition to differentiation or survival cues, $p75^{NTR}$ provides retrograde transport to neuronal cell types (28), triggering apoptosis in certain virally transformed neuronal cells (29) or survival when expressed in neutrophils (30). In this regard, $p75^{NTR}$ is analogous to other members of the tumor necrosis factor receptor superfamily (31). Interestingly $p75^{NTR}$ is overexpressed by aggressive malignant melanoma cells to the brain (11-13, 32).

We have made the following observations: 1) NT responsiveness, invasion and ECM degradation correlated with aberrant overexpression of $p75^{NTR}$ but not Trk, in brain-colonizing murine and human melanoma cells (11-13); 2) an inverse relationship was found between $p75^{NTR}$ and NT expression at the invasion front of brain-metastatic melanoma clinical specimens (33) and; 3) NT stimulated activation of the nonreceptor tyrosine kinase c-Yes, but not c-Src, in brain-metastatic melanoma cells (34).

These observations are important because they assert NT as paracrine growth factors involved in brain metastases formation. They also suggest that $p75^{NTR}$ can act as a molecular determinant for brain metastasis and can be modulated to affect the brain-metastatic process.

It is in fact known that the growth of tumors depends on cell proliferation, and this usually occurs in response to specific growth factors from either autocrine or paracrine sources (3, 4). During their growth, tumors frequently contain highly malignant cells that become clonally selected from the proliferating growth faction of cells and eventually become dominant cell subpopulations (35, 36). In brain metastasis, invasive tumor cells responsive to brain trophic factors such as NT and that possess aberrant levels of their receptor $p75^{NTR}$ may survive host selective pressures and proliferate, further diversify, and become clonally dominant.

3.1 p75 Neurotrophin Receptor ($p75^{NTR}$) Involvement in Tumor Invasion and Brain Metastasis Formation

During malignant progression, tumor cells undergo genomic changes and show differences in the expression of particular gene products (36). For example, malignant melanoma cells from patient brain metastases exhibit characteristic chromosomal alterations, such as a high frequency of translocation or deletion breakpoints at 11q23, terminal translocations at 17q25, or an isochromosome for the long arm of chromosome 17 (37). As mentioned, human melanoma cells also show progression-associated increases in the expression of $p75^{NTR}$ (8, 13, 38), whose gene is located at 17q21-22.

We have examined the roles of neurotrophin receptors in human melanoma invasion and brain colonization by using the MeWo human melanoma cellular system. It consists of parental MeWo cells that exhibit intermediate metastatic potential, nonmetastatic 3S5 cells and, notably, 70W variant cells that have the capacity to form brain colonies in nude mice as first example of human melanoma tumor capable of brain colonization *in vivo* (39). We found that overexpression of $p75^{NTR}$ is associated with brain colonization and neurotrophin-mediated enhancement of extracellular matrix invasion (11-13). The expression of NGF-binding $p75^{NTR}$, but not TrkA, on brain-metastatic 70W cells was determined either at the mRNA level (by Northern blotting) and at the protein level (by immunoprecipitation analysis of radioiodinated cell surface proteins) (13). Specific kinetic analyses confirmed the presence of low-affinity NT binding as NTR binding state employing a newly developed NT binding assay that, coupled with use of the LIGAND program, has a detection limit as low as 25 fmoles/assay (40). Anti-$p75^{NTR}$ monoclonal antibody specifically precipitated higher amounts of an appropriately sized, radioiodinated $p75^{NTR}$ component in lysates of the brain-metastatic melanoma cells (13). The expression of $p75^{NTR}$ on the surface of the MeWo parental cell line was low but detectable, and we could not detect $p75^{NTR}$ on nonmetastatic 3S5 cells. In addition, we could not detect the expression of the *trkA* gene or $p140^{TrkA}$ on the surfaces of any of the human MeWo melanoma cells (13).

We next examined whether NGF binding to $p75^{NTR}$ leads to the formation of NGF receptor complexes ($p75^{NTR}$ dimers) in MeWo cells. Immunoprecipitation was performed in the presence of excess exogenous NGF. Addition of excess NGF caused a significant increase in the amount of immunoprecipitate formed and an augmented formation of high molecular weight immunocomplexes. For example, an approximately 200 kDa band was found by prior treatment with NGF or in the presence of excess exogenous NGF (13). Complexes of similar molecular weight size and possessing high affinity binding values for NGF have been reported on the human A875 melanoma cells (derived from a patient brain with metastatic involvement) (41) after NGF treatment. In addition, we observed influences of cell-cell contact on $p75^{NTR}$ expression by immunofluorescence analyses using the anti-$p75^{NTR}$ monoclonal antibody ME-20.4 (38). High-density cultures of metastatic melanoma cells showed significant reductions in $p75^{NTR}$ expression, in inverse relation to the extent of cell-cell contact (42), possibly related to the formation and aggregation of receptor complexes (41) and their endocytosis following NGF treament.

3.2 Neurotrophin Production and Brain Invasion

If neurotrophins are important in brain invasion and colonization by metastatic tumor cells, then there should be evidences that they are present at the invasion front of brain metastases. Therefore, we examined whether brain-invading melanoma cells induced changes in NT concentration or NT distribution at the invading edge of melanoma tumors *in vivo* (33). Brain tissue samples from human melanoma metastases and uninvolved brain tissues progressively distant from the melanoma lesions were examined immunohistochemically (for protein assessment) and/or by employing nonradioactive *in situ* hybridization techniques (for mRNA assessment) for presence and distribution of NT/NTR. Firstly, using digoxigenin-labeled sense/antisense riboprobes against $p75^{NTR}$ and NGF for *in situ* hybridization, we determined whether the expression of $p75^{NTR}$ and NGF mRNA were related to brain metastasis in human melanoma. We detected $p75^{NTR}$ mRNA at the invasion front of human melanoma brain metastases, whereas $p75^{NTR}$ expression was not found in adjacent tissue. In contrast, human NGF mRNA levels were increased in tissues surrounding the melanoma lesions, supporting the notion that NGF and NT are important in determining melanoma brain-metastatic microenvironment. Secondly, using antibodies specific for $p75^{NTR}$, Trk, NGF and related NT, we found high but heterogeneous levels of $p75^{NTR}$ and TrkC (putative high-affinity binding receptor for the neurotrophin NT-3) expression in malignant melanoma metastatic to the brain. Lower levels of expression were found in primary melanoma as or in metastatic melanomas targeting sites other than brain. Finally, we found elevated levels of synthesis of NGF and NT-3, but not BDNF or NT-4/5, in brain tissues surrounding melanoma lesions. These studies which used clinical specimens further support critical roles for NT/NTR axis in the progression of melanoma and its capability to brain colonization.

4. HEPARANASE FUNCTIONALITY IN MELANOMA PROGRESSION TO THE BRAIN-METASTATIC PHENOTYPE

During metastases formation, migrating tumor cells are confronted by natural tissue barriers, such as basement membranes (BM) that surround the blood vessels (43-47) or ECM that is an integral part of the BBB. The ability of malignant cells to penetrate these barriers depends upon the production and activation of enzymes capable of ECM degradation (48-52). ECM/BM are rigid structures formed from such macromolecules as type IV collagen, laminin, entactin, nidogen, fibronectin, and proteoglycans (16, 45-47), one type being HSPGs. We know that HSPGs play a central role in embryonic morphogenesis, angiogenesis, neurite outgrowth, and tissue repair (53-58). ECM/BM HSPGs also provide a readily available storage depot for growth factors and cytokines (55, 56, 96). Since HSPGs are now recognized as active biological modulators, their degradation at the level of HS chains is expected to have significant regulatory consequences in cancer metastasis. Indeed, HSPG catabolism is observed in inflammation, wound repair, diabetes, and neoplasia, including melanoma (57-67). Melanoma heparanase responsible for HS degradation cleaves HS at specific intrachain sites resulting in the formation of fragments of discrete molecular weight size (62, 64). Therefore, heparanase was identified as belonging to the family of endo-β-D-glucuronidases (59-66). Heparanase differs from heparinases or HS-specific elimination enzymes, cleaving HS into characteristic large molecular weight fragments, approximately one-third of the original size (63). Heparanase activities have been also described in immune system and cancer cells other than melanoma (67-87). Increased levels of heparanase activity are associated with metastatic melanoma and other invasive tumors types and several evidences have demonstrated its role in tumor cell invasion into distant organs (65, 67, 68-71).

We have made the following observations: 1) highly invasive human melanoma cells degrade labeled ECM HS and purified HS cell-surface subpopulations faster than do sublines of lower metastatic potential (11, 12); 2) a heparanase activity is responsible for HS degradation (11, 12, 62); 3) in correlation with increased invasiveness, select NT augment heparanase production in brain-metastatic melanoma, making it a major candidate enzyme responsible for ECM degradation (11, 12) and; 4) heparanase recognizes specific motifs within HS chains associated with the binding domains to angiogenic factors and with the binding of an HS-interacting protein (HIP), recently cloned and characterized (66).

Of note, human heparanase has not been purified nor well characterized or cloned until very recently (14, 84-87). Therefore, molecular tools to explore the potentially important roles of heparanase in disease have been lacking for almost 20 years since the first reports describing the enzymatic activity. Interestingly, the newly discovered cDNA sequences of human heparanase derived from normal and tumor cells represent the same gene (14, 84-87).

We have postulated that heparanase play two critical roles in the biology of brain metastasis which are 1) in local invasive processes by degrading HS of HSPGs and, 2) in the release of HS-bound angiogenic factors at the metastatic site with the brain being the ideal environment because of its high levels of NT production. Heparanase can therefore dually be relevant in brain-metastatic melanoma in consideration of the strong angiogenic properties exhibited by melanoma cells in the brain. In fact, although metastasizing cancer cells may produce as many as 15 different matrix-digesting proteases, the new findings show there is only one heparanase. Heparanase inhibition not only can inhibit cancer cells' ability to invade, but also hinders the formation of new blood vessels that feed tumors, a process called angiogenesis. Another important aspect in the area of brain invasion and metastasis are therefore investigations aimed to assess heparanase contributions in angiogenic events. In the multistep process of tumorigenesis, angiogenesis is an early and essential requirement, characterized by uncontrolled endothelial cell proliferation (88, 92). Fibroblast growth factors (FGFs), a family of structurally related polypeptides, are highly mitogenic for vascular endothelial cells and among the most potent inducers of neovascularization (88). This is valid for acidic and basic FGF (bFGF), in particular (88-95). ECM HSPGs may serve as a reservoir for angiogenic factors that can be extracted from subendothelial ECM produced *in vitro* (88, 96). Despite the bFGF ubiquity in normal tissues, endothelial cell proliferation in these tissues is usually very low (97), suggesting that bFGF is sequestered from its site of action in such a reservoir. Reports have also asserted the importance of ECM HSPGs as promoting bFGF-receptor binding (56). Down-regulation of perlecan, the major BM HSPG, expression has specific effects on bFGF-mediated mitogenic response, basal growth, and bFGF binding (98). Displacement of bFGF from the ECM by heparanase can provide a novel mechanism for induction of neovascularization in normal and pathological conditions (69, 71, 99). Likewise, platelets and activated cells of the immune system that are often attracted by tumor cells may indirectly stimulate tumor angiogenesis via a heparanase activity (100-102). Several studies have indicated that heparin and HS inhibit the mitogenic activity of angiogenic bFGF and at the same time stabilize and protect the molecule from inactivation (91-93). bFGF is stored in ECM in a highly stable, inactive form. Its release from ECM as a complex with HS fragments may yield a form of bFGF that is more stable than free bFGF and capable of binding the high-affinity plasma-membrane receptors.

We have evaluated the ability of human melanoma heparanase, purified through chromatographic techniques, to modulate bFGF activity as controlled release from HSPGs by observing *in vivo* angiogenesis induction (103). We consider these studies of significance because they will provide knowledge of 1) direct participation of heparanase as purified enzyme in neovascularization and, 2) its interactions with the local environment by releasing angiogenic and active growth factors that are firmly HSPG-sequestered.

4.1 Neurotrophins Enhance Invasion and Heparanase Production in Brain-metastatic Melanoma Cells

Neurotrophins can enhance the invasive properties of certain melanoma cells (11-13). This may be required to penetrate the BBB. We have examined the effects of NT on invasion of brain-metastatic melanoma cells using filters (Transwell™) coated with Matrigel™ or purified ECM HSPGs in chemoinvasion assays. As a chemoattractant we placed brain microvessel endothelial cell-conditioned medium in the lower chamber; (endothelial cell motility factors from lung or liver did not substitute for brain endothelial cell motility factors). Treatment of the prototypic NT, nerve growth factor (NGF), resulted in a 7.9-fold increase in the extent of matrix invasion of the 70W cells, corresponding with increases in heparanase and, to a lesser extent, in type IV collagenase (MMP-2; 13) activities. NGF caused less matrix invasion and heparanase induction by MeWo parental cells, and there was no increase in invasion or heparanase NGF regulation in nonmetastatic 3S5 cells (11-13). Similar results were found replacing NGF for the other NT member NT-3, but not for BDNF or NT-4/5, consistent with presence of specific NTR on the cell-surface of brain-metastatic melanoma cells (12).

The ability to invade was only apparent if the human melanoma cells were grown on ECM or HSPGs and placed on an invasion substrate in the presence of brain microvessel endothelial cell-conditioned medium. This suggested that, in addition to their response to neurotrophic factors, melanoma cells must interact with the appropriate matrix and receive paracrine motility signals to be highly invasive (4, 5, 13).

5. BRAIN METASTASIS AND THE BRAIN ENVIRONMENT

Cellular and molecular passages into the brain are strictly regulated by the BBB. Metastatic cells must breach this barrier to invade and colonize the brain parenchyma. As discussed above, invasion into the brain requires that metastatic cells increase their expression of certain cell surface receptors, degradative enzymes, growth factors and cytokines. Additionally, they must respond to invasion-stimulating cytokines such as NT and other paracrine growth factors. To penetrate the BBB, brain-metastasizing melanoma cells express high levels of basement membrane hydrolytic enzymes, such as type IV collagenases, cathepsins, plasminogen activators, and, of importance, NT-regulated heparanase (11, 12, 62). Although highly metastatic cells generally expressed higher amounts of degradative enzymes than nonmetastatic cells, some of these enzymes like heparanase are induced to even higher levels by the presence/production of NT as paracrine invasion factors in the microenvironment. Additionally, heparanase can also be provided by certain normal cells, such as microvessel endothelial cells (62) or, as for the brain, astrocytes (104; see also section 5.1). If the appropriate paracrine signals are received by malignant cells, they can be stimulated to increase synthesis and release of heparanase/ECM degradative enzymes. The final result is BBB degradation. Neoplastic cells with the potential to colonize the brain may therefore express unique molecular determinants and may also respond to brain-derived factors important for their invasion in the brain (16).

Melanoma cells that have successfully penetrated the first two zones of the BBB, the brain microvessel endothelium and the thick basement membrane (BM), must then establish interactions with, and respond to, astroglial cells. Astrocytes, which are thought to provide structural support for CNS neurons, have been reported to play an important role in maintaining homeostasis. Astrocytes responses to extracellular stimuli have been widely studied (20, 108, 109). Among the different cytokines that are produced by astrocytes, NGF seems to play a specific role in brain-metastatic cell invasion. Exogenously supplied NGF, and possibly other NT, may stimulate melanoma cells to invade further into the brain parenchyma by binding to overexpressed cell surface $p75^{NTR}$, thus triggering heparanase production (33, 110). Interestingly, brain-metastatic melanoma cells express interleukin-1beta (IL-1β), transforming growth factor-beta1 (TGF-β1), and basic fibroblast growth factor (bFGF) (105), which have been shown to stimulate NGF synthesis by astrocytes (20, 109, 111-113). Therefore, brain metastasis outcome can depend on interactions with, and responsiveness to, astrocyte-released NGF which aid in the survival, growth, and invasion of $p75^{NTR}$-expressing melanoma cells in the CNS.

5.1 Astrocytes Contribute to the Brain-metastatic Specificity of Melanoma Cells by Producing Heparanase

Astrocytes are among the first brain cells encountered by extravasating melanoma cells and able to produce NGF, the prototypic NT (108). Furthermore, they are capable of binding NT since they express members of the Trk receptor family and the p75 low-affinity NT receptor ($p75^{NTR}$; 108). They are relevant because astrogliosis is a predictable pathologic trauma response to brain invasive events: injury-reacting astrocytes are frequently found in areas surrounding melanotic lesions. They may thus play important roles in the development of brain metastases. To test this hypothesis, we employed purified *in vitro* astrocyte cultures and investigated the presence of heparanase in astrocyte cultures. Primary glial cells were obtained from newborn rat and mouse cerebra using established purification methods (106). The majority of astrocytes appeared as large and flat cells with fibroblast-like morphology. Their identification as astrocytes was confirmed by positive immunoreactivity with an antibody against the astrocyte-specific intermediate filament glial fibrillary acidic protein (GFAP) (106). Neither vimentin, myelin basic protein, or gangliosides, as recognized by their respective monoclonal antibodies, were expressed by the astrocyte cultures, confirming purity of astrocytic preparations. We next examined the astrocyte cultures for expression of heparanase. A specific heparanase transcript was detected by semiquantitative reverse transcriptase-polymerase chain reaction (RT-PCR). This transcript was upregulated in astrocytes previously incubated with purified and biologically active NGF (104). Similar results were obtained using human brain-metastatic 70W cells. Heparanase activity was also detectable and NGF-regulated in cellular extracts from purified astrocytes. This was shown by the appearance of distinct HS degradation products detected in agarose gel shift assays or by HPLC analysis (104).

We also analyzed heparanase activity in brain-metastatic melanoma cells and astrocytic cell populations in logarithmic growth by obtaining HPLC-derived elution profiles of HS digested products at various incubation times (11). Cultures of highly brain-metastatic cells showed a gradual and time-dependent increase in heparanase activity for up to 72 hr. After this time interval, the levels of heparanase in the cultures plateaued. Similarly, cultures of astrocytes also produced heparanase in a time-dependent manner. Moreover, coincubation of brain-metastatic melanoma cells and astrocytes in equicellular numbers resulted in a super-additive increase of enzymatic activity above that expected by both cell types.

To determine if poorly metastatic melanoma cells acquire an increased metastatic potential following heparanase gene upregulation eukaryotic expression vectors were constructed that contained the full-length human heparanase cDNA, and used them to transfect melanoma cells. This was done by stably transfecting melanoma cells of low invasion potential and heparanase content (either of human 3S5, A375P or murine B16-F1 melanoma cells) (107). Heparanase preparations from stably transfected cell clones were able to fully digest HS chains from ECM-associated, purified HSPGs subpopulations. The heparanase gene transfected cells expressed increased heparanase, as detected by Western blotting analysis using a monoclonal antibody against human active heparanase (HEP-MAb) (104), and possessed high levels of heparanase activity (107). Transfection of the human heparanase gene into these cells resulted in functional enzymatic activity and also significantly increased (7 to 14-fold) invasion of the transfected cells using *in vitro* chemoinvasion assays with purified HSPG as a barrier (104, 107).

Finally, we incubated brain-metastatic melanoma cells with astrocyte-conditioned medium (ACM) and examined its effects on their invasive behaviors. Consistent increases in *in vitro* invasion were found following exposure of these cells to ACM. Invasion was most pronounced using ACM from NGF-treated astrocytes, and the invasion effects of ACM were completely abrogated in presence of HEP-MAb (104). The invasion enhancement caused by this NGF treatment was also abolished in presence of a neutralizing NGF MAb confirming the relevance of melanoma cells and astrocyte heparanases and their NT-regulation in invasion events.

5.2 Reciprocal Cytokine Regulation of Brain-metastatic Cell Growth

The induction of reactive astrocytes during tumor cell lodgment in the brain and its subsequent invasion may promote the secretion of trophic and growth factors (114). These in turn may cause differences in the organ specificity of metastatic cells by the requirement of response to these growth factors. Therefore, brain metastasis formation may depend on interactions with and responses to astrocyte-released molecules which aid in the survival, growth, and invasion of melanoma cells in the CNS (5, 114). This reciprocal cytokine regulation of growth probably also extends to parenchymal cell types as well as to the ECM as growth factors depot. The observation of metastatic growth explosion at certain organ sites can be therefore explained by the notion that reciprocal release of cytokines and other factors by tumor and host cells stimulate the motility, invasion, and growth of both tumor and host cells. Tumor cells that colonize the brain can have unusual responses to various cytokines. For example, melanoma cells that colonize the meninges and ventricles are growth stimulated by TGF-β, while others that colonize the brain parenchyma are growth inhibited by TGF-β (115). Other growth factors, such as epidermal growth factor (EGF), interleukin-1beta (IL-1β), tumor necrosis factor alpha (TNF-α), acidic FGF (aFGF), and bFGF, can stimulate NGF synthesis by reactive astrocytes, a response that can be potentiated by treatment with dibutyryl-cAMP (111-113, 116). Other cell types that increase the synthesis of NGF in response to bFGF treatment are the meningeal fibroblasts (112). As stated, stimulatory cytokines, also inhibitory cytokines can be equally valid. For example, an important inhibitory cytokine for melanoma cells is interleukin-6 (IL-6) (117), produced by astrocytes following TNF exposure (118).

We consider this reciprocal cytokine regulation of invasion and growth as an important aspect of understanding the basic mechanisms responsible for brain metastases formation (110, 114). Furthermore, astrocytes are likely to be involved in the development of brain metastases since they contribute to the invasive capabilities of brain-colonizing melanoma cells. Our laboratory is the first to provide evidence that astrocytes are a source *per se* of heparanase, able to produce active heparanase and modulate *in vitro* cell invasion. Thus, our data support the concept that melanoma brain invasion results from establishing reciprocal circuits between the tumor cells and the normal glial cells present in the CNS (119) with cellular responses to brain tissue injury to be considered as a paradigm for brain-metastatic processes. Following mechanical/chemical brain insults, increased NGF/p75^{NTR} presence is imperative for regenerational events of injured areas (20, 114). These properties may be paralleled by brain-invasive melanoma cells whose invasion can trigger NT production (i.e., NGF) and NT-regulated heparanase secretion by brain cells as response to the invasion event. Melanoma cells overexpressing p75^{NTR} can benefit from such a synergistic microenvironment and further invade into the brain parenchyma.

Heparanase activity has also been detected in tumor cells other than brain-metastatic melanoma cells where it has been shown to be related to metastatic potential (11, 14, 15, 60, 63, 85, 86). We have provided a convincing evidence for direct roles of heparanase in tumor metastasis by the conversion of melanoma cells from nonmetastatic or poorly metastatic to metastatic behavior after stable transfection and overexpression of the heparanase gene (107). Moreover, we found a preferential expression of the heparanase gene in specimens of human invasive melanomas compared to noninvasive primary melanomas or normal tissues (107). Together, these data support the notion that heparanase can be necessary and essential for the metastasis of tumor cells to sites such as brain.

6. CONCLUSIONS AND FUTURE PERSPECTIVES

The brain is a unique microenvironment and only specialized tumor cells with certain properties have the capacity to home to, invade, and colonize this organ. These cells must attach to brain microvessel endothelial cells, invade the BBB by expressing high concentrations of heparanase (and possibly other degradative enzymes), survive by responding to NT and other brain trophic factors produced by normal brain cells or stored in the ECM, and proliferate by responding to paracrine growth factors. Although we now know much more about the molecular determinants underlying brain metastasis formation than ever before, there remains much to learn.

One future goal is to test in *in vivo* settings the hypothesis that p75^{NTR} overexpression and NT responsiveness are truly determinants for invasion and establishment of clonal dominance by human melanoma metastatic to the brain. Our laboratory has recently selected both high- and low-p75^{NTR} (p75$^{NTR-H/L}$) variants by fluorescence-activated cell sorting (FACS) and related *in vitro* invasion and NT responsiveness to p75^{NTR} content (42). By using these sorts and p75$^{NTR-H/L}$ subclones obtained from transfection studies, we will investigate the extent of p75$^{NTR-H/L}$ expressors to colonize the brain of nude mice. We will also use a dominant negative strategy that will allow interference with p75^{NTR}-mediated signalling after infection of melanoma cells with recombinant adenovirus containing a p75^{NTR} dominant negative mutant. Brain tumor growth by dominant-negative inhibition of p75^{NTR} will be determined by delivering the viral-infected cells in nude mice. These investigations are of significance because they will provide rigorous and unequivocal proof of p75^{NTR} as key determinant responsible for brain metastases formation. Lack of such knowledge is an important problem, because, without it, acquiring the ability to therapeutically inhibit brain-metastases through p75^{NTR} mechanisms is unlikely.

Additional future aims will include the complete characterization of NT-regulated heparanase gene expression in melanoma progression and mechanistic studies (*in vitro* as well as *in vivo*) to clarify its involvement in both invasive and angiogenic processes. Now that the heparanase gene has been cloned and its cDNA sequence is known, more complete strategies can be formulated to provide direct evidences for heparanase roles in metastasis, brain metastasis in particular. To this end, a detailed study of the enzyme intracellular localization, membrane association, and secretory properties in cultured cells and tissue specimens has been initiated.

Questions related to the potential benefit of the therapeutic suppression of heparanase in brain-metastatic events remain unanswered. Further studies with astrocytic sublines, permissive or inhibitory to astrocyte boundary formation, as well as *in vivo* experiments using specific inhibitors of heparanase action will be useful to address these questions. Elucidating these mechanisms of action may some day result in the development of new and effective therapeutic approaches to the treatment of brain metastases.

ACKNOWLEDGMENTS

These studies were partially supported by grants from the U.S. National Institute of Health, the Skin Cancer Research Program at U.T. M.D. Anderson Cancer Center, and by the Vivian L. Smith Foundation for Neurological Research.

REFERENCES

1. Wright DC, Delaney TF. Treatment of metastatic cancer to the brain. In Cancer: Principals and Practice of Oncology, DeVita VT, Hellmann S, Rosenberg SA, eds., New York: J.B. Lippincott, 1989.

2. Steck PA, Nicolson GL. Metastasis to the central nervous system. In Molecular Genetics of Nervous System Tumors, Levine A, Schmidek H, eds., New York: Wiley, 1993.
3. Nicolson GL. Paracrine and autocrine growth mechanisms in tumor metastasis to specific sites with particular emphasis on brain and lung metastasis. Cancer Metastasis Rev 1993; 12: 325-43.
4. Nicolson GL. Cancer progression and growth: relationship of paracrine and autocrine growth mechanisms to organ preference of metastasis. Exp Cell Res 1993; 204: 171-80.
5. Nicolson GL, Menter D, Herrmann JL, Cavanaugh P, Jia L-B, Hamada J, Yun Z, Marchetti D. Tumor metastasis to brain: role of endothelial cells, neurotrophins and paracrine growth factors. Crit Rev Oncogenesis 1994; 15: 451-71.
6. Stehlin JS, Hills WJ, Rufino C. Disseminated melanoma: biologic behavior and treatment. Arch Surg 1967; 94: 495-501.
7. Sugarbaker EV. Patterns of metastases in human malignancies. Cancer Biol Rev 1981; 2: 235-78.
8. Herlyn M, Thurin J, Balaban G, Bennicelli JL, Herlyn D, Elder DE, Bondi E, Guerry D, Nowell P, Clark WH, Koprowski H. Characteristics of cultured human melanocytes isolated from different stages of tumor progression. Cancer Res 1985; 45: 5670-76.
9. Herlyn M, Kath R, Williams N, Valyi-Nagy I, Rodeck U. Growth regulatory factors for normal, premalignant and malignant human cells. Adv Cancer Res 1989; 54: 213-34.
10. Albino AP, Davis BM, Nanus DM. Induction of growth factor RNA expression in human malignant melanoma: markers of transformation. Cancer Res 1991; 51: 4815-20.
11. Marchetti D, Menter DG, Jin L, Nakajima M, Nicolson GL. Nerve growth factor effects on human and mouse melanoma cell invasion and heparanase production. Int J Cancer 1993; 55: 692-9.
12. Marchetti D, McQuillan DJ, Spohn WC, Carson DD, Nicolson GL. Neurotrophin stimulation of human melanoma cell invasion: selected enhancement of heparanase activity and heparanase degradation of specific heparan sulfate subpopulations. Cancer Res 1996; 56: 2856-63.
13. Herrmann JL, Menter DG, Hamada J, Marchetti D, Nakajima M, Nicolson, GL. Mediation of NGF-stimulated extracellular matrix invasion by the human melanoma low-affinity p75 neurotrophin receptor: melanoma p75 functions independently of *trkA*. Mol Biol Cell 1993; 4: 1205-16.
14. Toyoshima M, Nakajima M. Human heparanase. J Biol Chem 1999; 274: 24153-160.
15. Nakajima M, Irimura T, Nicolson GL. Heparanase and tumor metastasis. J Cell Biochem 1986; 36: 157-7.
16. Iozzo RV, Murdoch AD. Proteoglycans for the extracellular environment: clues from the gene and protein side offer new perspectives in molecular diversity and function. FASEB J 1996; 10: 598-614.
17. Raff MC, Barres BA, Burne JF, Coles HS, Ishizaki Y, Jacobson MD. Programmed cell death and the control of cell survival: lessons from the nervous system. Science 1993; 262: 695-700.
18. Snider WD. Functions of the neurotrophins during nervous system development: what knockouts are teaching us. Cell 1994; 77: 627-38.

19. Bradshaw RA, Blundell TL, Lapatto R, et al. Nerve growth factor revisited. Trends Biochem Sci 1993; 18: 48-52.
20. Snider WD. Functions of the neurotrophins during nervous system development: what knockouts are teaching us. Cell 1994; 77: 627-38.
21. Johnson D, Lanahan A, Buck CR, et al. Expression and structure of the human NGF receptor. Cell 1986; 47: 545-54.
22. Friedman WJ, Thakur S, Seidman L, Rabson AB. Regulation of nerve growth factor mRNA by interleukin-1 in rat hippocampal astrocytes is mediated by NFkappaB. J Biol Chem 1996; 271(49): 31115-20.
23. Meakin SO, Shooter EM. The nerve growth factor family of receptors. Trends Neurosci 1992; 15: 323-31.
24. Saltiel AR, Decker SJ. Cellular mechanisms of signal transduction for neurotrophins. BioEssays 1994; 16: 405-11.
25. Verdi JM, Birren SJ, Ibanez CF, et al. P75LNGFR regulates Trk signal transduction and NGF-induced neuronal differentiation in MAH cells. Neuron 1994; 12: 733-45.
26. Chao MV. Ceramide: a potential second messenger in the nervous system. Mol Cell Neurosci 1995; 6: 91-6.
27. Barrett GL, Bartlett PF. The p75 nerve growth factor receptor mediates survival or death depending on stage of sensory neuron development. Proc Natl Acad Sci USA 1994; 91: 6501-05.
28. Olshefski R, Taylor B, Heitger A, Hasegawa A, Ladish S. Induction of programmed cell death and immunosuppression by exogenous sphingolipids are separate processes. Eur J Biochem 1996; 241(1): 47-55.
29. Rabizadeh S, Oh J, Zhong LT, et al. Induction of apoptosis by the low-affinity NGF receptor. Science 1993; 261: 345-8.
30. Kannan Y, Usami K, Okada M, et al. Nerve growth factor suppresses apoptosis of murine neutrophils. Biochem & Biophys Res Comm 1992; 186: 1050-6.
31. Smith CA, Farrah T, Goodwin RG. The TNF receptor superfamily of cellular and viral proteins: activation, costimulation, and death. Cell 1994; 76: 959-62.
32. Menrad A, Herlyn, M. Tumor progression, biology, and host response in melanoma. Curr. Opinion in Oncol 1992; 4(2): 351-6.
33. Marchetti D, McCutcheon I, Ross JM, Nicolson GL. Inverse expression of neurotrophin receptor and neurotrophins at the invasion front of brain-metastatic human melanoma tissues. Int J Oncol 1995; 7: 87-94.
34. Marchetti D, Parikh N, Sudol M, Gallick GE. Enhancement of protein tyrosine kinase c-Yes, but not c-Src, by select neurotrophins in human brain-metastatic melanoma cells. Oncogene 1998; 16: 3253-60.
35. Nowell PC. The clonal evolution of tumor cell. Science 1976; 194: 23-8.
36. Nicolson GL. Tumor cell instability, diversification and progression to the metastatic phenotype: from oncogene to oncofetal expression. Cancer Res 1987; 47: 1473-87.
37. Morse HG, Gonzalez R, Moore GE, Robinson WA. Preferential chromosome 11q and/or 17q aberrations in short-term cultures of the metastatic melanoma resections from the brain. Cancer Genet Cytogenet 1992; 64: 118-26.
38. Ross AH, Grob P, Bothwell M, Elder DE, Ernst CS, Marano N, Ghrist BF, Slemp CC, Herlyn M, Atkinson B, Koprowski H. Characterization of nerve growth factor receptor in neural crest tumors using monoclonal antibodies. Proc Natl Acad Sci USA 1984; 81: 6681-85.

39. Ishikawa M, Fernandez B, Kerbel RS. Highly pigmented human melanoma variant which metastasizes widely in nude mice, including skin and brain. Cancer Res 1988; 48: 4897-4903.
40. Marchetti D, McManaman JL. Characterization of nerve growth factor binding to embryonic rat spinal cord motoneurons. J Neurosci Res 1990; 27: 211-8.
41. Buxser S, Puma P, Johnson GL. Properties of the nerve growth factor receptor: relationship between receptor structure and affinity. J Biol Chem 1985; 260:1917- 26.
42. Walch ET, Albino AP, Marchetti D. Correlation of overexpression of the low-affinity p75 neurotrophin receptor with augmented invasion and heparanase production in human malignant melanoma cells. Int J Cancer 1999; 82: 112-20.
43. Gladson CL. The extracellular matrix of gliomas: modulation of cell function. J Neuropath & Exp Neurol 1999; 58(10): 1029-40.
44. Gladson CL, Wilcox JN, Sanders L, et al. Cerebral microenvironment influences expression of the vitronectin gene in astrocytic tumors. J Cell Sci 1995; 108: 947-56.
45. Iida J, Meijne AM, Knutson JR, Furcht LT, McCarthy JB. Cell surface chondroitin sulfate proteoglycans in tumor cell adhesion, motility, and invasion. Sem Cancer Biol 1996; 7(3): 155-62.
45. Comito MA, Savell VH, Cohen MB. CD44 expression in neuroblastoma and related tumors. J Ped Hem/Oncol 1996; 19(4): 292-6.
46. Moscatello, DK, Santra M, Mann DM, McQuillan DJ, Wong AJ, Iozzo RV. Decorin suppresses tumor cell growth by activating the epidermal growth factor receptor. J Clin Inv 1998; 101(2): 406-12.
47. Liotta LA, Steeg PS, Stetler-Stevenson WG. Cancer metastasis and angiogenesis: an imbalance of positive and negative regulation. Cell 1991; 64(2): 327-36.
48. Himelstein BP, Canete-Soler R, Bernhard EJ, Dilks DW, Muschel RJ. Metalloproteinases in tumor progression: the contribution of MMP-9. Inv & Met 1994-95; 14(1-6): 246-58.
49. Powell WC, Matrisian LM. Complex roles of matrix metalloproteinases in tumor progression. Curr Topics in Micr & Immunol 1996; 213: 1-21.
50. Liotta LA, Thorgeirsson UP, Garbisa S. Role of collagenase in tumor cell invasion. Cancer Met Rev 1982; 1: 277-88.
51. Sloane BF, Honn KV. Cysteine proteinase and metastasis. Cancer Met Rev 1984; 3: 249-263.
52. Seftor RE, Seftor EA, Stetler-Stevenson WG, Hendrix MJ. The 72 kDa type IV collagenase is modulated via differential expression of alpha v beta 3 and alpha 5 beta 1 integrins during human melanoma cell invasion. Cancer Res 1993; 53(14): 3411-5.
53. Yanagishita M, Hascall VC. Cell surface heparan sulfate proteoglycans. J Biol Chem 1992; 267: 9451-54.
54. Iozzo RV. Cell surface heparan sulfate proteoglycan and the neoplastic phenotype. J Cell Biochem 1988; 37: 61-78.
55. Aviezer D, Levy E, Safran M, Svahn C, Buddeke E, Shmidt E, David G, Vlodavski I, Yayon A. Differential structural requirements of heparin and heparan sulfate proteoglycans that promote binding of bFGF to its receptor. J Biol Chem 1994; 269: 114-21.
56. McKeehan WL, Kan M. Heparan sulfate fibroblast growth factor receptor complex: structure-function relationships. Mol Repr & Dev 1994; 39(1): 69-81.
57. Ilhreke NS, Wrenshall LE, Lindman BJ, Platt JL. Role of heparan sulfate in immune system-blood vessel interactions. Immunol Today 1993; 14: 500-5.

58. Bernfield M, Kokenyesi R, Kato M, Hinkes MT, et al. Biology of syndecans: a family of transmembrane heparan sulfate proteoglycans. Ann Rev Cell Biol 1992; 8: 365-93.
59. Dong J, Kukula AK, Toyoshima M, Nakajima M. Genomic organization and chromosome localization of the newly identified human heparanase gene. Gene 2000; In Press.
60. Nakajima M, Irimura T, Di Ferrante N, Nicolson GL. Heparan sulfate degradation: relation to tumor invasive and metastatic properties of mouse B16 melanoma sublines. J Biol Chem 1984; 259: 2283-90.
61. Nakajima M, De Chavigny A, Johnson CE, Hamada J, Nicolson, GL. Suramin. J Biol Chem 1991; 266: 9661-66.
62. Marchetti D. Specific degradation of subendothelial matrix proteoglycans by human brain-metastatic melanoma/brain-endothelial cell heparanases. J Cell Physiol 1997; 173 (3): 334-42.
63. Nakajima M, Irimura T, Nicolson GL. Tumor metastasis-associated heparanase (heparan sulfate endoglycosidase) activity in human melanoma cells. Cancer Letters 1986; 31: 277-83.
64. Nakajima M., Irimura T., Nicolson GL. A solid phase substrate of heparanase: its application to assay of human melanoma for heparan sulfate degradative activity. Anal Biochem 1986; 157: 162-71.
65. Nakajima M, Irimura T, Nicolson GL. Heparanases and tumor metastasis. J Cell Biochem 1988; 36: 157-67.
66. Marchetti D, Shouchun L, Spohn WC, Carson DD. A synthetic peptide of heparan sulfate interacting protein (HIP) and heparanase recognize common sites on cell surface but not cell secreted heparan sulfate species. J Biol Chem 1997; 272(25): 15891-97.
67. Walch ET, Marchetti D. Roles of neurotrophins and neurotrophin receptors in the *in vitro* invasion and heparanase production of human prostate cancer cells. Clin & Exp Met 1999; 17: 307-14.
68. Becker M, Moczar M, Poupon MF, Moczar E. Solubilization and degradation of extracellular matrix by various metastatic cell lines derived from a rat rhabdomyosarcoma. J Natl Cancer Inst 1986; 77: 417-29.
69. Vlodavsky I, Fuks Z, Bar-Ner M, Ariav Y, Schirrmacher V. Lymphoma cell-mediated degradation of sulfated proteoglycans in the subendothelial extracellular matrix: relationship to tumor cell metastasis. Cancer Res 1983; 43: 2704-11.
70. Ricoveri W, Cappelletti R. Heparan sulfate endoglycosidase and metastatic potential in murine fibrosarcoma and melanoma. Cancer Res 1986; 46: 3855-62.
71. Bar-Ner M, Kramer MD, Schirrmacher V, Ishai-Michaeli R, Fuks Z, Vlodavsky I. Sequential degradation of heparan sulfate in the subendothelial extracellular matrix by highly metastatic lymphoma cells. Int J Cancer 1985; 35: 483-91.
72. Höök M, Wasteson A, Oldberg A. A heparan sulfate-degrading endoglycosidase from rat liver tissue. Biochem Biophys Res Comm 1975; 67: 1422-28.
73. Dempsey LA, Plummer TB, Coombes SL, Platt JL. Heparanase expression in invasive trophoblasts and acute vascular damage. Glycobiology 2000; 10: 467-75.
74. Klein U, von Figura K, Hoppe-Seyler Z. Substrate specificity of a heparan sulfate-degrading endoglucuronidase from human placenta. Physiol Chem 1979; 360: 1465-72.
75. Bame KJ, Robson K. Heparanases produce distinct populations of heparan sulfate glycosaminoglycans in chinese hamster ovary cells. J Biol Chem 1996; 272: 2245-52.

76. Bashkin P, Razin E, Eldor A, Vlodavski I. Degranulating mast cells secrete an endoglycosidase which degrades heparan sulfate in subendothelial matrix. Blood 1990; 75: 2204-12.
77. Goshen M, et al. Purification and characterization of placental heparanase and its expression by cultured cytotrophoblasts. Mol Hum Reprod 1996; 2: 679-84.
78. Naparstek Y, Cohen IR, Fuks Z. Activated T lymphocytes produce a matrix-degrading heparan sulphate endoglycosidase. Nature 1984; 310: 241-44.
79. Oosta GM, Favreau LV, Beeler DL, Rosenberg RD. Purification and properties of human platelet heparitinase. J Biol Chem 1982; 257: 11249-55.
80. Hoogewerf AJ, Leone JW, Reardon IM, Howe WJ, Asa D, Heinrikson RL, Ledbetter ST. CXC chemokines connective tissue activating peptide-III and neutrophil activating peptide-2 are heparin/heparan sulfate-degrading enzymes. J Biol Chem 1995; 270: 3268-77.
81. Thunberg L, Brackstrom G, Wasteson A, Robinson HC, Ogren S, Lindahl U. Enzymatic depolymerization of heparin-related polysaccharides. Substrates specificities of house mastocytoma and human platelet endo-β-D-glucuronidases. J Biol Chem 1982; 257: 10278-82.
82. Irimura T, Nakajima M, Nicolson GL. Chemically modified heparins as inhibitors of heparan sulfate specific endo-β-glucuronidase (heparanase) of metastatic melanoma cells. Biochemistry 1986; 25: 5322-28.
83. Pikas D, Li J-P, Vlodavski I, Lindhal U. Substrate specificity of heparanases from human hepatoma and platelets. J. Biol. Chem. 1998; 273: 18770-7.
84. Eccles S. Heparanase: breaking down barriers in tumors. Nat Med 1999; 7: 735-36.
85. Vlodavski I, Friedman, Y, Elkin M, et al. Mammalian heparanase: gene cloning, expression and function in tumor progression and metastasis. Nat Med 1999; 7: 793-802.
86. Hulett MD, Freeman C, Hamdorf BJ, et al. Cloning of mammalian heparanase, an important enzyme in tumor invasion and metastasis. Nat Med 1999; 7: 803-9.
87. Kussie PH, Hulmes JD, Ludwig DL, Patel S, et al. Cloning and functional expression of a human heparanase gene. Biochem Biophys Res Comm 1999; 261: 183-7.
88. Folkman J, Klagsbrun M. Angiogenic factors. Science 1987; 235: 442-7.
89. McKeehan WL, Wang F, Kan M. The heparan sulfate-fibroblast growth factor family: diversity of structure and function. Progr Nucleic Acid Res & Mol Biol 1998; 59: 135-76.
90. Zhou FY, Kan M, Owens RT, McKeehan WL, Thompson JA, Linhardt RJ, Hook M. Heparin-dependent fibroblast growth factor activities: effects of defined heparin oligosaccharides. Eur J Cell Biol 1997; 73(1):71080-6.
91. Gospodarowicz D, Cheng J. Heparin protects basic and acidic FGF from inactivation. J Cell Physiol 1986; 128: 475-84.
92. Goto F, Goto K, Weindel K, Folkman J. Synergistic effects of vascular endothelial growth factor and basic fibroblast growth factor on the proliferation and cord formation of bovine capillary cells within collagen gels. Lab Invest 1993; 69: 508-17.
93. Saksela O, Rifkin DB. Release of basic fibroblast growth factor-heparan sulfate complexes from endothelial cells by plasminogen activator-mediated proteolytic activity. J Cell Biol 1990; 110: 767-75.
94. Shively JE, Conrad HE. Formation of anhydrous sugars in the chemical depolymerization of heparin. Biochemistry 1976; 15: 3932-42.

95. Kimelman D, Kirschner M. Synergistic induction of mesoderm by FGF and TGF-beta and the identification of an mRNA coding for FGF in the early *Xenopus* embryo. Cell 1987; 51: 869-77.
96. Folkman J, Klagsbrun M, Sasse J, Wadinski M, Ingber D, Vlodavski I. A heparin-binding angiogenic protein, basic fibroblast growth factor, is stored within basement membrane. Am J Path 1988; 130: 393-400.
97. Denekamp J. Vasculature as a target for tumor therapy. Prog Appl Microcirc 1984; 4: 28-38.
98. Aviezer D, Iozzo RV, Noonan DM, Yayon A. Suppression of autocrine and paracrine functions of basic fibroblast growth factor by stable expression of perlecan antisense cDNA. Mol Cell Biol 1997; 17: 1938-46.
99. Rodeck U, Becker D, Herlyn M. Basic fibroblast growth factor in human melanoma. Cancer Cells 1991; 3(8): 308-11.
100. Naparstek Y, Cohen IR, Fuks Z. Activated T lymphocytes produce a matrix-degrading heparan sulphate endoglycosidase. Nature 1984; 310: 241-4.
101. Yahalom J, Eldor A, Fuks Z, Vlodavsky I. Degradation of sulfated proteoglycans in the subendothelial extracellular matrix by human platelet heparitinase. J Clin Invest 1984; 74: 1842-49.
102. Matzner Y, Bar-Ner M, Yahalom J, Ishay-Michaeli R, Fuks R, Vlodavsky I. Degradation of heparan sulfate in the subendothelial basement membrane by an readily released heparanase from human neutrophils. J Clin Invest 1985; 76: 1306-1313, 1985.
103. Marchetti D, et al. 2000; Unpublished observations.
104. Marchetti D, Li J, Shen RJ. Astrocytes contribute to the brain-metastatic specificity of melanoma cells by producing heparanase. Cancer Res 2000; In Press.
105. Menter DG, Herrmann, JL, Marchetti D, Nicolson, GL. Involvement of neurotrophins and paracrine growth factors in brain metastases formation. Inv & Met 1995; 14: 372-84.
106. McCarthy KD, DeVellis J. Preparation of separate astroglial and oligondedroglial cell cultures from rat cerebral tissue. J Cell Biol 1980; 85: 890-902.
107. Marchetti D, Nicolson GL. Human heparanase: a molecular determinant of brain metastasis. Adv Enz Reg 2000; In Press.
108. Yoshida K, Gage FH. Cooperative regulation of nerve growth factor synthesis and secretion in fibroblasts and astrocytes by fibroblast growth factor and other cytokines. Brain Res 1992; 569: 14-25.
109. Yoshida K, Kakihana M, Chen LS, Ong M, Baird A, Gage FH. Cytokine regulation of nerve growth factor-mediated cholinergic neurotrophic activity synthesized by astrocytes and fibroblasts. J Neurochem 1992; 59: 919-31.
110. Marchetti D, Nicolson GL. Neurotrophin stimulation of human melanoma cell invasion: selected enhancement of heparanase activity and heparanase degradation of specific heparan sulfate subpopulations. Adv Enz Reg 1997; 37: 111-34.
111. Yoshida K, Gage FH. Fibroblast growth factors stimulate nerve growth factor synthesis and secretion by astrocytes. Brain Res 1991; 538: 118-26.
112. Lindholm D, Hengerer B, Zafra F, Thoenen H. Transforming growth factor-beta1 stimulates expression of nerve growth factor in the rat CNS. Neuroreport 1990; 1: 9-12.
113. Lindholm D, Castren E, Kiefer R, Zafra F, Thoenen H. Transforming growth factor-beta 1 in the rat brain: increase after injury and inhibition of astrocyte proliferation. J Cell Biol 1992; 117: 395-400.

114. Nicolson GL, Menter DG, Herrmann JL, Yun Z, Cavanaugh PG, Marchetti D. Brain metastasis: role of trophic, autocrine and paracrine factors in tumor invasion and colonization of the central nervous system, In: Attempts to Understand Metastasis Formation II, Gunthert U, Birchmeier W, eds., Berlin: Springer-Verlag 1996.
115. Fujimaki T, Fan D, Staroselsky AH, Gohji K, Bucana CD, Fidler IJ. Critical factors regulating site-specific brain metastasis on murine melanomas. Int J Oncol 1993; 3: 789-99.
116. Ono T, Sato H, Kashimoto T, Okumoto T, Miyamoto K. Stimulation of biosynthesis of nerve growth factor by acidic fibroblast growth factor in cultured mouse astrocytes. Neurosci Lett 1991; 126: 18-20.
117. Lu C, Vickers MF, Kerbel R. Interleukin 6: a fibroblast-derived growth inhibitor of human melanoma cells from early but not advanced stages of tumor progression. Proc Natl Acad Sci USA 1992; 89: 9215-19.
118. Sawada M, Suzumura A, Marunochi T. TNF alpha induces IL-6 production by astrocytes but not by microglia. Brain Res 1992; 583: 296-9.
119. Nicolson GL. Autocrine and paracrine mechanisms in cancer progression and metastasis. In: Encyclopedia of Cancer, Bertino JR, ed., San Diego: Academic Press, 1996.

Chapter 6

AUTOCRINE MOTILITY FACTOR AND ITS RECEPTOR AS REGULATORS OF METASTASIS

Yasuharu Onishi[1,2], Arayo Haga[1,3], Avraham Raz[1]

[1]Karmanos Cancer Institute. Wayne State University, Detroit, MI, USA; [2]Toyama Medical and Pharmaceutical University, Japan; [3]Gifu Pharmaceutical University, Japan

INTRODUCTION

Despite the advances in diagnostic techniques for early detection of malignant tumors and the significant improvement in therapeutic procedures, the mortality rate of cancer has been increasing year after year, and the most frequent cause of death by cancer is metastasis. Therefore, elucidation of the metastasis mechanism is one of principal subjects in cancer study and would lead to development of new treatment methods. Cancer metastasis is a complex process which is composed of sequential events including tumor cell detachment from the primary neoplasm, invasion of the surrounding tissues, intravasation of the hematogenous or lymph system and transit in the circulation, arrest in a secondary organ, extravasation from the vessel, and growth in the newly defined site (1). The migration of tumor cells is one of the most significant constituents which contributes to the successful establishment of cancer metastasis (2). Many previous analyses (3-11) have indicated a positive relationship between metastatic potential and motile ability of tumor cells.

Tumor cell motility is thought to be controlled in an autocrine or paracrine fashion by soluble tactors which can dramatically affect cell's motile properties in the tumor environment (12). Among the various soluble factors, tumor cell autocrine motility factor (AMF) is a tumor-secreted cytokine which was originally distinguished from A2058 human melanoma cells and induces both directional (chemotactic) and random (chemokinetic) motility of the AMF-producing cells via binding to its receptor, a cell surface glycoprotein of 78kD (gp78, AMFR) (13-16). Therefore, in the light of the motility-regulating effects, AMF and its receptor have been proposed to play a significant role during the metastatic cascade of cancer (17). Thus, existence of the receptor-mediated AMF pathway correlates well with malignant tumor progression (18).

In this chapter, we describe AMF and its receptor as regulators of metastasis mainly on our studies.

D.R. Welch (ed.), Cancer Metastasis – Related Genes, 109–122.

Specificity of Autocrine Motility Factor

AMF was identified as a proteinaceous molecule secreted by A2058 human melanoma cells that stimulated the locomotion of the very cells that produced it, thereby providing its name (13). Liotta *et al.* demonstrated that although untransformed parental NIH-3T3 fibroblasts did not secrete appreciable AMF since their conditioned media was incapable of stimulating either the producer cells of the transformed lines, they were able to respond wilh an equivalent motile response when stimulated by AMF from three different ras-transformed NIH-3T3 clones, all of which were metastatic in animal tests. AMF has been shown to stimulate the motility of various kinds of cancer cell lines and untransformed fibroblasts, however it does not activate T lymphocyte or neutrophil motility (2, 13, 14, 16, 17, 19). These data suggest that the activity of the AMF is at least partially tumor cell specific factor with potential relevance to invasion and metastasis.

Identification of AMF

Our results obtained from partial amino acid sequencing of a purified murine AMF has shown a sequence identity to the previously cloned gene products, i.e., the cytokine neuroleukin (NLK) and the glycolytic enzyme phosphohexose isomerase (PHI) (20). Furthermore, molecular cloning and sequencing has revealed that the human gene encoding AMF cDNA is identical to NLK and PHI (21). In addition, myeloid leukemia maturation factor (MF) was also identified as NLK/PHI (22, 23). AMF/NLK/PHI/MF is a polypeptide of an apparent molecular mass of 55 and 64 kDa under non-reducing and reducing conditions, respectively. NLK is a neutrophic factor which promotes the survival of spinal and sensory neurons *in vitro* (24, 25) and is associated with terminal axonal sprouting (26), and its blockade may contribute to motor neuron diseases and AIDS-related dementia (27-29). Also, NLK is produced and secreted by lectin-stimulated T-cells, and induces immunoglobulin secretion by peripheral blood mononuclear cells (30). PHI is a member of the ecto/exoenzyme family and catalyzes the reversible isomerization of glucose-6-phosphate to fructose-6-phosphate, an essential reaction in glycolysis and gluconeogenesis (31-33). Enzyme deficiency in humans leads to congenital hemolytic anaemia which can be associated with neurological dysfunction (34-36). Elevated serum PHI levels serve as tumor progressior narkers in patients with malignant tumors including gastrointestinal, pulmonary, renal, and mammary carcinoma, and relative activities are correlated with the development of metastases (37-42). Moreover, PHI has been suggested as a urinary marker for bladder cancer (43).

MF was shown to mediate the differentiation of human myeloid leukemic HL-60 cells to terminal monocytic cells (22, 23). MF mediates reductions in proliferating S and G2M cells, mature monocytic cells acquired complement receptor, phagocytic capacity and adherence morphology. Furthermore, abnonnal elevated PHI enzymatic activities were detected in the plasma of acute myelogenous leukemia patients (23). Thus, AMF/NLK/PHI/MF exhibits multifunctional activities in various biological processes.

Several recent studies (44-46) reported the crystal structure of the AMF/NLK/PHI/MF. The crystal structure of the bacterial PHI was resolved as an open twisted and structural motif consisting of two global domains and two protruding parts and was shown to exhibit AMF and NLK activities. In addition, it was postulated that the substrate-binding site for PHI and AMF is located within the domains' interface (44). Jeffery *et al.* (45) have demonstrated the three-dimensional structure of rabbit muscle PHI bound with the competitive inhibitor glucose-6-phosphate by X-ray crystallography. Also, Chou *et al.* have shown that the crystal structures of PHI complexed with the transition state analogue 5-phospho-D-arabinonate (5PA) and N-bromoacetylethanolamine (BAP) phosphate at 2.5- and 2.3-Å resolution, respectively (46). The inhibitors, 5PA and BAP, not only affect the enzymatic activity of PHI but also inhibit the AMF-induced cell motility of colon-26 murine colon carcinoma cells. The locations of the substrate binding sites for PHI and the receptor binding sites for AMF are overlapped.

Mechanism of AMF Secretion

We have recently shown that AMF mRNA expression level is significantly enhanced in tumor cells as compared with normal cells and that although the intracellular protein levels are similar, 57 kDa AMF protein is selectively secreted into the conditioned media only by neoplastic cells such as A31 murine angiosarcoma and HT-1080 human fibrosarcoma cells (21). Thus, the overexpression and secretion of AMF are restricted to malignant cells. AMF lacks a consensus secretary signal sequence (24), which regulates protein secretion via the classical endoplasmic reticulum (ER)-Golgi route while functioning, at least in part, as a secreted soluble cytokine. Interestingly, the selective release of 57 kDa AMF into the conditioned media is thought to be actively performed via an alternative, nonclassical pathway, which has been indicated in the secretion of the proteins devoid of secretary signal sequence, such as iterleukin (IL)-1 (47, 48), fibroblast growth factor (49), galectin-1(50), galectin-3 (51, 52), and others. A consensus sequence (if any) shared by proteins using the alternative secretary pathway is not yet fully resolved. Therefore, there must be certain cascades of regulation steps which may be controlled by various molecules and mechanisms. Until now the studies demonstrating direct, receptor-ligand interaction relied mainly on contingent evidence, due to an insufficient quantity of AMF necessary to perform a detailed biochemical analysis. We have recently generated, expressed and isolated a recombinant human AMF (rhAMF) (53). To test whether the purified rhAMF is properly folded and biologically active, three migration assays were employed, each focusing on a different cellular migratory aspect i.e., chemotaxis (Fig. A), wound healing (Fig. B), and random motility (Fig. C). rhAMF exhibited a highly potent motogenic activity on human fibrosarcoma HT-

1080 cells over an average of 10 pg/ml-10 ng/ml, that was independent of the assay used and was similar to the activity of the native AMF (16, 20). Thus, rhAMF retained the biological activities of the native AMF, i.e., catalyzes PHI activity and stimulated cell motility. lsolation of rhAMF which is identical to the endogenous species provides an opportunity to study the structural-functional relationship of the molecule.

The deduced amino acid sequence of human AMF protein 12 includes potential phosphorylation site, consisting of nine casein kinase II (CK II) and three protein kinase C (PKC) consensus sequences (21). Our recent study has shown that AMF possesses three potential CK II phosphorylation sites containing serines, i.e., Ser185-, Ser441- and Ser455 but a single serine residue at position 185 of human AMF is phosphorylated by CK II. CK II is a ubiquitous protein kinase, responsible for serine and threonine phosphorylation specified by acidic side chains of many proteins. Furthermore, higher CK II activity was detected in metastatic melanoma compared with dermal nevus (54), suggesting that the overall phosphorylation activities of CK II may contribute to the cancer metastasis. AMF/PHI/NLK/MF protein undergoes CK II-mediated serine phosphorylation and propose that conformational structural changes induced by phosphorylation may play a role in its nonclassical secretory pathway.

AUTOCRINE MOTILITY FACTOR RECEPTOR

The cell surface receptor for the cytokine such as AMF is implicated in regulation of cellular motility (55). The receptor for AMF has been identified as 78 kDa cell surface glycoprotein designated gp78. A role for AMFR in the metastatic process was first suggested by its increased O-linked glycosylation in B16-F1 mouse melanoma cells grown in a spherical configuration on a non-adhesive substrate which exhibited enhanced lung-colonizing ability in mice (56-58). The growth pattern is comparable to tumor cell interaclions in the tumor mass or as aggregates in the circulalion prior to invasion through the endothelium (59-61) and the ability to grow in such a manner is distinct to the neoplastic phenotype (8). The causal involvement of gp78 in metastasis was further demonstrated by the observation that treatment of B16-F1 cells grown in monolayer with Fab fragments from a polyclonal anti-gp78 antibody prior to intravenous inoculation into the tail vein of syngenic mice resulted in an enhanced lung-colonization similar to that seen with cells grown in a spherical configuration (57). Similarly, cells treated with a mAb against gp78 displayed increased motility in vitro as evidenced by phagokinetic clearing of particles from gold-coated substrates and enhanced *in vivo* lung-colonizing capability compared to stimulation with control antibodies (14, 62). The induction of motility *in vitro* by either conditioned media from these B16-F1 cells or the anti-gp78 mAb was found to be of similar magnitude and each was specifically inhibited by PT (14), indicating use of a common signalling pathway.

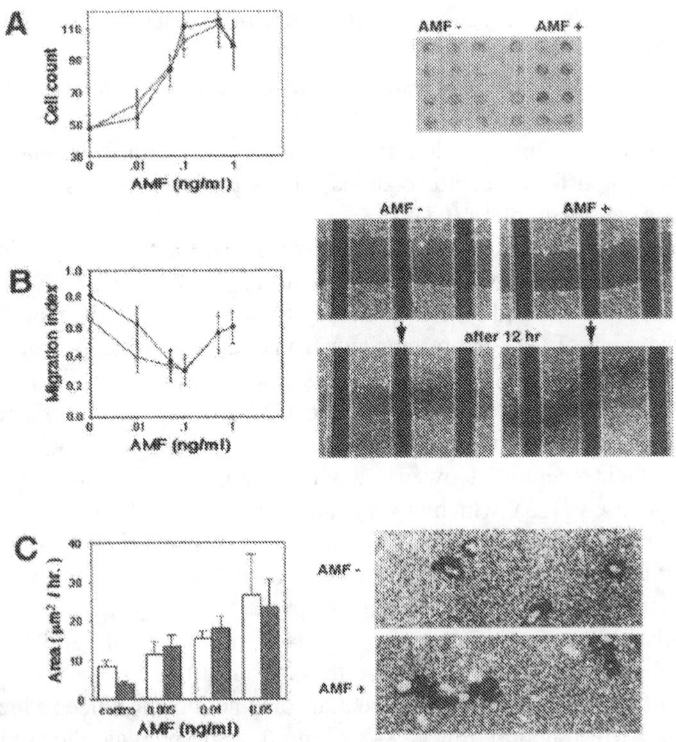

Figure 1. Cell motility-stimulating activity of purified recombinant human AMF. Motility of human fibrosarcoma HT-1080 (open circles and bars) and fibroblast IMR-90 cells (closed circles and bars) was assayed by Boyden chamber assay (A), wound healing assay (B) and phagokinetic assay (C). (Adapted from Biochimica Biophysica Acta 2000; 1480: 235-244)

We have recently cloned the full-length cDNAs for both human and mouse AMFR genes in order to elucidate the functional role of AMFR in cancer invasion and metastasis (63). Structural analysis demonstrated that both genes encode a protein of 643 amino acids containing a seven transmembrane domain, a RING-H2 motif which is a zinc finger variant (64), and a leucine zipper motif (65) and showed a 94.7% amino acid sequence identity to each other. Furthermore, AMFR protein included a potential N-glycosylation site and several potential O-glycosylation sites, supporting the previous finding that AMFR is glycosylated with both – and O-glycosaccharides (59). Protein sequence database analysis showed that the AMFR protein had no significant homology with registered protein sequences including the ones of seven transmembrane protein, but a significant structural similarity to a *Caenorhabditis elegans* hypothelical protein, F26E4.11. Thus, the AMFR gene is an evolutionarily conserved gene which encodes a novel type of seven tansmembrane protein having the RING-H2 and leucine zipper motifs.

Since most of the seven transmembrane proteins are recognized as a member of guanine nucleotide binding protein (G-protein)-coupled receptors (66), our observation is in accordance with the previous findings that AMF interacts with a cell surface receptor coupled with a pertussis toxin-sensitive G-protein to initiate cell locomotion (67).

In addition, the identification of AMFR as a seven transmembrane domain receptor induced a question whether AMF belongs to the C-C, C-X-C gene family of chemoattractants and AMFR to the CXC-receptor (CXCR) gene family (68, 69) because most of CXCR proteins signal via heterotrimeric G-proteins. A homology search has revealed that AMF does not contain the C-C or C-X-C motif and AMFR is not homologous with the CXCR gene family. Therefore, AMF and its receptor are thought to be different from the C-C, C-X-C chemokines and their receptors. However, AMF contains a similar motif, C-X-X-C (Cys330-Phe-Glu-Cys333), which is very rare and present only in the disulfide isomerase protein family as defined as the thioredoxin-box motif (70) and in one cytokine known as the macrophage migration inhibitor (MIF) (71, 72). AMF and MIF share several structural and functional properties, may be secreted by activated T cells via a non-classical pathway, lack signal peptide and have an intramolecular disulfide (21, 71, 72). Since the MIF disulfide bridge was shown to be in the motif where its integrity is essential for function and since reducing a gents inhibit the AMF function, this C-X-X-C motif might regulate the structural integrity and function of AMF.

Cell Surface Expression of AMFR

The AMFR protein is localized to the leading edge, trailing edge, and a region proximal to the nucleus on the surface of motile cells (14). This cell surface distribution may be characteristic of proteins involved directly in the locomotory process which require receptor recycling to the leading edge for the maintenance of forward movement.

Different high- and low-metastatic variant sublines of B16 and K-1735 murine melanomas were examined for AMFR surface expression and localization (62). The highly metastatic variants exhibited an increased lung-colonization *in vivo* and motility *in vitro* by the stimulation with a mAb against AMFR. Fluorescence-activated cell sorter (FACS) analysis revealed that the high-metastatic sublines expressed less AMFR on their cell surface than the low-metastatic counterparts. Immunofluorescent examination showed that less motile, low-metastatic cells had multiple focal areas of AMFR on the cel surface, while most of the high-metastatic and more motile cells displayed a single intensity labeled area at the cell periphery. These observations suggest that the directed endocytosis of AMFR may be associated with the metastatic ability.

AMF-Signaling of Cell Motility

The motility-stimulating signal transduction pathway of AMF in melanoma and fibrosarcoma cells is specifically sensitive to pertussis toxin (PT) as well (17, 67). AMF stimulated the incorporation of inositol into cellular lipids and inositol phosphates, especially inositol triphosphates (73, 74). This AMF-stimulated production of inositol phosphates was dose dependent, correlated with induction of motility, and was in part inhibited by pretreatment of the cells with PT. Stimulation of cells with AMF caused the extension of pseudopodia which exhibited prominent axial actin filament bundles. Isolated pseudopodia were highly enriched for laminin and integrin receptors after stimulation, containing over 20 times the laminin and fibronectin receptors of plasma memhranes of untreated cells (75). Exogenous 12-(S)-hydroxyeicosatetraenoic acid [12-(S)-HETE], a lipoxygenase metabolite of arachidonic acid, stimulated the motility of high-metastatic melanoma cells in a manner analogous to AMF without affecting the migration of the low-metastatic variant (19, 76). This effect was associated with enhanced expression of the AMFR and AMF treatment stimulated production of exogenous 12-(S)-HETE as well as upregulation of the 12-lipoxygenase enzyme in the migration-responsive cells exclusively. Although PT inhibits AMF action entirely, the effects of 12-(S)-HETE are unaffected by pretreatment with PT, and this eicosanoid appears to act downstream of the PT-sensitive G-protein in the AMF signaling pathway. However, the motility effects of both AMF and 12-(S)-HETE are abolished by inhibitors of protein kinase C (76, 77). Furthermore, it has been shown that both arachidonic acid metabolism and protein kinase C activity are related to locomotory induced by AMF and not its proliferative effects. During the course of tumor cell invasion and metastasis, malignant cells must interact with components of the extracellular matrix (ECM) encountered in the various tissue microenvironments through which they move. AMF differentially affects the adhesion, spreading and migration of high- and low-metastatic murine melanoma cells on the ECM components fibronectin, laminin and collagen IV via alterations of focal contact architecture (78). Moreover, the unique migration patterns in these cells which reflect differences in degradation and/or remodeling of the cellular substratum were observed (79). These profiles of matrix interaction were influenced distinctly by AMF and dictated by both substrate composition and cellular phenotype. Our findings show the dynamic interplay that exists between cellular signals from the insoluble ECM and secreted cytokines like AMF. With regard to the relationship between these signals and the metastatic phenotype of the cells, it appears that differential responses dependent on the available substrate proteins as well as signals from circulating molecules might mediate the invasion and migration of metastasizing tumor cells by regulating cellular responses to either degrade and invade the ECM or remodel the ECM and arrest.

Regulation of AMFR

It has been demonstrated that metabolites and synthetic analogues of vitamin A (retinoids) can inhibit carcinogenesis and suppress the tumorigenicity of several tumor cell types by inducing terminal differentiation (80). Our previous study showed that beta-all-trans-retinoic acid (RA) inhibited the invasiveness of human melanoma cells with different metastatic propensities in time- and dose-dependent manner and that AMFR expression decreased 13- to 50-fold compared with control cells (81). This suggests that RA has either a direct or indirect effect on AMFR expression which may be directly related to decreased tumor cell invasion. The inhibitory effect is indirect, since characterization of AMFR promoter failed to distinguish an RA-response element (82). Furthermore, we indicated that reduced migratory response of RA-treated melanoma cells to AMF and the AMF-mimicking anti-AMFR mAb correlated with decreased AMFR expression (83), suggesting that the two events are related and that previously reported observations of RA-mediated suppression of tumor cell invasion metastasis may be related, at least in part, to suppression of cell motility resulting from diminished levels of AMFR. Cell density-dependent downregulation of AMFR appears to be a hallmark of lower malignant potential, since normal murine BALB/c 3T3-A31 fibroblasts as well as normal human bladder epithelial cells show complete downregulation of both the surface and total expression of AMFR when grown to confluence (17, 84). Cells which display progressively malignant and especially metastatic phenotypes appear to be capable of maintaining steady state expression of AMFR in high density conditions (85). The display of AMFR in a polarized fashion on the malignant cell surface coupled with the motile responses to AMF, irrespective of cell density, suggests a mechanism for the exploitation of continued expression for invasion.

Clinical Utilization of AMF and its Receptor System

AMF and its receptor have been postulated to play an important role in the most devastating aspects of cancer, i.e. tumor cell metastasis, based on their motility-regulating effects (86). The presence of AMF in the urine is a marker of transitional cell carcinoma of the bladder (87). Recently, it was shown that expression of the AMFR correlates with malignant potential in bladder cell lines (84). Moreover, immunohistochemical examination in specimens from human bladder carcinoma patients revealed a strict negative correlation between AMFR expression and prognosis (85). Importantly, upregulation of AMFR in the more severe bladder cancer were associated with a concomitant downregulation of the calcium-dependent homotypic cell adhesion molecule E-cadherin. These results suggest that enhanced migration of bladder carcinoma cells with invasion of the suburothelial lining is due to both the acquisition of motility-promoting molecules such as AMFR as well as the loss of the molecules which promotes the sedentary phenotype (e.g., E-cadherin) and indicate the potential utility of a dual-antigen approach for improving early diagnosis of high risk bladder cancer patients, especially in the cases where the normal predictors of stage and grading fail. The possible utility of AMF and its receptor system as prognostic indicators was previously observed by many other reports (88-94). These findings implicate the AMF system as a very useful prognostic indicator of malignancy.

CONCLUSION

Here, we have shown the importance of autocrine motility factor and its receptor that function as regulator during the metastatic steps. Further study about characterization of AMF and AMFR will be needed to increase our understanding of the fundamental processes which control motility in mammalian cells and provide the basis for the development of more effective clinical treatments and specific modalities which inhibit tumor cell motility in vitro and invasion and metastasis *in vivo*.

ACKNOWLEDGEMENTS

The authors thank Ms. Vivian Powell for careful editing. This work was supported in part by grant CA51714 from NIH.

REFERENCES

1. Fidler IJ. The evolution of biological heterogeneity in metastatic neoplasms. In Cancer Invasion and Metastases; Biologic and Therapeutic Aspects. Edited by Nicolson GL and Milas L, pp 5-26, Raven Press, New York 1984.
2. Nabi IR. Watanabe H, Raz A. Autocrine motility factor and its receptor: role in cell locomotion and metastasis. Cancer Metastasis Rev 1992; 11: 5-20.
3. Orr FW, Varani J, Delikatny J, Jain N, Ward PA. The chemotactic response of tumor cells. A model for cancer metastasis. Am J Pathol 1981; 102: 160-167.
4. Raz A, Geiger B. Altered organization of cell-substrate contacts and membrane-associated cytoskeleton in tumor cell variants exhibiting different metastatic capabilities. Cancer Res 1982; 42: 5183-5190.
5. Volk T, Geiger B, Raz A. Motility and adhesion properties of high- and low-metastatic murine neoplastic cells. Cancer Res 1984: 44: 811-824.
6. Geiger B, Volk T, Raz A. Cell contacts and cytoskeletal organization of metastatic cell variants. Exp Biol Med 1985; 10: 39-53.
7. Zvibel I, Raz A. The establishment and characterization of a new Balb/c angiosarcoma tumor system. Int J Cancer 1985; 36: 261-272.
8. Raz, A. Ben-Ze'ev A. Cell-contact and architecture of malignant cells and their relationship to metastasis. Cancer Metastasis Rev 1987; 6: 3-21.
9. Mohler JL. Partin AW, Isaacs JT, Coffey DS. Time lapse video microscopic identification of Dunning R-2237 adenocarcinoma and normal rat prostatic cells. J Uro 1987; 137: 544-547.

10. Mohler JL, Partin AW, lsaacs JT, Coffey DS. Metastatic potential prediction by a visual grading system of cell motility: prospective validation in the Dunning R-3327 prostatic adenocarcinoma model. Cancer Res 1988; 48: 4312-4317.
11. Partin AW, Schoeniger JS, Mohler JL, Coffey DS. Fourier analysis of cell motility: correlation of motility with metastatic potential. Proc Natl Acad. Sci USA 1989; 86: 1254-1258.
12. Stracke ML, Murata J. Aznavoorian S, Liotta LA. The role of the extracellular matrix in tumor cell metastasis. In vivo 1994: 8: 49-58.
13. Liotta LA, Mandler R, Murano G, Katz DA, Gordon RK, Chiang PK, Schiffman E. Tumor autocrine motility factor. Proc Natl Acad Sci USA 1986; 83: 3302-3306.
14. Nabi IR, Watanabe H, Raz A. Identification of B16-F I melanoma autocrine motility-like factor receptor. Cancer Res 1990; 50: 409-414.
15. Silletti S, Watanabe H, Carmi P, Hogan V, Nabi IR, Raz A. Purification of B16-F1 melanoma autocrine motility factor and its receptor. Cancer Res 1991; 51: 3507-3511.
16. Watanabe H, Carmi P, Hogan V, Raz T, Silletti S, Nabi IR, Raz A. Purification of human tumor cell autocrine motility factor and molecular cloning of its receptor. J Bio Chem 1991; 266: 13442-13448.
17. Silletti S, Raz A. Autocrine motility factor is a growth factor. Biochem Biophys Res Commun 1193; 194: 446-457.
18. Silletti S, Raz A. Regulation of autocrine motility factor receptor expression in tumor cell locomotion and metastasis. Curr Topics Microbiol Immunol 1996: 213/II: 137-169.
19. Silletti S, Timar J, Honn KV, Raz A. Autocrine motility factor induces differential 12-lipoxygenase expression and activity in high and low-metastasis K-1735 melanoma cell variants. Cancer Res 1994; 54: 5752-5756.
20. Watanabe H, Takehara K, Date M, Shinozaki T, Raz A. Tumor cell autocrine motilility factor is the neuroleukin/phosphohexose isomerase polypeptide. Cancer Res 1996; 56: 2960-2963.
21. Niinaka Y, Paku S, Haga A, Watanabe H, Raz A. Expression and secretion of neuroleukin/phosphohexose isomerase/maturation factor as autocrine motility factor by tumor cells. Cancer Res 1998; 58: 2667-2674.
22. Xu W, Seiter K, Feldman E, Ahmed T, Chiao JW. The differentiation and maturation mediator for human myeloid leukemia cells shares homology with neuroleukin or phosphohexose isomerase. Blood 1996; 87: 4502-4506.
23. Chiao JW, Xu W, Seiter K, Feldman E. Neuroleukin mediated differentiation induction of myelogenous leukemia cells. Leuk. Res 1999; 23:13-18.
24. Gurney ME, Heinrich SP, Lee MR, Yin H-S. Molecular cloning and expression of neuroleukin, a neutrophic factor for spinal and sensory neurons. Science (Washington DC) 1986; 234: 566-574.
25. Chaput M, Claes V, Portetelle D, Cludts I, Cravador A, Burny A, Gras H. Tartar A. The neutrophic factor neuroleukin in 90% homologous with phosphohexose isomerase. Nature (Lond.) 1988; 332: 454-455.
26. Gurney ME. Suppression of sprouting at the neuromuscular junction by immune sera. Nature 1984; 307: 546-548.
27. Gurney ME, Belton AC, Cashman N, Antel JP. Inhibition of terminal axonal sprouting by serum from patients amyotrophic lateral sclerosis. N Engl J Med 1984; 311: 933-939.

28. Lee MR, Ho DD, Gurney ME. Functional interaction and partial homology between human immunodeficiency virus and neuroleukin. Science 1987; 237: 1047-1051.
29. Ho DD, Pomerantz RJ, Kaplan JC. Pathogenesis of infection with human immunodeficiency virus. N Engl J Med 1987; 317: 278-286.
30. Gurney ME, Apatoff BR, Spear GT, Baumel MJ, Antel JP, Bania MB, Reder AT. Neuroleukin a lymphokine product of lectin-stimulated T cells Science 1986; 234: 574-581.
31. DeLorenzo RJ, Ruddle FH. Genetic control of two electrophoretic variants of glucosephosphate isomerase in the mouse (Mus musculus). Biochem Genet 1969; 3: 151-162.
32. Harrison RAP. The detection of hexokinase, glucosephosphate isomerase and phosphoglucomutase activities in polyacrylamide gel after electrophoresis: a novel method using immobilized glucose 6-phosphate dehydrogenase. Anal Biochem 1974; 61: 500-507.
33. Gracy RW, Tilley BE. Phosphoglucose isomerase of human erythrocytes and cardiac tissue. Methods Enzymol 1975; 41: 392-400.
34. Whitelaw AG, Rogers PA, Hopkinson DA, Gordon H, Emerson PM, Darley JH, Reid C, Crawfurd MA. Congenital hemolytic anaemia rcsulting from glucose phosphate isomerase deficiency: genetics, clinical picture, and prenatal diagnosis. J Med Genet 1979: 16: 189-196.
35. Kugler W, Breme K, Lapse P, Muirhead H, Davis C, Winkler H, Schroter W, Lakome M. Molecular basis of neurological dysfunction coupled with hemolytic anemia in human glucose-6-phosphate isomerase (GPI) deficiency. Hum Genet 1998; 103: 450-455..
36. Baronciani L, Zanella A, Bianchi P, Zappa M, Alfinito F, Iolascon A, Tannoia N, Beutler E, Sirchia G. Study of the molecular defects in glucose phosphate isomerase-deficient patients affected by chronic hemolytic anemia. Blood 1996 88: 2306-2310.
37. Baumann M, Brand K. Purification and characterization of phosphohexose isomeras from human gastrointestinal carcinoma and its potential relationship to neuroleukin. Cancer Res 1988; 48: 7018-7021.
38. Baumann M, Kappel A, Brand K, Siegfield W, Paterok E. The diagnostic validity of the serum tumor marker phosphohexose isomerase (PHI) in patients with gastrointestinal, kidney, and breast cancer. Cancer Invest 1990; 8: 351-356.
39. Filella X, Molina R, Jo J, Mas E, Ballesta A. Serum phosphohexose isomerase activities in patients with colorectal cancer. Tumor Biol 1991; 12: 360-367.
40. Patel PS, Rawal GN, Rawal RM, Patel GH, Balar DB, Shah PM, Patel DC. Comparisol between serum levels of carcinoembryonic antigen, sialic acid and phosphohexose isomerase in lung cancer. Neoplasia 1995; 42: 271-274.
41. Bodansky O. Serum phosphohexose isomerase in cancer. II. As an index of tumor growth in metastatic carcinoma of the breast. Cancer (Phila.) 1954; 7: 1200-1226.
42. Schwarz MK. Laboratory aids to diagnosis: enzymes. Cancer (Phila.) 1976: 37: 542-544.
43. Guirguis R, Javadpour N, Sharareh S, Biswas C, el-Amin W, Mansur I, Kim JS. A new method for evaluation of urinary autocrine motility factor and tumor cell collagenase stimulating factor as markers for urinary tract cancer. J Occup Med 1990; 32: 846-853.

44. Sun YJ, Chou CC, Chen WS, Wu RT, Meng M, Hsiao CD. The crystal structure of a multifunctional protein: Phosphohexose isomerase/autocrine motility factor/neuroleukin. Proc Natl Acad Sci USA 1999, 96, 5412-5417.
45. Jeggery CJ, Bahnson B, Chien W, Ringe D, Petsko GA. Crystal structure of rabbit phosphohexose isomerase, a glycolytic enzyme that moonlights as neuroleukin, autocrine motility factor, and differentiation mediator. Biochemistry 2000; 39: 955-964.
46. Chou CC, Sun YJ, Meng M, Hsiao CD. The crystal structure of phosphohexose isomcrase/autocrine motility factor/neuroleukin complexed with its carbohydrate phosphate inhibitors suggests its substrate/receptor recognition. J Biol Chem 2000; 275: 23154-23160.
47. Rubartelli A, Colzolino F, Talio M, Sitia R. A novel secretary pathway for interleukin-l f@, a protein lacking a signal sequence. EMBO J 1990; 9: 1503-1510.
48. Suttles J, Giri JG, Mizel S. IL-1 secretion by macrophages. Enhancement of IL-1 secretion and processing by calcium inophorcs. J Immunol 1990; 144: 175-182.
49. Mignatti P, Morimolo T, Rifkin DB. Basic fibroblast growth factor, a protein devoid of secretary signal sequence, is released by cells via a pathway independent of the endoplasmic reticulum-Golgi complex. J Cell Physiol 1992; 151: 81-93.
50. Cooper DNW, Barondes SH. Evidence for export of a muscle lectin from cytosol to extracellular matrix and for a novel secretory mechanism. J Cell Biol 1990; 110: 1681-1691.
51. Sato S, Burdett I, Hughes RC. Secretion of the baby hamster kidney 30-kDa galactose-binding lectin from polarized and nonpolarized cells: a pathway independent of the endoplasmic reticulum-Golgi complex. Exp Cell Res 1993; 207: 8-18.
52. Lindstedt R, Apodaca G, Barondes SH, Mostov KE, Leffler H. Apical secretion of a cytosolic protein by madin-darby canine kidney cells. J Biol Chem 1993; 268: 11750-11755.
53. Haga A, Niinaka Y, Raz A. Phosphohexose isomerase/autocrine motility factor/neutroleukin/maturation factor is a multifunctional phosphoprotein. Biochimica Biophysica Acta 2000; 1480: 235-244.
54. Meggio F, Shugar D, Pinna LA. Ribofuranosyl-benzimidazole derivatives as inhibitors of casein kinase-2 and casein kinase-1. Eur J Biochem 1990; 187: 89-94.
55. Stracke ML, Engel JD, Wilson LW, Rechler MM, Liotta LA, Schiffman E. The type I insulin-like growth factor receptor is a motility receptor in human melanoma cells. J Bio Chem 1989; 264: 21544-21549.
56. Raz A, Ben-Ze'ev A. Modulation of the metastasis capability in B16 melanoma by cell shape. Science 1983; 221: 1307-1310.
57. Nabi IR, Raz A. Cell shape modulation alters glycosylation of a metastatic melanoma cell surface antigen. Int J Cancer 1987; 40: 396-401.
58. Nabi IR, Raz A. Loss of metastatic responsivcness to cell shape modulation in a newly characterized B16 melanoma adhesive variant. Cancer Res 1988; 48: 1258-1264.
59. Nicolson GL. Organ specificity of tumor metastasis: role of preferential adhesion, invasion and growth of malignant cells at secondary sites. Cancer Metastasis Rev I988; 7: 143-188.
60. Fidler IJ, Gersten DM, Hart IR. The biology of cancer invasion and metastasis. Adv Cancer Res 1988; 28: 149-250.

61. Poste G. Experimental systems for analysis of the malignant phenotype. Cancer Metastasis Rev 1982; 1; 141-199.
62. Watanabe H, Nabi JR, Raz A. The relationship between motility factor receptor internalization and the lung colonization capability of murine melanoma cells. Cancer Res 1991; 51: 2699-2705.
63. Shimizu K, Tani M, Watanabe H, Nagamachi Y, Niinaka Y, Shiroishi T, Ohwada S, Raz A, Yokota J. The autocrine motility factor gene encodes a novel type of seven transmembrane protein. FEBS Letters 1999; 456: 295-300.
64. Freemont PS. The RING finger. A novel protein sequence motif related to the zinc finger. Ann NY Acad Sci 1993; 684: 174-192.
65. Landschulz WH, Johnson PF, McKnight SL. The leucine zipper: a hypothetical structure common to a new class of DNA binding proteins. Science 1988; 240: 1759-1764.
66. Dohlman HG, Thorner J, Caron MG, Lefkowitz RJ. Model systems for the study of seven-transmembrane-segment recwptors. Annu Rev Biochem 1991; 60: 653-688.
67. Stracke ML, Guirguis R, Liotta LA, Schiffman E. Pertussis toxin inhibits stimulated motility independently of the adenylate cyclase pathway in human melanoma cells Biochem Biophys Res Commun 1987; 164: 339-345.
68. Arai H, Charo IF. Differentiation rcgulation of G-protein-mediated signaling by chemokine receptors. J Biol Chem 1996; 271: 21814-21819.
69. Gutkind JS. The pathways connecting G protein-coupled receptors to the nucleus through divergent mitogen-activated protein kinase cascade. J Biol Chem 1998; 273:1839-1842.
70. Kemmink J, Darby NJ, Dijkstra K, Nilges M, Creighton TE. The folding catalyst protein disulfide isomerase is constructed of active and inactive thioredoxin modules. Curr Bio 1997; 7: 239-245.
71. Kleemann R. Kapurniotu A, Frank RW, Gessner A, Mischke R. Flieger O, Juttner S, Brunner H, Bernhagcn J. Disulfide analysis reveals a role for macrophage migration inhibitory factor (MIF) as thiol-protein oxidoreductase. J Mol Biol 1998; 280: 85-102.
72. Swope M, Sun HW, Blade PR, Lolis E. Direct link between cytokine activity and a catalytic site for macrophage migration inhibitory factor. EMBO J 1998; 17: 3534-3541.
73. Smith CD, Cox CC, Snyderman R. Receptor coupled activation of phosphoinositide-specific phospholipase C by a G protein. Science 1986; 232: 97-100.
74. Kohn EC, Liotta LA, Schiffman E. Autocrine motility factor stimulates a three-fold increase in inositol trisphosphate in human melanoma cells. Biochem Biophys Res Commun 1990; 166: 757-764.
75. Guirguis R, Margulies I, Taraboletti G, Schiffman E, Liotta LA. Cytokine-induced pseudopodial protrusion is coupled to tumor cell migration. Nature 1987; 329: 261-263.
76. Timar J, Silletti S, Bazaz R, Raz A. Honn KV. Regulation of melanoma cell molility by the lipoxygenase metabolite 12-(S)-HETE. Int J Cancer 1993; 55: 1003-1010.
77. Kanbe K, Chigira M, Watanabe H. Effects of protein kinase inhibitors on the cell motility stimulated by autocrine motility factor. Biochimica Biophysica Acta 1994; 1222: 395-399.
78. Silletti S, Paku S, Raz A. Autocrine motility factor and the extracellular matrix. I. Coordinate regulation of melanoma cell adhesion, spreading and migration involves focal contact reorganization. Int J Canccr 1998; 76: 120-128.

79. Silletti S, Paku S, Raz A. Autocrine motility factor and the extracellular matrix. II. Degradation od remodeling od substratum components directs the motile response of tumor cells. Int J Cancer 1998; 76: 129-135.
80. Lotan R. Effects of vitamin A and its analogs (retinoids) on normal and neoplastic cells. Biochim Biophys Acta 1980; 605: 33-91.
81. Hendrix MJC, Wood WR, Seftor EA, Lolan D, Nakiyama M, Misiorowski RL, Seftor REB, Stetler-Stevenson WG, Bevacqua SJ, Liotta LA, Sobel ME, Raz A, Lotan R. Retinoic acid inhibition of human melanoma cell invasion through a rcconstituted basement membrane and its relation to decrease in the expression of proteolytic enzymes and motility factor receptor. Canccr Res 1990; 50: 4121-4130.
82. Huang B, Xie Y, Raz A. Identification of an upstream region that controls the transcription of the human autocrine motility factor receptor. Biochem Biophys Res Commun 1995; 212: 727-742.
83. Lotan R, Amos B, Watanabe H, Raz A. Suppression of motility factor response expression by retinoic acid. Cancer Res 1992: 52: 4878-4884.
84. Silletti S, Yao J, Sanford J, Mohammed AN, Otto T, Wolman SR, Raz A. Autocrine motility factor receptor in human bladder carcinoma: gene expression, loss of cell-contact regulation and chromosomal mapping. Int J Oncology 1993; 3: 801-807.
85. Silletti S, Paku S, Raz A. Tumor autocrine motility factor responses are mediated through cell contact and focal adhesion rearrangement in the absence of new tyrosine phosphorylation in metastatic cells. Am J Pathology 1996: 148: 1649-1660.
86. Van Roy FV, Mareel M. Tumor invasion: effects of cell adhesion and motility. Trends Cell Biol 1992; 2: 163-169.
87. Guirguis R, Schiffman E, Lui B, Birkbeck D, Engel J, Liotta LA. Detection of autocrine motility factor in urine as a marker of bladder cancer. J Natl Cancer Inst 1988; 80: 1203-1211.
88. Otto T, Birchmeier W, Schmidt U, Hinke A, Schipper J, Rubben H, Raz A. Inverse relation of E-cadherin and autocrine motility factor receptor expression as prognostic factor in patients with bladder carcinoma. Cancer Res 1994; 54: 3120-3123.
89. Nakarnori S, Watanabe H, Kameyama M, Imaoka S, Furukawa H, Ishikawa O, Sasaki Y, Kabuto T, Raz A. Expression of autocrine motility factor receptor in colorectal cancer as a predictor for disease recurrence. Cancer (Phila.) 1994; 74: 1855-1862.
90. Maruyama K, Watanabe H, Shiozaki H, Takayama T, Gofuku J, Yano H, Inoue M, Tamura S, Raz A, Monden M. Expression of autocrine motility factor receptor in human esophageal squamous cell carcinoma. Int J Cancer 1995; 64: 316-321.
91. Hirono Y, Fushida S, Yonemura Y, Yamamoto H, Watanabe H, Raz A. Expression of autocrine motility factor correlates with disease progression in human gastric canccr. Br J Cancer 1996; 74: 2003-2007.
92. Naga Y, Ishikawa O, Miyachi Y, Watanabe H. Expression of autocrine motility factor receptor in cutaneous malignant. Dermatology 1996; 192: 8-11.
93. Taniguchi K, Yonemura Y, Nojima N, Hirono Y, Fushida S, Fujimura T, Miwa K, Endo, Y, Yamamoto H, Watanabe H. The relation between the growth patterns of gastric carcinoma and the expression of hepatocyte growth factor receptor (c-met), autocrine motility factor receptor, and urokinase-type plasminogen activator receptor. Cancer 1998: 82: 2112-2122.
94. Ohta Y, Minato H, Tanaka Y, Go T, Oda M, Watanabe Y. Autocrine motility factor receptor expression associates with tumor progression in thyoma. Int J Oncol 2000; 17: 259-264.

Chapter 7

NM23 METASTASIS SUPPRESSOR GENE

Patricia S. Steeg, Taoufik Ouatas, Michael Mair, Susan E. Clare, Melanie T. Hartsough

Women's Cancers Section, Laboratory of Pathology, National Cancer Institute, Bethesda, MD; Division of Surgical Oncology, Department of Surgery, Northwestern University, Chicago, Il.

Abstract

The *nm23* gene family was identified on the basis of its reduced expression in highly metastatic melanoma cell lines, as compared to related cell lines of low metastatic potential. Transfection of *nm23* into metastatically competent melanoma and breast, oral squamous cell and colon carcinoma cell lines reduced tumor metastatic potential upon *in vivo* injection. No effect on primary tumor size was observed, making *nm23* a metastasis suppressor gene. Research in Drosophila, human breast carcinoma and rat neural cells show *nm23* to regulate differentiation, which may contribute to its metastatic suppressive phenotype. The biochemical mechanism of action of Nm23 in metastasis suppression is under investigation, and is hypothesized to involve a histidine protein kinase activity. Elevation of *nm23* expression in micrometastatic cells of cancer patients is hypothesized as a translational strategy to limit further colonization, invasion and induce differentiation, potentially with a clinical benefit.

Key words: Nm23, NDP kinase, breast, melanoma, differentiation, histidine protein kinase

1. CLONING AND EXPRESSION

The *nm23* metastasis suppressor gene was discovered in a murine K-1735 melanoma model system, using a differential expression strategy. Twenty four cDNA clones exhibiting different amounts of hybridization to labeled mRNA from the highly metastatic K-1735 TK line and low metastatic K-1735 Clone 19 cell line were tested for their expression patterns in seven related K-1735 cell lines of varying metastatic potentials. One clone, *nm23* (nonmetastatic 23), exhibited higher mRNA expression levels in the two low metastatic potential cell lines than in five more highly metastatic cell lines (1). Reduced Nm23 protein levels were found in the same model system using an anti-peptide antibody (2).

A family of *nm23* genes has been identified. In the human, *nm23-H1*, -H2, -DR, -H4, -H5 and –H6 share partial homology (2-8). The degree of amino acid

D.R. Welch (ed.), Cancer Metastasis – Related Genes, 123–143.

homology, subcellular localization, enzymatic activities and biological functions of the family members vary. A side-by-side comparison to test which genes have metastasis suppressor activity has not been conducted, and most research concerns *nm23-H1* and –H2. Homologs in other species are named as ndp kinases (nucleoside diphosphate kinases or ndk) or *awd* (abnormal wing discs).

An important question in the evaluation of genes identified in a model system is whether similar expression patterns are found in human tumors. In these experiments the *nm23* expression of a cohort of primary tumors was measured using northerns or in situ hybridizations for RNA levels or immunohistochemistry for protein levels. These data were compared to patient survival (disease free or overall) or histopathological characteristics of the primary tumor which are indicative of metastatic aggressiveness, such as the presence of metastases in the draining lymph nodes, Stage or grade of tumor, etc. Table 1 summarizes those studies performed with breast cancer cohorts. The various correlations reported are listed, including those which were nonsignificant by statistical analysis. Overall, 19/25 studies found a significant association between low *nm23* expression and an aspect of aggressive disease course, confirming the trend observed in the model system. The fact that these results were obtained using multiple antibodies and varying scoring systems adds strength to the overall conclusion. Similar trends are evident in cohort studies using melanoma (9-16), ovarian (17-21) and hepatocellular carcinoma (22-25) and several other cancer cell types.

Table 1. Trends in Nm23 expression in human breast carcinoma cohort studies.

Methodology:	N	Association:	Ref.:
Significant Associations between Reduced Expression and Increased Aggressiveness:			
IHC	71	Grade <0.05	(26)
IHC	82	Nodal Status 0.001	(27)
IHC (CAD)	168	MFS 0.012	
		OS, NS	(28)
IHC	72	DFS < 0.05	(29)
(+ MT)			
IHC	163	DFS 0.008	
	N-	Multivariate 0.03	(30)
IHC	102	DFS <0.01	
(+ Sialyl LX)		Nodal Status <0.01	
		Multivariate 0.032	(31)
IHC	112	Nodal Status 0.007	
		Grade 0.011	(32)
IHC	101	OS 0.05	
		Nodal Status 0.05	(33)
IHC	100	DFS 0.0433	
		Nodal Status 0.0036	
		Multivariate 0.051	(34)
IHC	33	Nodal Status 0.00001	(35)
Western 59		OS 0.001	

Table 1. *Continued.*

		MFS 0.001	
		ER 0.001	(36)
IHC	144	AX Nodal Status 0.035	
		IM Nodal Status 0.0146	(37)
IHC	130	OS 0.014	
		Nodal Status <0.01	(38)
IHC	39	OS 0.03	(39)
IHC	191	Stage NS	
		PgR 0.001	(40)
IHC	76	Differentiation <0.02	
		DFS, NS	(41)
RNA Northern	128	Nodal Status 0.007	(42)
RNA Northern	71	OS 0.003	
		DFS 0.002	
		Nodal Status 0.02	(43)
RNA In Situ Hyb.	17	Nodal Status 0.05	(44)
No Significant Associations:			
IHC	40 N-	DFS, NS	(45)
IHC	197	OS, Grade, NS	(46)
IHC	124	Nodal Status, NS	(37)
IHC	44 N-	DFS, NS	(47)
RNA	47	Nodal Status, NS	(48)
RNA	153	Nodal Status, NS	(49)

a. Abbreviations: N, Number of tumors analyzed. Also noted are any special characteristics of the tumor cohort; N-, Lymph node metastasis negative tumors only; Grade, Tumor grade; Nodal Status, Lymph node metastases; OS, Overall survival; DFS, Disease free survival; MFS, Metastasis free survival; Multivariate, Cox's proportional hazards model of survival; ER, Estrogen receptor expression; PgR, Progesterone receptor expression; NS, Not significant; IHC, Immunohistochemistry; CAD, computer assisted diagnosis; Sialyl LX, Conducted with Nm23-H1 and Sialyl Lewis X antigen; AX, Axillary; IM, Internal mammary.

Multivariate analysis, such as Cox's proportional hazards model, quantifies the strength of a predictor versus other available data, such as nodal status, tumor grade, etc. In only one case was a large study conducted using node-negative breast tumors, where prognostic information is needed for clinical decisions, and *nm23* was a significant predictor of outcome by multivariate analysis (30). In two other large cohort studies multivariate analysis showed Nm23 expression to be a significant independent predictor or nearly so. Although intriguing, *nm23* expression is not yet considered an independent prognostic factor.

The group which conducted a multivariate analysis of Nm23 in the node-negative breast cancer population has recently compiled data for several metastasis associated events, including PCNA and Ki67 for proliferation, microvessel count for angiogenesis, E-cadherin, Nm23, Vimentin and p53 (50). Of these, microvessel count ($P = 0.04$), Nm23 ($P = 0.004$), E-cadherin ($P < 0.0001$), p53 ($P = 0.0002$) and Vimentin ($P = 0.05$) were significant correlates of patient survival. Angiogenesis and loss of Nm23 were found to be early events in the malignant progression of this tumor type, with p53 and E-cadherin overexpression occuring later in progression

Other cancer types do not exhibit reduced *nm23* expression coincident with aggressive disease course. The most interesting is neuroblastoma, in which increased *nm23* expression and mutations occur in advanced (Stage IV) cancer (51). A mutation of serine 120 to a glycine was found in 6/18 advanced tumors (52), an amino acid that is thought to be a site of phosphorylation and which has functional effects on the motility phenotype (53). Alterations in the expression level of a metastasis suppressor gene may be irrelevant to the progression of some cancer cell types, or alternatively, the gene may be de-regulated by different means, such as mutation.

2. TRANSFECTION EXPERIMENTS

The hallmark of a metastasis suppressor gene is its reduction of metastatic potential *in vivo*, without a reduction in primary tumor size, upon transfection into a low expressing, metastatically competent tumor cell line. Table 2 summarizes the results of *nm23* transfection experiments reported to date. *In vivo* metastasis data was reported in 10 studies, showing a 40-98% reduction by the *nm23* transfectants, as compared to control transfectants. These experiments form the basis for the designation of *nm23* as a metastasis suppressor gene. In seven studies spontaneous tumor metastatic potential was quantitated, in which cells are injected to form a primary tumor, which then seeds out metastases. In each case, the primary tumor size between the *nm23*- and control- transfectants was comparable (54-60).

In vitro assays were also reported in transfection experiments, in order to test for cellular function in aspects of the metastatic process, and are summarized on Table 2. It is apparent that *nm23* can act at several stages of the metastatic process, including invasion, motility and colonization. The inhibition of motility to a variety of chemoattractants, including extracellular matrix proteins, growth factors, motility factors, etc. suggests that *nm23* serves to block the signal transduction process downstream of a specific receptor. In colonization, *nm23* suppressed the aberrant stimulatory effect of TGF-β on metastatic cells, also consistent with a signal transduction modulatory effect. Where tested, *nm23* did not inhibit proliferation (54, 55, 59, 67), again consistent with its lack of tumor suppressor activity.

Table 2. Comparison of Nm23- and Control transfectants for metastatic potential *in vivo* and metastasis-related function *in vitro*.

Cell Line	Decrease in metastasis	*In vitro* effects	Ref.
MDA-MB-435 Hu. Breast Ca.	50-90%	Dec. Motility Dec. Colonization Differentiation	(55, 61-64)
MDA-MB-435 Hu. Breast Ca.	90-100%	Phospholipid Metabolism	(56)
MDA-MB-231 Hu. Breast Ca.	44-46%	Dec. Motility	(57)
MTLN3 Rat Mammary Ca.	48%		(58)
K-1735 TK Mu. Melanoma	52-96%	Dec. Colonization	(54)
B16-FE7 Mu. Melanoma	83%		(59)
B16F10 Mu. Melanoma	93%	Dec. Invasion Dec. CAM	(65)
MelJuSo Mu. Melanoma	40-80%		(60)
Colon 26 Rat Colon Ca.	94%		(66)
DU145 Hu. Prostate Ca.	Not Reported	Dec. Colonization Dec. Invasion	(67)
LMF4 Hu. Oral Squamous Cell Ca.	73-98%		(68)

Abbreviations: Hu., Human, Mu., Murine, Dec., Decreased; CAM, Cell adhesion molecules, Ca., Carcinoma.

3. DIFFERENTIATION

A link between metastasis and differentiation is suggested daily in pathology laboratories. Pathologists characterize tumor grade, one component of which is the extent of tissue differentiation. In general, those tumors exhibiting poor differentiation, and consequently poor grade, often have an aggressive disease course. It has been proposed that metastasis may involve the aberrant re-expression of gene cascades operative in embryonic development , in which cells move, invade and grow albeit in an organized manner . One function of metastasis suppressor genes therefore may be to induce a differentiated state.

The *nm23* homolog in Drosophila is *awd*, which controls differentiation post-metamorphosis as presumptive adult epithelial cells in the imaginal discs attempt to differentiate (69, 70). Other organs affected include the brain and ovary. Mutations resulting in loss of *awd* expression were lethal and exhibited aberrant cellular morphology, poor architectural differentiation and a pronounced instability in phenotype. A mutation of *awd* proline 96 to serine (based on the *nm23-H1* numbering system) caused no developmental defects alone, but exhibited the aberrant differentiation of null *awd* when co-expressed with a prune (pn) eye color mutation (71). This mutation, known as killer-of-prune (k-pn), sits on a distinct loop structure of the Nm23/Awd protein by x-ray crystallography , modulates enzymatic activities (72) and is involved in the control of the tumor cell motility phenotype (62).

One of the most important contributions to understanding the possible mechanism of action of Nm23/Awd came from experiments in which the null *awd* germline was transfected with *nm23* or *awd* genes, and the extent of differentiation observed (73). Two conclusions were apparent: (1) the histidine 118 of Nm23/Awd was necessary but not sufficient for normal development and differentiation; and (2) only a minor fraction (4%) of total Awd enzymatic activity was needed for full differentiation. The latter data suggest that a minor subpopulation of Nm23/Awd proteins may exist with discreet biochemical functions.

An immunohistochemical survey of Nm23 expression through mouse embryonic development found increased expression coincident with functional differentiation in many epithelial tissues (74). Increased *nm23* expression has been linked functionally to the differentiation process in transfection studies using two mammalian cell types. PC12 phaeochromocytoma cells transfected with *nm23* exhibited differentiation *in vitro* in response to nerve growth factor, including neurite outgrowth and increased expression of microfilament and microtubule proteins (75, 76). Neurite extension was also observed in the absence of any differentiation inducer (75). Histidine 118 of Nm23 was required, in agreement with Drosophila data. A similar pattern of differentiation was observed when *nm23-DR* was transfected into neuroblastomas cell lines (77), and the corresponding histidine residue was required (78).

In the breast, a three-dimensional culture system in basement membrane proteins has been used to analyze morphological and biochemical aspects of differentiation. Normal human mammary epithelial cells produce duct-like circular structures, synthesize and directionally secrete basement membrane proteins to the outside of the "duct", and synthesize and directionally secrete sialomucin to the inside of the "duct", a process compromised in many breast tumor cell lines (79). The *nm23-H1* transfectant of the MDA-MB-435 human breast carcinoma cell line recapitulated most aspects of differentiation in this model system, including morphological duct-like colony formation, synthesis and directional secretion of basement membrane and sialomucin synthesis, in contrast to either control-transfectants or the parent cell line (63).

In contrast to the above studies in multicellular tissues, Nm23 has been linked to the de-differentiation of hematopoietic cells. The *nm23-DR* gene inhibited granulocytic differentiation and induced apoptosis (4). As an extracellular protein, Nm23 has been found to inhibit the differentiation of myeloid cells (80, 81). Neither the receptor nor the signaling process for this phenotype has been identified.

4. BIOCHEMICAL ACTIVITIES OF Nm23

4.1 Putative Biochemical Activities and Associations

The list of biochemical activities and protein:protein interactions ascribed to Nm23, which represent hypothetical mechanisms of metastasis suppression, is enormous. Table 3 presents an overview of the literature.

Several generalizations are possible. First, the nucleoside diphosphate kinase activity of Nm23, which is the basis for the discovery of the protein, is not the only biochemical activity that this protein possesses. This activity is thought to contribute to the regulation of nucleotide triphosphate pool concentrations, and therefore to impact DNA synthesis, G-protein function and other events. Most, if not all, of the reports concerning Nm23 association and function with small and heterotrimeric G proteins cannot rule out the possibility that Nm23 is simply regenerating GTP in these systems.

Second, Nm23 exhibits a protein kinase activity. Where studied, it can involve the phosphorylation of a serine on the substrate, and may have an intermediate step of Nm23 autophosphorylation on Histidine 118. If so, this protein kinase activity can be considered a type of histidine protein kinase activity, discussed below. It has been difficult to separate the NDP kinase activity of Nm23 from a protein kinase activity; in many cases extensive experimentation has been carried out to show that Nm23 is not regenerating ATP from contaminating nucleotide pools, which then permits autophosphorylation of the "substrate". Phosphatases have been reported to associate with Nm23, suggesting that kinase activity is regulated.

Table 3. Enzymatic activities and protein: protein associations attributed to Nm23.

Activity	Reference
Histidine autophosphorylation	
Autophosphorylation on His 118	(82-84)
Serine autophosphorylation	(53, 85-87)
Autophosphorylation on two peptides containing serine	
May be downstream of Histidine	
Nucleoside diphosphate kinase	(88)
Transfer of γ-phosphate from NTP to NDP	
Protein kinase	
General	(89)
Serine protein kinase	(90, 91)
Histidine protein kinase	
Histidine – Histidine transfer	(92)
Two-component	(93-95)
Transcriptional regulation	(96-100)
DNA Cleavage	(101)
Interactions:	
Small G-proteins:	
Ras, Arf, Rad, etc.	(102-104)
Heterotrimeric G-proteins	(105-107)
ROR/RZR Nuclear Orphan Receptors	(108)
GAPDH	(109, 110)
MER-5	(111)
Cytoskeleton	
Vimentin	(111)
Microtubules	(112-115)
Heat shock protein	(116, 117)
Prune	(118)
Phytochromes	(119-121)
Phosphatases	(122)

The most intriguing kinase activity of Nm23 is as a histidine protein kinase. This is a little known class of kinases in mammalian cells, but forms the primary signal transduction system in prokaryotes, called the two-component system. When a signal interacts with a receptor, a linked histidine protein kinase is autophosphorylated. The histidine kinase then passes the phosphate from its histidine to the aspartate of a receiver protein. The receiver protein has a DNA binding domain, which activates transcription of the genes required for the bacterial response. A Nm23 homolog was cloned into a two-component system, and functioned as the histidine protein kinase portion (95). Two-component pathways have been identified in eukaryotes such as yeast and plants; in this case the response receiver, when phosphorylated on an aspartate residue, sets into motion a poorly defined signal transduction process which results in the inhibition of MAP kinase. Do histidine kinases occur and function in mammalian cells? Nm23-H1 was found to phosphorylate an aspartate residue of a bovine brain lysate protein, aldolase. In this case, the protein does not autophosphorylate, providing concrete evidence of a protein kinase activity. Other possibilities include histidine-histidine, and histidine-serine phosphorylations, based on published "ice bucket" reactions. Our ability to observe these transphosphorylations is severely limited by the pH under which SDS-PAGE is run, which destroys acid-labile phosphohistidine. It has been estimated that up to 6% of cellular phosphorylation is phosphohistidine.

Third, the list of associations of Nm23 with cellular proteins is extensive. Nm23 is an abundant protein, with a long half-life. The two best studied forms, Nm23-H1 and –H2, have acidic and basic pIs, and can nonspecifically bind proteins via ionic interactions. Perhaps the best indicator of specificity is whether the association with Nm23 alters the biochemical activity of the target protein, for which incomplete data is available.

Fourth, the data from Drosophila development, which indicate that only a small fraction of total enzymatic (NDP kinase) activity of Nm23 is needed for biological activity, suggest that a relatively minor subpopulation of Nm23 with a particular function or association may be relevant. In this regard, we may need to consider a structure of Nm23 other than the published crystallography (123-125). Mutation of serine 120 of Nm23 has been reported to display a novel structure upon renaturation, that of a molten globule (126). Such a form may have a different active site or association profile, etc. and may be relevant to the biology of metastasis or differentation.

Fifth, if Nm23 functions in metastasis suppression by the limitation of signal transduction, as suggested in the transfection experiments, then several mechanistic possibilities are found on Table 3. Function as a histidine protein kinase in an as-yet unknown two-component like signal transduction reaction is postulated. Binding to cytoskeletal elements, if their function is impacted, appears relevant to the motile and invasive phenotypes. The participation of Nm23 homologs in plant signal transduction in the phytochromes, is incompletely understood, but may she light on conserved pathways.

4.2 Structure: Function Correlations

We have used site directed mutagenesis to change amino acids of Nm23 thought to be critical to development, found to be mutated in cancers, or identified as sites of phosphorylation. The constructs were expressed in bacteria, and the recombinant Nm23 protein was purified and characterized biochemically (72). Wild type and mutant *nm23* constructs, tethered to a constitutive promoter, were transfected into MDA-MB-435 human breast carcinoma cells and assays for metastasis suppressor function in one aspect of the process *in vitro*, motility in Boyden chamber assays (62). In general, wild type *nm23-H1* inhibited the motility of MDA-MB-435 cells to serum, which represents an amalgamation of factors, or to a defined chemoattractant, autotaxin. Mutation of either proline 96 to serine (P96S, the killer of prune mutation) or serine 120 to glycine or alanine

(S120G, S120A, a site of phosphorylation and mutation in neuroblastomas) significantly abrogated the motility inhibitory phenotype. Mutation of serine 44 to alanine (S44A), another site of autophosphorylation, left Nm23 with motility suppressive capacity, confirming the specificity of the other mutations in this model system. The wild type and mutated *nm23-H1* transfected cell lines exhibited comparable *in vitro* proliferation rates, again providing evidence of relevance to the metastatic process. It was impossible to test the impact of histidine 118 mutation, as transfectants could not be produced.

Recombinant wild type or site directed mutated Nm23-H1 proteins have been analyzed for several of the biochemical activities listed on Table 3. Do the P96S and S120G mutations, which have altered biological function, exhibit any differences in biochemical function? The data are summarized below:

1. The NDP kinase activity of the wild type and mutant proteins are comparable.
2. S120G exhibits decreased serine autophosphorylation, in some but not all assays.
3. Both S120G and P96S exhibit reduced function in assays of histidine protein kinase activity, including the transphosphorylation of succinyl thiokinase and GST-Nm23 on histidine residues. The P96S is a true kinase-defective mutant, whereas the S120G deficiency may result from poor autophosphorylation.
4. Both S120G and P96S are deficient in histidine kinase activity for a bovine brain protein on an aspartate residue, later identified as aldolase (93). This is an example of histidine-aspartate histidine protein kinase activity, typical of two-component pathways.

These data support the hypothesis that a histidine protein kinase activity underlies the metastasis suppressive effect of Nm23. The NDP kinase activity failed to correlate with suppression of motility. Other activities or associations, as yet untested, may also vary among these mutants, and have contributory activity to metastasis as well.

Structure:function comparisons in differentiation have reached similar conclusions. The histidine residue of Nm23 was required for the induction of PC12 or neuroblastomas differentiation (78). Site directed mutagenesis of three *nm23-DR* motifs was performed and tested for differentiation induction in neuroblastomas cells, the histidine phosphorylation residue (His 134), the homolog of serine 44 (Ser 61) and an RGD (Integrin) postulated binding site (78). All recombinant proteins with the exception of the histidine mutant exhibited NDPK activity. Both the histidine and serine mutant forms were deficient in the induction of differentiation, on morphological or gene expression grounds. The RGD mutant was comparable to wild type *nm23-DR*. On the assumption that serine 44 (61) phosphorylation is downstream of histidine autophosphorylation, the data are compatible with a histidine protein phosphotransferase activity underlying a differentiation effect.

5. TRANSLATIONAL RESEARCH CONCERNING Nm23

5.1 Potential Clinical Utility of Metastasis Suppressor Genes

The identification and characterization of metastasis suppressor genes is expected to lead to the development of new clinical strategies to control the metastatic process in patients. This premise is based on several concepts:

1. Primary tumors can often be successfully managed by surgery and radiation therapy. Metastatic disease is a major contributor to patient morbidity and mortality.
2. For many breast and other cancer patients, only a minor percentage of patients (~6% in breast cancer) have detectable distant metastases at the time of diagnosis and surgery. A greater percentage have evidence of some metastatic spread, for instance to the regional lymph nodes (40-50%). Another percentage of patients with no evidence of spread will eventually go on to develop distant disease. Based on these data, it can be concluded that the metastatic process has not been completed in many patients.
3. Even if detectable metastases have not been found, it cannot be proven that the initial parts of the metastatic process have not been completed. For instance, cells may have invaded from the primary tumor, intravasated and extravasated the circulatory system, and be sitting as occult micrometases in a distant organ. What remains is angiogenesis and metastatic colonization at the distant site.
4. Metastatic colonization is known to differ from primary tumor growth or proliferation *in vitro*, but is as yet incompletely characterized. For instance, many nonmetastatic tumor cells are inhibited by TGF-β, while metastatically competent cells are stimulated. Thus, reactivity to cytokines may distinguish the metastatically competent cells from the primary tumor mass.
5. Many metastasis suppressor genes affect, at least in part, aspects of metastatic colonization. Nm23 appears to influence colonization, particularly in response to TGF-β. It also influences differentiation, which may impact outgrowth at a distant site.
6. Data compiled to data in human tumor cohorts indicate that mutations in metastasis suppressor genes are relatively rare. Thus, one may only have to "turn on" expression of the wild type gene, rather than eliminate any number of gain-of-function mutational effects.

5.2 Re-expression of Nm23-H1

We have hypothesized that *nm23-H1* expression may represent a valuable therapeutic target based on: (a) its reduced expression in many breast tumors from patients with aggressive disease in cohort studies, and (b) its suppression of *in vivo* metastasis upon transfection into a number of tumor cell lines. Several aspects of *nm23-H1* expression noted in the transfection experiments may affect colonization at the distant site, including soft agar colonization, responsiveness to TGF-b, and induction of differentiation. A cohort study in which side-by-side analysis of Nm23-H1 expression by immunohistochemistry, and allelic deletion, shed light on the mechanism of Nm23-H1 de-regulation (127). Allelic deletion (or loss of heterozygosity, LOH) was observed frequently at the *nm23-H1* locus, but was accompanied by varying levels of protein expression. Apparently, the remaining allele could be highly or poorly transcribed and translated. Poor patient survival was not correlated with allelic deletion, but was correlated with low Nm23-H1 protein expression. Twenty tumors were sequenced for mutations in the Nm23-H1 coding region, and none found. Thus, down-regulation of expression, rather than the deletion-mutation model often found for tumor suppressor genes, is likely the case in breast cancer.

The promoter regions for several *nm23* homologs have been identified (128). We have used a model system of four unrelated human breast carcinoma cell lines to characterize the *nm23-H1* promoter. These lines include MCF-7 and ZR-75, which are relatively high in *nm23-H1* expression and virtually nonmetastatic *in vivo*, and MDA-MB-435 and MDA-MB-231, which are low in *nm23-H1* expression and metastatically competent in vivo. A 2.1 kb portion of the *nm23-H1* promoter, generally 5' of its transcription initiation site, was found to induce differential reporter gene expression when transfected into the four cell lines in a manner consistent with the known Nm23 expression levels (i.e., several fold higher in the MCF-7 and ZR-75 cells than in the MDA-MB-435 and MDA-MB-231 cells) (129). This region was therefore thought to contain the information needed to determine high versus low levels of *nm23-H1*. Deletion of restriction fragments from the 2.1 kb promoter narrowed down the region of interest to a 250 bp fragment. Investigation of the transcription factor binding sites within this region, their pattern of usage among the various cell lines, and their functional effect on Nm23-H1 expression is ongoing.

REFERENCES

1. Steeg, P. S., Bevilacqua, G., Kopper, L., Thorgeirsson, U. P., Talmadge, J. E., Liotta, L. A., and Sobel, M. E. Evidence for a novel gene associated with low tumor metastatic potential., J. Nat'l. Cancer Inst. 80: 200-204, 1988.
2. Rosengard, A. M., Krutzsch, H. C., Shearn, A., Biggs, J. R., Barker, E., Margulies, I. M. K., King, C. R., Liotta, L. A., and Steeg, P. S. Reduced Nm23/Awd protein in tumor metastasis and aberrant Drosophila development, Nature. 342: 177-180, 1989.

3. Stahl, J. A., Leone, A., Rosengard, A. M., Porter, L., King, C. R., and Steeg, P. S. Identification of a second human nm23 gene, nm23-H2., Cancer Res. 51: 445-449, 1991.
4. Venturelli, D., Martinez, R., Melotti, P., Casella, I., Peschile, C., Cucco, C., Spampinato, G., Darzynkiewicz, Z., and Calabretta, B. Overexpression of DR-nm23, a protein encoded by a member of the nm23 gene family, inhibits granulocyte differentiation and induces apoptosis in 32Dc13 myeloid cells., Proc. Nat'l Acad. Sci. USA. 92: 7435-7439, 1995.
5. Milon, L., Rosseau-Merck, M.-F., Munier, A., Erent, M., Lascu, I., Capeau, J., and Lacombe, M.-L. nm23-H4, a new member of the family of human nm23/nucleoside diphosphate kinase genes localised on chromosome 16p13., Hum. Genet. 99: 550-557, 1997.
6. Munier, A., Feral, C., Milon, L., Pinon, V., Gyapay, G., Capeau, J., Guellaen, G., and Laconbe, M. A new human nm23 homologue (nm23-H5) specifically expressed in testis germinal cells., FEBS Letters. 434: 289-294, 1998.
7. Ysuki, H., Nitta, M., Furuya, A., Hanai, N., Fujiwara, T., Inagaki, M., Kochi, M., Ushio, Y., Saya, H., and Nakamura, H. A novel nucleoside diphosphate (NDP) kinase, Nm23-H6, localizes in mitochondria and affects cytokinesis., J. Cellular Biochem. 76: 254-269, 1999.
8. Mehus, J., Deloukas, P., and Lambeth, D. NME6: a new member of the nm23/nucleoside diphosphate kinase gene family located on human chromosome 3p21.3., Hum. Genet. 104: 454-459, 1999.
9. Florenes, V. A., Aamdal, S., Myklebost, O., Maelandsmo, G. M., Bruland, O. S., and Fodstad, O. Levels of nm23 messenger RNA in metastatic malignant melanomas: inverse correlation to disease progression, Cancer Res. 52: 6088-6091, 1992.
10. Betke, H., Korabiowska, M., Brinck, U., Trybus-Galuszka, H., Kellner, S., Okon, K., Stachura, J., and Droese, M. The role of nm23 in melanoma progression and its prognostic significance, Pol. J. Pathol. 49: 93-96, 1998.
11. Bodey, B., Kaiser, H., and Goldfarb, R. Immunophenotypically varied cell subpopulations in primaty and metastatic human melanomas. Monoclonal antibodies for diagnosis, detection of neoplastic progression and receptor directed immunotherapy., Anticancer Res. 16: 517-532, 1996.
12. Caligo, M. A., Grammatico, P., Cipollini, G., Varesco, L., Porto, G. D., and Bevilacqua, G. A low NM23.H1 gene expression identifying high malignancy human melanomas., Melanoma Res. 4: 179-184., 1994.
13. Greco, I., Calvisi, G., Ventura, L., and Cerrito, F. An immunohistochemical analysis of nm23 gene product expression in uveal melanoma., Melanoma Res. 7: 231-236, 1997.
14. McDermott, N. C., Milburn, C., Curran, B., Kay, E. W., Walsh, C. B., and Leader, M. B. Immunohistochemical expression of nm23 in primary invasive malignant melanoma is predictive of survival outcome, J. Pathol. 190: 157-162, 2000.
15. Xerri, L., Grob, J. J., Battyani, Z., Gouvernet, J., Hassoun, J., and Bonerandi, J. J. NM23 expression in metastasis of malignant melanoma is a predictive prognostic paramenter correlated with survivial., Br. J. Cancer. 70: 1224-1228., 1994.
16. Easty, D. J., Maung, K., Lascu, I., Veron, M., Fallowfield, M. E., Hart, I. R., and Bennett, D. C. Expression of NM23 in human melanoma progression and metastasis., Br. J. Cancer. 74: 109-114., 1996.

17. Scambia, G., Ferrandina, G., Marone, M., Panici, P. B., Giannitelli, C., Piantelli, M., Leone, A., and Manusco, S. Nm23 in ovarian cancer: Correlationwith clinical outcome and other clinico-pathological and biochemical prognositic markers., J. Clin. Oncol. 14: 334-342., 1996.
18. Mandai, M., Konishi, I., Koshiyama, M., Mori, T., Arao, S., Tashiro, H., Okamura, H., Nomura, H., Hiai, H., and Fukumoto, M. Expression of metastasis-related nm23-H1 and nm23-H2 genes in ovarian carcinomas: Correlation to clinicopathology, EGFR, c-erb-B-2 and c-erb-B-3, and sex steroid receptor expression., Cancer Res. 54: 1825-1830., 1994.
19. Qian, M., Feng, Y., Xu, L., Zheng, S., and Zhou, X. Expression of antimetastatic gene nm23-H1 in epithelial ovarian cancer, Chin. Med. J. 110: 142-144, 1997.
20. Srivatsa, P. J., Cliby, W. A., Keeney, G. L., Dodson, M. K., Suman, V. J., Roche, P. C., and Podratz, K. C. Elevated nm23 protein expression is correlated with diminished progression-free survival in patients with epithelial ovarian carcinoma, Gynecol. Oncol. 60: 363-372, 1996.
21. Veil, A., Dall'Agnese, L., Canzonieri, V., Sopracordevole, F., Capozzi, E., Carbone, A., Visentin, M., and Boiocchi, M. Suppressive role of the metastasis-related nm23-H1 genen in human ovarian carcinomas: Association of high messenger RNA expression with lack of lymph node metastasis., Cancer Res. 55: 2645-2650, 1995.
22. Iizuka, N., Oka, M., Noma, T., Nakazawa, A., Hirose, K., and Suzuki, T. Nm23-H1 and N+m23-H2 messenger RNA abundance in human hepatocellular carcinoma., Cancer Res. 55: 652-657, 1995.
23. Nakayama, T., Ohtsuru, A., Nakao, K., Shima, M., Nakata, K., Watanabe, K., Ishii, N., Kimura, N., and Nagataki, S. Expression in human hepatocellular carcinoma of nucleoside diphosphate kinase, a homolog of the nm23 gene product., J. Nat'l. Cancer Inst. 84: 1349-1354., 1992.
24. Shimada, M., Taguchi, K., Hasegawa, H., Gion, T., Shirabe, K., Tsuneyoshi, M., and Sugimachi, K. Nm23-H1 expression in intrahepatic or extrahepatic metastases of hepatocellular carcinoma, Liver. 18: 337-342, 1998.
25. Yamaguchi, A., Urano, T., Goi, T., Takeuchi, K., Niimoto, S., Nakagawara, G., Furukawa, K., and Shiku, H. Expression of human nm23-H1 and nm23-H2 proteins in hepatocellular carcinoma., Cancer. 73: 2280-2284., 1994.
26. Midulla, C., Iorio, P. D., Nagar, C., Pisani, T., Cenci, M., Valli, C., nofroni, I., and Vecchione, A. Immunohistochemical expression of p53, nm23-H1, Ki67 and DNA ploidy: Correlation with lymph node status and other histopahtologic parameters in breast cancer., Anticancer Res. 19: 4033-4037, 1999.
27. Yoshida, H., Kijima, H., Terasaki, Y., Takeshita, T., Omiya, H., Shimojima, K., Shimbori, M., Sato, T., Sato, S., Onoda, N., Yamazaki, H., Tamaoki, N., Ueyama, Y., and Nakamura, M. Heterogenous expression of nm23 gene product as a predictor of lymph nodal status in human breast cancer, Int. J. Oncol. 13: 1141-6, 1998.
28. Charpin, C., Bouvier, C., Garcia, S., Martini, F., Andrac, L., Lavaut, M., and Allasia, C. Automated and quantitative immunocytochemical assays of Nm23/NDPK protein in breast carcinomas., Int. J. Cancer. 74: 416-420, 1997.
29. Vaquea-Ramirez, F., Gonzalez-Campora, J., Hevia-Alvarez, E., Fernandez-Santos, J., Rios-Martin, J., Otal-Salaverri, C., and Gonzalez-Campora, R. P-glycoprotein, metallothionein and NM23 protein expressions in breast carcinoma., Pathology Res. and Practice. 196: 553-559, 2000.

30. Heimann, R., Ferguson, D., and Hellman, S. The relationship between nm23, angiogenesis, and the metastatic proclivity of node-negative breast cancer., Cancer Res. 58: 2766-2771, 1998.
31. Yamaguchi, A., Ding, K., Maehara, M., Goi, T., and Nakagawara, G. Expression of nm23-H1 gene and Sialyl Lewis X antigen in breast cancer, Oncology. 55: 357-362, 1998.
32. Bertheau, P., Steinberg, S., and Merino, M. c-erbB-2, p53 and nm23 gene product expression in breast cancer in young women: Immunohistochemical analysis and clinicopathologic correlation., Human Pathol. 29: 323-329, 1998.
33. Mao, H., Liu, H., Li, S., Fu, X., and Feng, Y. The relationship between nm23 gene expression and distal metastasis and prognosis in breast cancer, Eur. J. Cancer. 34 Suppl 5: 487, 1998.
34. Han, S., Yun, I.-K., Noh, D.-Y., Choe, K.-J., Song, S.-Y., and Chi, J. Abnormal expression of four novel molecular markers represents a highly aggressive phenotype in breast cancer. Immunohistochemical assay of p53, nm23, erbB-2, Cathepsin D protein., J. Surg. Oncol. 65: 22-27, 1997.
35. Duenas_Gonzalez, A., Abad-Hernandez, M. M., Garcia-Mata, J., Paz-Bouza, J. I., Cruz-Hernandez, J. J., and Gonzalez-Sarmiento, R. Analysis of nm23-H1 expression in breast cancer. Correlation with p53 expression and clinicopathologic findings., Cancer Lett. 101: 137-142., 1996.
36. Toulas, C., Mihura, J., Balincourt, C. d., Marques, B., Marek, E., Soula, G., Roche, H., and Favre, G. Potential prognostic value in human breast cancer of cytosolic Nme1 protein detection using an original hen specific antibody., Br. J. Cancer. 73: 630-635., 1996.
37. Noguchi, M., Earashi, M., Ohnishi, I., Kinoshita, K., Thomas, M., Fusida, S., Miyazaki, I., and Mizukami, Y. Nm23 expression versus Helix pomatia lectin binding in human breast cancer metastases., Int. J. Oncol. 4: 1353-1358., 1994.
38. Tokunaga, Y., Urano, T., Furakawa, K., Kondo, H., Kanematsu, T., and Shiku, H. Reduced expression of Nm23-H1, but not of Nm23-H2, is concordant with the frequency of lymph-node metastasis of human breast cancer., Int. J. Cancer. 55: 66-71, 1993.
39. Barnes, R., Masood, S., Barker, E., Rosengard, A. M., Coggin, D. L., Crowell, T., King, C. R., Porter-Jordan, K., Wargotz, E. S., Liotta, L. A., and Steeg, P. S. Low Nm23 protein expression in infiltrating ductal breast carcinomas correlates with reduced patient survival., American Journal of Pathology. 139: 245-250, 1991.
40. Nakopoulou, L. L., Tsitsimelis, D., Lazaris, A. C., Tzonou, A., Gakiopoulou, H., Dicglou, C. C., and Davaris, P. S. Nm-23, c-erB-2, and progesterone resceptor expression in invasive breast cancer: correlation with clinicopathologic parameters, Cancer Detect. Prev. 23: 297-308, 1999.
41. Russo, A., Bazan, V., Morello, V., Valli, C., Giarnieri, E., Dardanoni, G., Cucciarre, S., Carreca, I., Bazan, P., Tomasino, R., and Vecchione, A. Nm23-H1 protein immunohistochemical expression in human breast cancer-Relationship to prognostic factors and risk of relapse., Oncol. Reports. 3: 183-189., 1996.
42. Caligo, M., Cipollini, G., Berti, A., Viacava, P., Collecchi, P., and Bevilacqua, G. NM23 gene-expression in human breast carcinomas - Loss of correlation with cell-proliferation in the advanced phase of tumor porgression., Int. J. Cancer. 74: 102-111., 1997.

43. Hennessy, C., Henry, J., May, F. E. B., Westly, B., Angus, B., and Lennard., T. W. J. Expression of the antimetastatic gene nm23 in human breast cancer: An association with good prognosis., J. Nat'l Cancer Inst. 83: 281-285, 1991.
44. Bevilacqua, G., Sobel, M. E., Liotta, L. A., and Steeg, P. S. Association of low nm23 RNA levels in human primary infiltrating ductal breast carcinomas with lymph node involvement and other histopatholgical indicators of high metastatic potential., Cancer Research. 49: 5185-5190, 1989.
45. Russell, R., Geisinger, K., Mehta, R., White, W., Shelton, B., and Kute, T. nm23-Relationship to the metastatic potential of breast cancerinoma cell lines, primary human xenografts and lymph node negative breast carcinoma patients., Cancer. 79: 1158-1165, 1997.
46. Sawan, A., Lascu, I., Veron, M., Anderson, J. J., Wright, C., Horne, C., and Angus, B. NDP-K/nm23 expression in human breast cancer in relation to relapse, survival and other prognostic factors: An immunohistochemical study., J. Pathol. 172: 27-34., 1994.
47. Kapranos, N., Karaiossifidi, H., Kouri, E., and Vasilaros, S. Nm23 expression in breast ductal carcionmas: a ten year follow-up study in a uniform group of node-negative breast cancer patients, Anticancer. 16: 2987-2990, 1996.
48. Albertazzi, E., Cajone, F., Leone, B. E., Naguib, R. N. G., Lakshmi, M. S., and Sherbert, G. V. Expression of metastasis-associated genes h-mts1 (S100A4) and nm23 in carcinoma of breast is related to disease progression, DNA and Cell Biology. 17: 335-342, 1998.
49. Goodall, R., Dawkins, H., Robbins, P., Hahnel, E., Sarna, M., Hahnel, R., Papadimitriou, J., Harvey, J., and Sterrett, G. Evaluation of the expression levels of nm23-H1 RNA in primary breast cancer, benign breast disease, axillary lymph nodes and normal breast tissue., Pathology. 26: 423-428., 1994.
50. Heimann, R. and Hellman, S. Individual characterisation of the metastatic capacity of human breast carcinoma., Eur. J. Cancer. 36: 1631-1639, 2000.
51. Leone, A., Seeger, R. C., Hong, C. M., Hu, Y. Y., Arboleda, M. J., Brodeur, G. M., Stram, D., Slamon, D. J., and Steeg, P. S. Evidence for nm23 RNA overexpression, DNA amplification and mutation in aggressive childhood neuroblastomas., Oncogene. 8: 855-865, 1993.
52. Hailat, N., Keim, D. R., Melhem, R. F., Zhu, X.-X., Eckerstrom, C., Brodeur, G. M., Reynolds, C. P., Seeger, R. C., Lottspeich, F., Strahler, J. R., and Hanash, S. M. High levels of p19/nm23 protein in neuroblastoma are associated with advanced disease and N-myc gene amplification., Journal of Clinical Investigation. 88: 341-345, 1991.
53. MacDonald, N. J., DeLaRosa, A., Benedict, M. A., Freije, J. M. P., Krutsch, H., and Steeg, P. S. A novel phosphorylation of Nm23, and not its nucleoside diphosphate kinase activity, correlates with suppression of tumor metastasis., J. Biol. Chem. 269: 25780-25789, 1993.
54. Leone, A., Flatow, U., King, C. R., Sandeen, M. A., Margulies, I. M. K., Liotta, L. A., and Steeg, P. S. Reduced tumor incidence, metastatic potential, and cytokine responsiveness of nm23-transfected melanoma cells., Cell. 65: 25-35, 1991.
55. Leone, A., Flatow, U., VanHoutte, K., and Steeg, P. S. Transfection of human nm23-H1 into the human MDA-MB-435 breast carcinoma cell line: Effects on tumor metastatic potential, colonization, and enzymatic activity., Oncogene. 8: 2325-2333, 1993.

56. Bhujwalla, Z., Aboagye, E., Gilles, R., Chack, V., Mendola, C., and Backer, J. Nm23-transfected MDA-MB-435 human breast carcinoma cells form tumors with altered phospholipid metabolism and pH: A 31P nuclear magnetic resonance study in vivo and in vitro., Magnetic Res. Med. 41: 897-903, 1999.
57. Russell, R., Pedersen, A., Kantor, J., Geisinger, K., Long, R., Zbieranski, N., Townsend, A., Shelton, B., Brunner, N., and Kute, T. Relationship of nm23 to proteolytic factors, proliferation and motility in breast cancer tissues and cell lines., Br. J. Cancer. 78: 710-717, 1998.
58. Fukuda, M., Ishii, A., Yasutomo, Y., Shimada, N., Ishikawa, N., Hanai, N., Nagara, N., Irimura, T., Nicolson, G., and Kimura, N. Metastatic potential of rat mammary adenocarcinoma cells associated with decreased expression of nucleoside diphosphate kinase/nm23: Reduction by transfection of NDP Kinase a isoform, an nm23-H2 gene homolog., Int. J. Cancer. 65: 531-537., 1996.
59. Baba, H., Urano, T., Okada, K., Furukawa, K., Nakayama, E., Tanaka, H., Iwasaki, K., and Shiku, H. Two isotypes of murine nm23/Nucleoside Diphosphate Kinse, nm23-M1 and nm23-M2, are involved in metastatic suppression of a murine melanoma line., Cancer Res. 55: 1977-1981., 1995.
60. Miele, M. E., Rosa, A. D. L., Lee, J. H., Hicks, D. J., Dennis, J. U., Steeg, P. S., and Welch, D. R. Suppression of human melanoma metastasis following introduction of chromosome 6 is independent of NME1 (nm23), Clin. Exp. Metastasis. 15: 259-265, 1997.
61. Kantor, J. D., McCormick, B., Steeg, P. S., and Zetter, B. R. Inhibition of cell motility after nm23 transfection of human and murine tumor cells., Cancer Res. 53: 1971-1973, 1993.
62. MacDonald, N., Freije, J., Stracke, M., Manrow, R., and Steeg, P. Site directed mutagenesis of nm23-H1: Mutation of proline 96 or serine 120 abrogates its motility inhibitory activity upon transfection into human breast carcinoma cells., J. Biol. Chem. 271: 25107-25116, 1996.
63. Howlett, A. R., Petersen, O. W., Steeg, P. S., and Bissell, M. J. A novel function for Nm23: Overexpression in human breast carcinoma cells leads to the formation of basement membrane and growth arrest., J. Nat'l. Cancer Inst. 86: 1838-1844, 1994.
64. Bemis, L. and Schedin, P. Reproductive state of rat mammary gland stroma modulates human breast cancer cell migration and invasion., Cancer Res. 60: 3414-3418, 2000.
65. Parhar, R. S., Shi, Y., Zou, M., Farid, N. R., Ernst, P., and Al-Sedairy, S. Effects of cytokine mediated modulation of Nm23 expression on the invasion and metastatic behavior of B16F10 melanoma cells., Int. J. Cancer. 60: 204-210, 1995.
66. Tagashira, H., Hamazaki, K., Tanaka, N., Gao, C., and Namba, M. Reduced metastatic potential and c-myc overexpression of colon adenocarcinoma cells (Colon 26 line) transfected with nm23-R2 rat nucleoside diphosphate kinase a isoform., Int. J. Mol. Med. 2: 65-68, 1998.
67. Lim, S., Lee, H.-Y., and Lee, H. Inhibition of colonization and cell-matrix adhesion after nm23-H1 transfection of human prostrate carcinoma cells, Cancers Lett. 133: 143-149, 1998.
68. Miyazaki, H., Fukuda, M., Ishijima, Y., Negishi, A., Hirayama, R., Ishikawa, N., Amagasa, T., and Kimura, N. Overexpression of nm23-H2/NDP Kinase B in a human oral squamous cell carcinoma cell line results in reduced metastsis, differentiated phenotype in the metastatic site, and growth factor-independent proliferative activity in culture, Clin.Cancer Res. 5: 4301-4307, 1999.

69. Dearolf, C., Hersperger, E., and Shearn, A. Developmental consequences of awdb3, a cell autonomous lethal mutation of Drosophila induced by hybrid dysgenesis., Developmental Biology. 129: 159-168, 1988a.
70. Dearolf, C., Tripoulas, N., Biggs, J., and Shearn, A. Molecular consequences of awdb3, A cell autonomous lethal mutation of Drosophila induced by hybrid dysgenesis., Developmental Biology. 129: 169-178, 1988b.
71. Biggs, J., Hersperger, E., Dearolf, C., and Shearn, A. Analysis of the lethal interaction between the prune and Killer of prune mutations of Drosophila., Genes and Development. 2: 1333-1343, 1988.
72. Freije, J. M. P., Blay, P., MacDonald, N. J., Manrow, R. E., and Steeg, P. S. Site-directed mutation of Nm23-H1. Mutations lacking motility suppressive capacity upon transfection are deficient in histidine-dependent protein phosphotransferase pathways in vitro., J. Biol. Chem. 272: 5525-5532, 1997.
73. Xu, J., Liu, L., Deng, F., Timmons, L., Hersperger, E., Steeg, P., Veron, M., and Shearn, A. Rescue of the awd mutant phenotype by expression of human Nm23/NDP kinase in Drosophila., Develp. Biol. 177: 544-557, 1996.
74. Lakso, M., Steeg, P. S., and Westphal, H. Embryonic expression of Nm23 during mouse organogenesis., Cell Growth Diff. 3: 873-879, 1992.
75. Ishijima, Y., Shimada, N., Fukada, M., Miyazaki, H., Orlov, N., Orlova, T., Yamada, T., and Kimura, N. Overexpression of nucleoside diphosphate kinases induced neurite outgrowth and their substitution to inactive forms leads to suppression of nerve growth factor- and dibutryl cAMP-induced effects in PC12D cells., FEBS Lett. 445: 155-159, 1999.
76. Gervasi, F., D'Agnano, I., Vossio, S., Zupi, G., Sacchi, A., and Lombardi, D. nm23 influences proliferation and differentiation of PC12 cells in response to nerve growth factor., Cell Growth Diff. 7: 1689-1695, 1996.
77. Amendola, R., Martinez, R., Negroni, A., Venturelli, D., Tanno, B., Calabretta, B., and Raschella, G. DR-nm23 gene expression in neuroblastoma cells: Relationship to integrin expression, adhesion characteristics, and differentiation., J. Nat'l Cancer Inst. 89: 1300-1310, 1997.
78. Negroni, A., Venturelli, D., Tanno, B., Amendola, R., Ransac, S., Cesi, V., Calabretta, B., and Raschelia, G. Neuroblastoma specific effects of DR-nm23 and its mutant forms on differentiation and apoptosis., Cell Death and Diff. 7: 843-850, 2000.
79. Barcellos-Hoff, M. H., Aggeler, J., Ram, T. G., and Bissell, M. J. Functional differentiation and alveolar morphogenesis of primary mamary cultures on reconstituted basement membrane., Development. 105: 223-235., 1989.
80. Okabe-Kado, J., Kasukabe, T., Honma, Y., Hayashi, M., Henzel, W. J., and Hozumi, M. Identity of a differentiation inhibiting factor for mouse myeloid leukemia cells with Nm23/nucleoside diphosphate kinase., Biochem. Biophys. Res. Comm. 182: 987-994, 1992.
81. Okabe-Kado, J., Kasukabe, T., Hozumi, M., Honma, Y., Kimura, N., Baba, H., Urano, T., and Shiku, H. A new function of Nm23/NDP kinase as a differentiation inhibitory factor, which does not require its kinase activity., FEBS Lett. 363: 311-315, 1995.
82. Walinder, O. Evidence of the presence of 1-Phosphohistidine as the main phosphorylated component at the active site of bovine nucleoside diphosphate kinase., Acta. Chem. Scand. 23: 339-341, 1969.

83. Edlund, B., Rask, L., Olsson, P., Wålinder, O., Zetterqvist, Ö., and Engström, L. Preparation of crystalline nucleoside diphosphate kinase from baker's yeast and identification of 1-[32P]phosphohistidine as the main phosphorylated product of an alkaline hydrolysate of enzyme incubated with adenosine [32P] triphosphate., Eur. J. Biochem. 9: 451-455, 1969.
84. Parks, R. E. and Agarwal, R. P. Nucleoside diphosphokinases., Vol. 8, p. 307-333. New York: Academic Press, 1973.
85. Bominaar, A., Tepper, A., and Veron, M. Autophosphorylation of nucleoside diphosphate kinase on non-histidine residues., FEBS Lett. 353: 5-8., 1994.
86. Muñoz-Dorado, J., Almaula, N., Inouye, S., and Inouye, M. Autophosphorylation of nucleoside diphosphate kinase from Myxococcus xanthus, J. Bacteriol. 175: 1176-1181, 1993.
87. Hemmerich, S. and Pecht, I. Oligomeric structures and autophosphorylation of nucleoside diphosphate kinase from rat mucosal mast cells, Biochemistry. 31: 4580-4587, 1992.
88. Wallet, V., Mutzel, R., Troll, H., Barzu, O., Wurster, B., Veron, M., and Lacombe, M. L. Dictyostelium Nucleoside Diphosphate Kinase Highly Homologous to Nm23 and Awd Proteins Involved in Mammalian Tumor Metastasis and Drosophila Development, J. Nat'l. Cancer Inst. 82: 1199-1202, 1990.
89. Ann, K.-S. and Nelson, D. A nucleoside diphosphate kinase from Paramecium tetraurelia with protein kinase activity., J. Eukaryotic Microbiol. 43: 365-372, 1996.
90. Engel, M., Veron, M., Theisinger, B., Lacombe, M.-L., Seib, T., Dooley, S., and Welter, C. A novel serine/threonine-specific protein phosphoransferase activity of Nm23/nucleoside-diphosphate kinase., Eur. J. Biochem. 234: 200-207, 1995.
91. Inoue, H., Takahashi, M., Oomori, A., Sekiguchi, M., and Yoshioka, T. A novel function for nucleoside diphosphate kinase in Drosophila., Biochem. Biophys. Res. Comm. 218: 887-892, 1996.
92. Wagner, P. and Vu, N.-D. Phosphorylation of ATP-Citrate Lyase by Nucleoside diphosphate kinase., J. Biol. Chem. 270: 21758-21764, 1995.
93. Wagner, P. D., Steeg, P. S., and Vu, N.-D. Two-component kinase like activity of Nm23 correlates with its motility suppressing activity, Proc. Nat'l. Acad. Sci. USA. 94: 9000-9005, 1997.
94. Wagner, P. and Vu, N.-D. Histidine to aspartate phosphotransferase activity of nm23 protein: Phosphorylation of Aldolase C on Asp 319., Biochem. J. 346: 623-630, 2000.
95. Lu, Q., Park, H., Egger, L., and Inouye, M. Nucleoside-diphosphate kinase-mediated signal transduction via histidyl-aspartyl phosphorelay systems in Escherichia coli., J. Biol. Chem. 271: 32886-32893, 1996.
96. Postel, E. H., Berberich, S. J., Flint, S. J., and Ferrone, C. A. Human c-myc transcription factor PuF identified as Nm23-H2 nucleoside diphosphate kinase, a candidate suppressor of tumor metastasis., Science. 261: 478-480, 1993.
97. Chae, S.-K., Lee, N.-S., lee, K.-J., and Kim, E. Transactivation potential of the C-terminus of human Nm23-H1, FEBS Lett. 423: 235-238, 1998.
98. Hildebrandt, M., Lacombe, M., Mesnildrey, S., and Veron, M. A human NDP-Kinase-B specifically binds single-stranded poly-pyrimidine sequences., Nucl. Acids Res. 23: 3858-3864., 1995.
99. Ji, L., Arcinas, M., and Boxer, L. The transcription factor, NM23H2, binds to and activates the translocated c-myc allele in Burkitt's lymphoma., J. Biol. Chem. 270: 13392-13398., 1995.

100. Michelotti, E., Sanford, S., Freije, J., MacDonald, N., Steeg, P., and Levens, D. Nm23/PuF does not directly stimulate transcription through the CT element in vivo., J. Biol. Chem. 272: 22526-22530, 1997.
101. Postel, E. Cleavage of DNA by human NM23-H2/Nucleoside diphosphate kinase involves formation of a covalent protein-DNA complex., J. Biol. Chem. 274: 22821-22829, 1999.
102. Zhu, J., Tseng, Y.-H., Kantor, J., Rhodes, C., Zetter, B., Moyers, J., and Kahn, C. Interaction of the Ras-related protein associated with diabetes Rad and the putative tumor metastasis suppressor NM23 provides a novel mechanism of GTPase regulation., Proc. Nat'l. Acad. Sci. USA. 96: 14911-14918, 1999.
103. Randazzo, P., Northup, J. K., and Kahn, R. A. Activation of a small GTP-binding protein by nucleoside diphosphate kinase., Science. 254: 850-853, 1991.
104. Randazzo, P. A., Northup, J. K., and Kahn, R. A. Regulatory GTP binding proteins (ARF, Gt and RAS) are not activated directly by nucleoside diphosphate kinase., J. Biol. Chem. 267: 18182-18189., 1992.
105. Klinker, J. and Seifert, R. Nucleoside diphosphate kinase activity in soluble transducin preparations, Eur. J. Biochem. 261: 72-80, 1999.
106. Kikkawa, S., Takahanshi, K., Takahashi, K. I., Shimada, N., Michio, M., Kimura, N., and Katada, T. Conversion of GDP into GTP by nucleoside diphosphate kinase on the GTP-binding proteins., J. Biol. Chem. 265: 21536-21540, 1990.
107. Kimura, N. and Shimada, N. Evidence for complex formation between GTP binding protein (Gs) and membrane-associated nucleoside diphosphate kinase., Biochem. Biophys. Res. Comm. 168: 99-106., 1990.
108. Paravicini, G., Steinmayr, M., André, E., and Becker-André, M. The metastasis suppressor candidate nucleoside diphosphate kinase Nm23 specifically interacts with members of the ROR/RZR nuclear orphan receptor subfamily, Biochem. Biophys. Res. Comm. 227: 82-87, 1996.
109. Engel, M., Seifert, M., Theisinger, B., Seyfert, U., and Welter, C. Glyceraldehyde-3-phosphate dehydrogenase and Nm23-H1/Nucleoside diphosphate kinase A. Two old enzymes combine for the novel Nm23 protein phosphotransferase function., J. Biol. Chem. 273: 20058-20065, 1998.
110. Weber, B., Weber, W., Buck, F., and Hilz, H. Isolation of the myc transcription factor nucleoside diphosphate kinase and the multifunctional enzyme glyceraldephyde-3-phosphate dehydrogenase by cAMP affinity chromatography., Int. J. Biochem. Cell Biol. 27: 215-224., 1995.
111. Otero, A. Copurification of vimentin, energy metabolism ensymes and a MER5 homolog with nucleoside diphosphate kinase. Identification of tissue-specific interactions., J. Biol. Chem. 272: 14690-14694, 1997.
112. Nickerson, J. A. and Wells, W. W. The microtubule-associated nucleoside diphosphate kinase., J. Biol. Chem. 259: 11297-11304, 1984.
113. Melki, R., Lascu, I., Carlier, M. F., and Veron, M. Nucleoside diphosphate kinase does not directly interact with tubulin or microtubules., Biochem. Biophys. Res. Comm. 187: 65-72, 1992.
114. Lombardi, D., Sacchi, A., D'Agostino, G., and Tibursi, G. The association of the Nm23-M1 protein and beta-tubulin correlates with cell differentiation., Exp. Cell Res. 217: 267-271., 1995.
115. Biggs, J., Hersperger, E., Steeg, P. S., Liotta, L. A., and Shearn, A. A Drosophila gene that is homologous to a mammalian gene associated with tumor metastasis codes for a nucleoside diphosphate kinase., Cell. 63: 933-940, 1990.

116. Barthel, T. and Walker, G. Inferences concerning the ATPase properties of DnaK and other HSP70s are affected by the ADP kinase activity of copurifying nucleoside diphosphate kinase., J. Biol. Chem. 274: 36670-36678, 1999.
117. Leung, S.-M. and Hightower, L. E. A 16-kDa protein functions as a new regulatory protein for Hsc70 molecular chaperone and is identified as a member of the Nm23/nucleoside diphosphate kinase family, J. Biol. Chem. 272: 2607-2614, 1997.
118. Reymond, A., Volorio, S., Merla, G., Al-Maghtheh, M., Zuffardi, O., Bulfone, A., Ballabio, A., and Zollo, M. Evidence for interaction between human Prune and nm23-H1 NDPKinase., Oncogene. 18: 7244-7252, 1999.
119. Bovet, L., Meylan_Bettex, M., Eggman, T., Martinoia, E., and Siegenthaler, P.-A. CDP phosphotransferase activity in spinach intact chloroplasts: Possible involvement of nucleoside diphosphate kinase II,, Plant Physiol. Biochem. 37: 645-652, 1999.
120. Choi, G., Yi, H., Lee, J., Kwon, Y.-K., Soh, M., Shin, B., Luka, Z., Hahn, T.-R., and Song, P.-S. Phytochrome signalling is mediated through nucleoside diphosphate kinase 2., Nature. 401: 610-613, 1999.
121. Ogura, T., Tanaka, N., Yabe, N., Komatsu, S., and Hasunuma, K. Characterization of protein complexes containing nucleoside diphosphate kinase with characteristics of light signal transduction through phytochrome in etiolated pea seedlings., Photochem. and Photbiol. 69: 397-403, 1999.
122. Shankar, S., Kavanaugh-Black, A., Kamath, S., and Chakabarty, A. Characterization of a phosphoprotein phosphatase for the phosphorylated form of nucleoside-diphosphate kinase from Pseudomonas aeruginosa., J. Biol. Chem. 270: 28246-28250., 1995.
123. Cherfils, J., Morera, S., Lascu, I., Veron, M., and Janin, J. X-ray structure of nucleoside diphosphate kinase complexed with thymidine diphosphate and Mg2+ at 2-A resolution., Biochemistry. 33: 9062-9069, 1994.
124. Chiadmi, M., Morera, S., Lascu, I., Dumas, C., LeBras, G., Veron, M., and Janin, J. Crystal structure of the Awd nucleotide diphosphate kinase from Drosophila., Structure. 1: 283-293, 1993.
125. Dumas, C., Lascu, I., Morera, S., Glaser, P., Fourme, R., Wallet, V., Lacombe, M. L., Veron, M., and Janin, J. X-ray structure of nucleoside diphosphate kinase., EMBO J. 11: 3203-3208, 1992.
126. Lascu, I., Schaertl, S., Wang, C., Sarger, C., Giartosio, A., Briand, G., Lacombe, M.-L., and Knorad, M. A point mutation of human nucleoside diphosphate kinase A found in aggressive neuroblastoma affects protein folding., J. Biol. Chem. 272: 15599-15602, 1997.
127. Cropp, C., Lidereau, R., Leone, A., Liscia, D., Cappa, A., Campbell, G., Barker, E., Doussal, V. L., Steeg, P., and Callahan, R. NME1 protein expression and loss of heterozygosity mutations in primary human breast tumors., J. Nat'l Cancer Inst. 86: 1167-1169, 1994.
128. DeLaRosa, A., Mikhak, B., and Steeg, P. Identification and characterization of the promoter for the human metastasis suppressor gene nm23-H1., Archives Med. Res. 27: 395-401, 1996.
129. Ouatas, T., Clare, S., Hartsough, M., DeLaRosa, A., and Steeg, P. A cassette of mammary-specific transcription factor binding sites contributes to nm23-H1 promoter activity in human breast carcinoma cell lines., J. Biol. Chem. Submitted for publication.

Chapter 8

GENE REGULATION IN MELANOMA METASTASIS

Menashe Bar-eli

University of Texas M. D. Anderson Cancer Center, Department of Cancer Biology, Houston, Texas

Abstract

The molecular changes associated with the transition of melanoma cells from radial growth phase (RGP) to vertical growth phase (VGP, metastatic phenotype) are not very well defined. We previously demonstrated that expression of the cell surface adhesion molecule MCAM/MUC18 correlates directly with the metastatic potential of human melanoma cells. In addition, the progression of human melanoma towards the metastatic phenotype is associated with loss of expression of the tyrosine-kinase receptor c-KIT. We found that both genes MCAM/MUC18 and c-KIT, are regulated by the transcription factor AP-2 and that metastatic melanoma cells do not express AP-2. Re-expression of AP-2 in highly metastatic cells decreased their tumorigenicity and inhibited their metastatic potential in nude mice, while expression of dominant-negative AP-2 gene (AP-2B) augmented their tumor growth *in vivo*. The AP-2 transfected cells displayed down regulation of MCAM/MUC18 and MMP-2 and re-expression of the c-KIT receptor. Because AP-2 also regulates other genes that are involved in the progression of human melanoma, such as E-cadherin, p21/WAF-1, HER-2, Bcl-2, IGF-R1, FAS/APO-1 and the thrombin receptor (PAR-1), we propose that loss of AP-2 is a critical event in the development of malignant melanoma. The progression of human melanoma from RGF to VGP is also associated with over expression of the transcription factors CREB and ATF-1. We found that CREB/ATF-1 may act as survival factors for melanoma cells. In addition, some of the genes regulated by AP-2 such as MCAM/MUC18, MMP-2 and FAS/APO-1 are also regulated by CREB/ATF-1. It is therefore feasible that the balance between AP-2 and CREB/ATF-1 expression is among the factors determining the acquisition of the metastatic phenotype in human melanoma.

Key words: malignant melanoma, metastasis, gene regulation, transcription factor

D.R. Welch (ed.), Cancer Metastasis – Related Genes, 145–168.

INTRODUCTION

Malignant melanoma is a common human cancer with high mortality rates. The incidence of this disease is currently increasing faster than that of any other malignancy. The primary cause of melanoma is thought to be exposure to ultraviolet (UV) radiation (1). Other risk factors include fair skin, a hereditary predisposition toward the disease, increased age, and race. In the white population for example, the disease is increasing rapidly and is projected to effect approximately 1 in 75 persons by the turn of the century (2). The steady overall increase has been attributed to various causes ranging from the reduction of the stratospheric ozone to the changing lifestyles and attitudes that affect sun exposure. Although it is easily diagnosed and wholly curable if detected early, the high mortality rate (6,700 deaths/year in the United States) is due to the propensity for melanoma to metastasize if left untreated (3).

As it progress through a multistep process, malignant melanoma in humans switches from melanocytes to nevi, to radial growth phase (RGP) and subsequently to vertical growth phase (VGP, metastatic phenotype, *Figure 1*). These switches are associated with molecular and genetic changes (4-6). *Figure 1* outlines some of the molecular changes occurring during the different stages of melanoma progression. The mechanisms, however, that mediate this differential expression of genes during melanoma progression remain largely unknown. There has been, however, some progress in our understanding the development of malignant melanoma. Of particular relevance are recent results concerning abnormalities in the expression of transcription factors which in turn modulate and regulate several genes controlling tumor growth and metastasis of melanoma.

The prevalent working model for melanoma metastasis development defines it as series of interrelated, sequential steps involving the survival and growth of unique subpopulation of cells with metastatic properties that pre-existed within the parental neoplasm (7). By their ability to regulate the expression of several genes, transcription factors could play an important role during melanoma progression. Indeed, it has been recently demonstrated that the progression of human melanoma is associated with loss of expression of the AP-2 transcription factor and overexpression of the CREB (cAMP response element-binding protein) and the ATF-1 (activator transcription factor-1) transcription factors. Here we will summarize our recent data concerning the contribution of various genes to the metastatic phenotype and will describe how these genes are regulated during the progression of human melanoma.

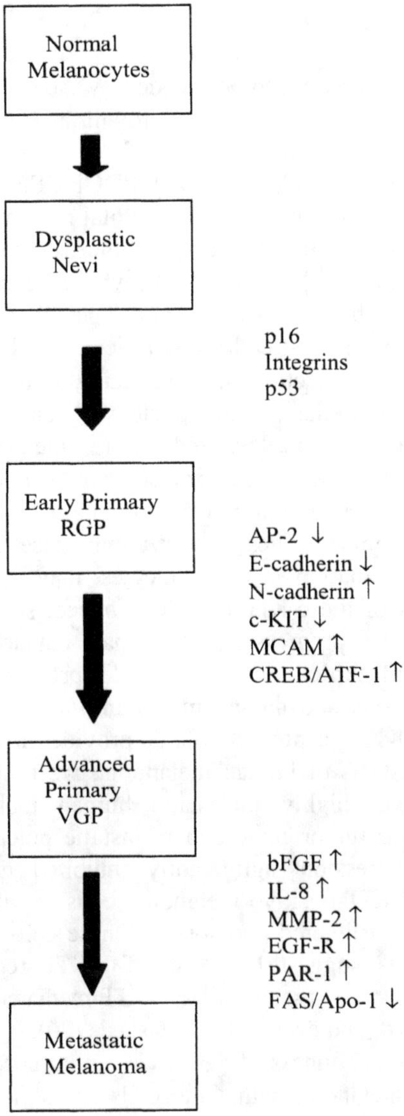

Figure 1. Molecular changes associated with the progression of human melanoma. Abnormalities in the p16/CDKN2 gene are usually an early event. Mutations in the p53 gene are infrequent, but were observed in early stages. Abnormal functons of w.t. p53 were also observed. Alterations in c-KIT, MCAM/MUC18, E- and N-cadherins and integrins occur in the transition from RGP to VGP. The AP-2 transcription factor is not expressed in metastatic cells while the CREB and ATF-1 transcription factors are upregulated in these cells. The expression of genes involved in angiogenesis, invasion and apoptosis such as bFGF, IL-8, MMP-2 EGF-R,PAR-1, and FAS/APO-1 correlates with higher metastatic potential of human melanoma cells.

ROLE OF C-KIT IN THE PROGRESSION OF MELANOMA

Expression of the tyrosine kinase receptor encoded by the c-KIT proto-oncogene progressively decreases during local tumor growth and invasion of human melanomas (8-10). The proto-oncogene c-KIT encodes a transmembrane tyrosine-protein kinase receptor related to the PDGF/CSF-1 (c-fms) receptor subfamily (11). c-KIT has been found to play a pivotal role in the normal growth and differentiation of embryonic melanoblasts. In mice, c-KIT has been mapped to the dominant white spotting (w) locus (12-13), whose ligand is the product of the sl locus (Steel) (14) which encodes the stem cell factor, SCF (also known as KIT-ligand, KL, steel factor or mast cell growth factor, MCF). Mutations in the (w) locus or (sl) (15, 16), or injection of neutralizing anti-KIT antibodies into pregnant mice (17) results in the piebald phenotype, characterized by white spotting of the fur and attributed to a local reduction in the number of cutaneous melanocytes. Mutations in the c-KIT receptor also have been identified in human piebald patients (18,19), suggesting that normal function of c-KIT is required for human melanocyte development. These observations raise the question as to whether malignant transformation of melanocytes may be associated with changes in the expression of the c-KIT receptor. Indeed, several recent studies have demonstrated that the progression of human cutaneous melanoma is associated with the loss of expression of the c-KIT proto-oncogene. The vast majority of metastatic lesions and human melanoma cell lines do not express detectable levels of the c-KIT receptor (8-10). To provide direct evidence that c-KIT plays a role in metastasis of human melanoma, we transfected the c-KIT gene into c-KIT negative, highly metastatic human melanoma cells and subsequently analyzed their tumorigenic and metastatic potential in nude mice (20). Enforced c-KIT expression significantly inhibited tumor growth and metastasis. Exposure of c-KIT-positive melanoma cells in vitro and in vivo to SCF, the ligand for c-KIT, triggered apoptosis of these cells but not of normal melanocytes. These results suggest that loss of c-KIT receptor may allow malignant melanoma cells to escape SCF/c-KIT-mediated apoptosis, thus contributing to tumor growth and eventually metastasis (20).

Our data point to the usefulness of SCF as an antitumor and antimetastatic agent in early stages of melanoma while the cells are still expressing c-KIT. Furthermore, c-KIT might be used in gene therapy for advanced tumors and established melanoma metastases. Specific expression of c-KIT in melanoma cells *in vivo* could be achieved by ligating the tyrosinase promoter upstream of the c-KIT gene in the retroviral expression vectors used in our studies. The mechanism(s) for the loss of c-KIT gene expression are unknown and will be the subject of this review.

ROLE OF MCAM/MUC18 IN MELANOMA PROGRESSION

Although c-KIT expression in malignant melanoma cells consists of loss of function, recent accumulating experimental evidence has demonstrated that the progression of human melanoma is associated with gain of expression of the adhesion molecule MCAM (previously known as MUC18 or Mel-CAM). One tumor cell property that is essential for metastasis is the expression of cell adhesion molecules, which mediate cell-to-cell or cell-to-matrix interactions. MCAM/MUC18 is a member of the immunoglobin superfamily that shares homology with gicerin (33% amino AA identity), a molecule mediating intercellular adhesion in the developing nervous system. It contains five immunoglobin-like domains, and its cytoplasmic domain contains several protein kinase recognition motifs, suggesting the involvement of MCAM/MUC18 in cell signaling (21). MCAM/MUC18 can mediate homotypic interaction either through the MCAM antigen (22, 23) or via heterophilic ligand (24). MCAM/MUC18 is strongly expressed by advanced primary and metastatic melanomas but is weaker and less frequent in nevus cells (25, 26). Approximately 70% of melanoma metastases express this molecule, and among primary tumors, expression increases with increasing vertical thickness, an important predicator of metastatic disease (27). Indeed, we have demonstrated that expression of MCAM/MUC18 by human melanoma cell lines correlates with their ability to grow and to produce metastases in nude mice (28), suggesting that MCAM/MUC18 may play a pivotal role in the development of malignant melanoma. This hypothesis is supported by the observation that the production of tumorigenic variants from a non-tumorigenic melanoma cell line is accompanied by MCAM/MUC18 upregulation (29) and by our recent observation that enforced MCAM expression in primary cutaneous melanoma leads to increased tumor growth and metastasis in vivo (22). The transfected cells displayed increased homotypic adhesion, increased attachment to human endothelial cells, decreased ability to adhere to laminin, and increased invasiveness through Matrigel-coated filters due to upregulation in the expression of MMP-2 (22). The above changes in function attributed to the expression of MCAM may underlie the contribution of MCAM/MUC18 to the malignant phenotype.

ROLE OF TRANSCRIPTION FACTOR AP-2 IN MELANOMA PROGRESSION

The mechanisms for upregulation of MCAM/MUC18 and loss of c-KIT gene expression during melanoma progression are unknown. We have previously reported that overexpression of MCAM/MUC18 in highly metastatic melanoma cells is not due to amplification or to rearrangement of the gene (28). The promoter of MCAM/MUC18 has been cloned and sequenced (21). It is a G + C rich promoter lacking the conventional TATA and CAAT boxes, but strikingly, it contains four putative AP-2 binding elements (21). Similarly, the human c-KIT promoter region lacks a typical "TATA box" but has a relatively high G + C content and three putative AP-2 binding sites. These observations coupled with our previous finding that highly metastatic human melanoma cells do not express the AP-2 transcription factor (5, 30), led us to hypothesize that AP-2 may regulate c-KIT and MCAM/MUC18 gene expression in human melanoma cells.

AP-2, a 52-kDa protein, was first purified from HeLa cells. Partial peptide sequences led to the isolation of the cDNA from a HeLa cell library (31), and the gene was mapped to a region on the short arm of chromosome 6 near the HLA locus (32, 33). The AP-2 protein binds to a consensus palindromic core recognition element with the sequence 5N-GCCNNNGGC-3N (31). Functional AP-2 binding sites have been identified in the enhancer regions of viral genes such as simian virus 40(SV40) (34), human T-cell leukemia virus type I, and cellular genes such as those for murine a major histocompatibility complex (H-$2K^b$), human metallothionein-IIa (huMTIIa), human proenkephalin, human keratin K14, c-ERB-2, plasminogen activator type I (PAI-I), and insulin-like growth factor-binding protein-5. The DNA-binding domain is located within the C-terminal half of the 52 kDa protein and consists of two putative amphiphatic α-helices separated by a 92-amino acid intervening span that is both necessary and sufficient for homodimer formation (35). An alternatively spliced AP-2 protein, AP-2B, which differs in its C terminus and acts as dominant-negative to AP-2 has been recently cloned (36).

AP-2 activity is regulated through a number of signal transduction pathways. Phorbol esters and signals that enhance cAMP levels induce AP-2 activity independently of protein synthesis, whereas retinoic acid treatment of teratocarcinoma cell lines result in a transient induction of AP-2 mRNA levels on a transcription level (36, 37). AP-2 is involved in mediating programmed gene expression both during embryonic morphogenesis and adult cell differentiation. By using *in situ* hybridization, a restricted spatial and temporal expression pattern has been observed during murine embryogenesis. In particular, regulated AP-2 expression was observed in neural crest-derived cell lineages (from which melanocytes are derived) and in facial and limb bud mesenchyme (33). Two recent reports of AP-2 null mutant mice have demonstrated that AP-2 is important for development of the cranial region and for midline fusions. The AP-2-null mice died at birth (38, 39).

Here, I will summarize our recent results demonstrating that (i) there is a direct correlation between expression of the AP-2 transcription factor and expression of c-KIT and MCAM/MUC18 in human melanoma cells; (ii) transfection of highly metastatic cells (c-KIT-negative, MCAM-positive and AP-2-negative) with the AP-2 gene resulted in a downregulation of MCAM/MUC18 and re-expression of c-KIT; (iii) transfection of AP-2 into highly metastatic melanoma cells inhibited their tumor growth and metastatic potential in nude mice, possibly through the regulation of c-KIT and MCAM/MUC18 and; (iv) expression of dominant-negative AP-2 (AP-2B) in primary cutaneous melanoma cells augmented their tumorigenicity in nude mice.

In an effort to determine the molecular basis for c-KIT's lack of expression in highly metastatic melanoma cells, we found that the c-KIT gene and its promoter in c-KIT-negative melanoma cells had no abnormalities (deletions, rearrangements or mutations) that can account for the lack of c-KIT expression (20, 30). These observations suggest that c-KIT expression might be regulated at the transcriptional level. To test this hypothesis, we subcloned the promoter of the c-KIT gene (–125 to +1) (40) in front of the luciferase reporter gene. Using the Dual-Luciferase Reporter System, we analyzed the luciferase activity driven by the c-KIT promoter in c-KIT-positive and c-KIT-negative melanoma cell lines. c-KIT-luciferase activity was higher in the human melanoma cell lines Mel-888 and Mel-501 (both express high levels of c-KIT mRNA) (10), as compared with the activity in c-KIT-negative A375SM cells which was given the reference value of 1. We observed low luciferase activity in the WM-2664 cell line which does not express c-KIT mRNA and protein. These results suggest that c-KIT expression is regulated at the transcriptional level in these melanoma cells.

The 1.2 kb c-KIT promoter lacks TATA or CAAT boxes, is highly G + C rich, and contains binding sites for SP-1, myb, GATA-1 and three putative AP-2 sites (40). Deletion of the proximal 185 bp of the promoter abolished the transcription of c-KIT in HEL cells, suggesting that only the proximal 185 bp of the promoter are necessary and sufficient for c-KIT expression. As we closely examined the proximal 185 bp promoter region, two AP-2 binding sites were identified within this region with a total of three putative AP-2 binding motifs within the 1.2 kb c-KIT promoter. The presence of two AP-2-binding sites within the essential region of c-KIT promoter suggested that AP-2 might regulate expression of the c-KIT gene. Indeed, during mouse development, AP-2 and c-KIT are co-expressed in several tissues including hindbrain, kidney and heart, and in neural crest-derived lineages from which melanocytes originate.

To assess the effect of AP-2 on c-KIT transcription, the c-KIT promoter-luciferase construct, pKLuc, was co-transfected into A375SM cells with increasing concentrations of an expression vector encoding for wild-type AP-2 (AP-2A, pSG5-AP-2) or with the control vector lacking AP-2A (pSG5). Using the β-actin-Renilla luciferase plasmid (pB-Actin-RL) vector as a control to normalize for transfection efficiency, we found that the luciferase activity driven by the c-KIT promoter was activated by AP-2A in A375SM cells in a dose-dependent manner. A 9.5-fold stimulation was observed in cells co-transfected with 6 μg of the plasmid expressing the AP-2A protein, which was not detected in transfections with the parent vector. Conversely, when the pKLuc construct was co-transfected into Mel-501 cells (which express high levels of c-KIT) and increasing concentrations of AP-2B, the dominant-negative form of AP-2, the luciferase activity was inhibited by 50% (41).

To assess the contribution of the AP-2 transcription factor on c-KIT expression and to the acquisition of the metastatic phenotype in human melanoma cells, we decided to re-express AP-2 in A375SM cells. A375SM cells are highly metastatic in nude mice (42) and also c-KIT-negative, and they express negligible levels of endogenous AP-2. We were able to isolate two clones of A375SM cells (designated as T1 and T2) that express high levels of functional AP-2 as was determined by Northern, Western blot and EMSA gel analyses (41). The expression of the 145 kDa c-KIT receptor was analyzed in the two AP-2 transfectants, T1 and T2. To determine c-KIT protein expression, whole-cell lysates were prepared and reacted with specific polyclonal anti-c-KIT antibody, the immunocomplexes were resolved by sodium dodecyl sulfate-polyacrylamide gel electrophoresis (SDS-PAGE), and c-KIT expression was determined by Western blot analysis. We found that c-KIT was expressed in the two AP-2 transfectants A375SM-AP-2.T1 and A375SM-AP-2.T2 but not in parental or control neo-transfected cells (41). Collectively, these data demonstrate a direct correlation between AP-2 and c-KIT expression.

In an effort to determine the molecular basis for MCAM upregulation in metastatic cells, we found that MCAM overexpression is not due to gene amplification or rearrangement (28). Moreover, sequencing of the entire MCAM promoter revealed no abnormalities that could account for MCAM overexpression in metastatic cells (data not shown). These observations suggest that MCAM expression might be regulated at the transcriptional level.

To test this hypothesis, the activity of the CAT reporter gene driven by the MCAM promoter (–642 to +26) was analyzed in MCAM-positive and MCAM-negative melanoma cells. Strong CAT activity was observed in the highly metastatic cells A375SM that express abundant levels of MCAM (22, 28), as compared with a background level of CAT activity in the non-metastatic MCAM-negative SB-2 melanoma cells (43).

The 0.9-kilobase pair MCAM promoter lacks TATA or CAAT boxes, is highly G + C-rich, and contains binding sites for SP-1, CREB, MYB, and four AP-2-binding elements. The presence of four AP-2-binding sites within the essential region of the MCAM promoter suggested that AP-2 might regulate MCAM gene expression. Competition experiments on the EMSA gels and the addition of human-recombinant AP-2 to nuclear extracts from A375SM cells have revealed lack of AP-2 expression in these cells (43).

To assess further the effect of AP-2 on MCAM transcription, the MCAM promoter-CAT construct was co-transfected into A375SM cells with increasing concentrations of an expression vector encoding AP-2 (pSG5-AP2) or with a control vector lacking the AP-2 cDNA (pSG5). These experiments showed that CAT activity driven by the MCAM promoter was inhibited by AP-2 in A375SM cells in a dose-dependent manner. The down regulation of the MCAM promoter by AP-2 was due to a direct interaction of AP-2 to a region of the MCAM promoter that is required for its transcription (43). These experiments indicate the presence of functional AP-2 elements within the MCAM promoter, which regulate MCAM expression in melanoma cells.

Since our promoter analysis indicated that AP-2 is an important regulator of MCAM gene expression, we therefore next examined the effect of AP-2 transfection into A375SM cells on MCAM expression. MCAM expression was determined by fluorescence-activated cell sorter (FACS) analysis on the cells surface of parental A375SM, neo-transfected, and in AP-2-transfected cells

(A375SM-AP-2.T1). By using specific monoclonal anti-MCAM antibody that recognizes the extracellular domain of MCAM (25,30), we were able to demonstrate that MCAM expression occurred in 98.4% of A375SM parental cells, in 90.1% of control, neo-transfected cells, but in only 17.9% of the AP-2-transfected A375SM-AP-2.T1 cells. Downregulation of MCAM expression in the AP-2 transfected cells was also confirmed by western blot analysis (43). Collectively, these results indicate that expression of AP-2 in highly metastatic melanoma cells caused downregulation of MCAM gene expression.

To determine the tumorigenicity of the AP-2-transfected A375SM cells which exhibited re-expression of c-KIT and downregulation in MCAM expression, we injected 1 X 10^6 cells subcutaneously into BALB/c nude mice and monitored tumor growth once a week for 50 days. Both A375SM parental and neo-transfected control cells grew in all mice (100% tumor uptake) and reached 1.0-1.4 cm in mean diameter within 6 weeks. In contrast, A375SM AP-2-transfected cells did not begin to form palpable tumors until 3 weeks (T2) or 7 weeks (T1) after injection and produced smaller tumors (0.1 and 0.45 cm in mean diameter, Table 1).

In the next set of experiments, the metastatic potential of AP-2-transfected A375SM cells was determined in an experimental lung metastasis assay. To that end, BALB/c nude mice were injected intravenously with 1 x 10^6 A375SM AP-2-transfected, parental, or neo-control cells, and 60 days later, the number of lung metastases was counted. As shown in Table 1, A375SM and A375SM-Neo (Neo.a and Neo.b) produced a high number of lung tumor colonies in all mice injected. In contrast, the A375SM-AP-2-transfected cells did not metastasize to lungs (T1) or produced only a few lung metastases in some mice (T2).

Table 1. Tumor growth and metastatic potential of AP-2-transfected cells.

	Tumorigenicity*	Experimental Lung Metastasis†	
Cells	Size (cm)	Median	Range
A375SM-P	1.8	121	74 > 200
A375SM-Neo.a	2.0	153	111 > 200
A375SM-Neo.b	1.7	87	54 -143
A375SM-AP-2.T1	0.3	0	All 0
A375SM-AP-2.T2	0.5	19	0-47

*Tumor cells ($1x10^6$) were injected subcutaneously, and tumor growth was measured twice weekly for 56 days.

†Tumor cells ($1x10^6$) were injected intravenously, and the number of lung colonies was determined after 60 days.

DOMINANT-NEGATIVE AP-2 AUGMENTS MELANOMA TUMOR GROWTH *IN VIVO*: REGULATION OF MMP-2

An alternatively spliced AP-2 protein, AP-2B, that differs in its C-terminus and acts as dominant-negative to AP-2, has been recently cloned (36). This protein contains the activation domain and part of the DNA-binding domain, but lacks the dimerization domain, which is necessary for DNA binding. Both transient and stable transfection experiments show that AP-2B inhibits AP-2 trans-activator function. Furthermore, constitutive AP-2B expression in the human teratocarcinoma cell line PA-2 causes a retinoic acid-resistant phenotype, anchorage-independent growth in soft agar, and tumorigenicity in nude mice (36). AP-2B has also been detected by RT-PCR in HeLa cells and the human prostate carcinoma cell line LNCaP (36).

To further investigate the role of AP-2 in the progression of human melanoma, we attempted to inactivate AP-2 in primary cutaneous melanoma by using the dominant-negative AP-2B gene. Stable transfection of primary cutaneous melanoma SB-2 cells with the dominant-negative AP-2B gene was confirmed by RT-PCR and Northern blot analyses. Electrophoretic mobility shift assay using nuclear extracts from these cell lines demonstrated decreased functional binding of AP-2B-transfected cells to the AP-2 consensus binding sequence compared with neo-transfected controls (44). In addition, CAT activity driven by a construct containing the AP-2 consensus binding sequence was downregulated in AP-2B transfected cells, indicating AP-2 activity was quenched in the transfected cells. Orthotopic (subcutaneous) injection of the dominant-negative AP-2B-transfected cell lines into nude mice increased their tumorigenicity compared to control neo-transfected cells. The AP-2B-transfected cells displayed an increase in MMP-2 expression (by Northern blot) and MMP-2 activity (by zymography), which resulted in an increase in invasiveness through Matrigel-coated filters (44). The AP-2B-transfected tumors also displayed an increase in microvessel density and angiogenesis. In the process of angiogenesis, endothelial cells must breech the extra cellular matrix (ECM) and extend towards the source of the angiogenic stimulus (45,46). Proteolysis of the ECM by MMP-2 permits endothelial cell migration and may also release sequestered angiogenic molecules such as vascular endothelial growth factor (VEGF) and basic fibroblast growth factor (bFGF). MMPs are implicated in this process by their expression in and around forming blood vessels (47) and by their ability to modulate endothelial cell proliferation and microtubule formation in vitro (48). Moreover, MMP-2 activity has been shown to correlate with metastatic potential in melanoma (49-52), and its promoter contains at least one AP-2 binding element (49, 53, 54). Activation of the type IV collagenase (MMP-2) may account for the increased tumorigenicity of the AP-2B-transfected cells. These results demonstrate that inactivation of AP-2 contributes to the progression of melanoma by its ability to regulate the MMP-2 gene.

HOW LOSS OF AP-2 EXERTS ITS METASTATIC EFFECT

Here we provide evidence that lack of expression of AP-2 in metastatic melanoma cells can result in deregulation of c-KIT, MCAM/MUC18 and the MMP-2 genes. All three genes are known to be involved in the progression of human melanoma (41,43,55,56).

In addition, other genes that are involved in the progression of human melanoma, such as E-cadherin (57), p21^{WAF-1} (58,59), HER-2 (60), plasminogen activator inhibitor type I (61) Bcl-2 (62), VEGF, PAR-1, and insulin like growth factor receptor-1, have either already been shown to be regulated by AP-2 (63,64) or represent likely targets for AP-2 gene regulation based on the existence of AP-2 elements in their promoters. Loss of AP-2 expression was also observed in advanced primary and metastatic melanoma lesions (65). Therefore, we propose that loss of AP-2 expression is a crucial event in the development of malignant melanoma. As such, AP-2 may act as a "major regulator" in the progression of human melanoma. In our proposed model, (*Figure 2*), AP-2 plays a pivotal role in regulating the expression of several genes whose products are involved in tumor growth and metastasis of melanoma. AP-2 regulates genes that are involved in proliferation, cell cycle regulation (HER-2, p21/WAF-1), apoptosis (c-KIT, Bcl-2, FAS/APO-1), adhesion (MCAM/MUC18, E-cadherin), and invasion/angiogenesis (MMP-2, plasminogen activator inhibitor type I, VEGF and PAR-1). These functional changes attributed to one transcription factor may underlie the contribution of AP-2 loss of expression to the malignant phenotype.

ROLE OF CREB/ATF-1 IN MELANOMA PROGRESSION

A large body of data concerning the molecular control of melanoma progression has come from studies using mitogens. In culture, melanocytes synergistically respond to a number of growth factors, which in combination with each other or with 12-0-tetradecanoylphorbol-13-acetate or cAMP stimulate not only proliferation but also pigmentation (66). These growth factors include several fibroblast growth factors, hepatocyte growth factor, and stem cell factor (also known as KIT ligand, MGF, and steel factor), all of which stimulate tyrosine kinase receptors. As melanocyte proliferation and differentiation are positively regulated by agents that increase cAMP (67-68), we have, therefore, focused on the transcription factors ATF-1 (Activating transcription factor 1) and CREB (cAMP-responsive element-binding protein), which are known to be activated by cAMP, as possible mediators of tumor growth and metastasis of human melanoma.

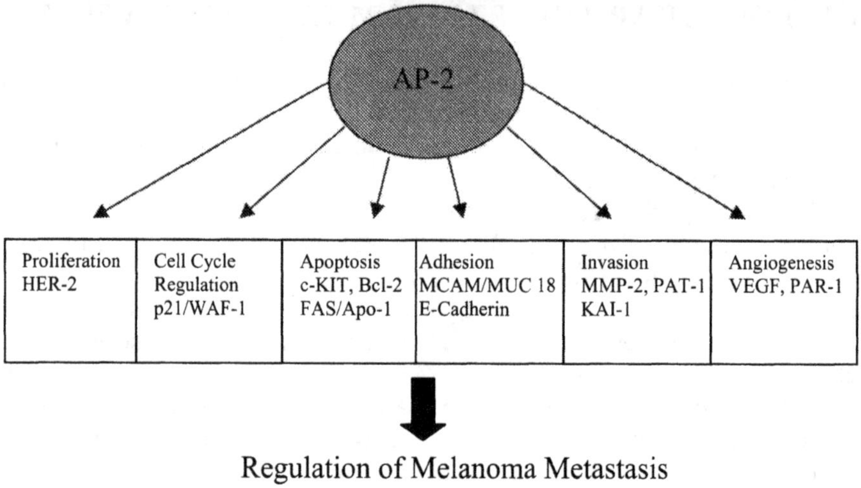

Figure 2. A model for the role of AP-2 in the progression of human melanoma. AP-2 may act as a major regulator of several genes involved in tumor growth and metastasis of melanoma.

ATF-1 and CREB are members or the large bZIP superfamily of transcription factors. Members of the CREB/ATF family bind to cAMP-responsive elements (CREs) within the promoter and enhancer sequences of many genes. ATF-1, CREB, and the cAMP-responsive element modulator protein (CREM) constitute the CREB/ATF subfamily within the bZIP superfamily, whose members are defined by their ability to heterodimerize with each other but not with members of other subfamilies (69). ATF-1, CREB, and CREM have similar structures and are highly homologous at the amino acid sequence level, especially within the bZIP region. Despite these similarities, members of the CREB multigene subfamily have distinct biological activities.

ATF-1, CREB, and CREM may act as either positive or negative regulators of transcription. Alternative mRNA splicing produces numerous isoforms of ATF-1 and CREB, that can account for the variability in transcriptional regulatory activities (70,71). In addition, each protein or isoform also possesses differing patterns of phosphorylation, and the specific patterns contribute to their ability as regulators of transcription (69). CREB and CREM have been shown to play important roles in basal and hormone-regulated transcription and differentiation while the role of ATF-1 is less well defined. ATF-1 homodimers appear to be weaker transcriptional activators than either CREB or certain forms of CREM, since ATF-1-mediated activation is enhanced by heterodimerization with either CREB or CREM (70). In addition, it has been demonstrated that CREB can efficiently form heterodimers with ATF-1 rather than form the CREB homodimer (72,73).

Previous studies have demonstrated that CREB expression correlates directly with the metastatic potential of murine melanoma cells (74), and that

ATF-1 is not detected in normal melanocytes but is easily found in metastatic melanoma cells (75). Whether these observations are causally related to tumorigenicity and metastasis of melanoma cells was not clear. In this review, we will summarize our current results describing how overexpression of the transcription factors CREB and ATF-1 contributes to the acquisition of the metastatic phenotype in human melanoma.

EFFECT OF CREB/ATF-1 ON TUMORIGENICITY AND METASTASIS OF HUMAN MELANOMA CELLS

To study the contribution of CREB and its associated proteins (mainly ATF-1) to tumor growth and metastasis of human melanoma cells, we have previously taken the approach of using a dominant-negative form of CREB, KCREB. KCREB has a single base pair substitution in its DNA-binding domain that causes a change at position 287 from Arg to Leu. This mutation prevents the binding of KCREB to the CRE-element (76); KCREB can associate with CREB and other proteins that interact with CREB, including members of the ATF and AP-1 transcription factor families and the coactivator CBP. But, because of the mutation in the DNA-binding domain, heterodimers formed with KCREB do not possess the same degree of transcriptional activity as they would in the case of the wild-type CREB. We reasoned that overexpression of KCREB in cells should quench CREB and its associated proteins. Indeed, the ability of KCREB to inhibit CRE-dependent transcription was demonstrated previously in F9 tetratocarcinoma and CA-77 thyroid carcinoma cells (76,77).

To analyze the role of CREB in tumor growth and metastasis of melanoma cells, we have used MeWo human melanoma cells that ectopically express a dominant-negative of CREB, KCREB, that had been mutated in the DNA-binding domain and had lost its ability to bind the CRE element (76, 77). The isolation and the characterization of the neo-transfected cells and the two KCREB-transfected clones, K-2 and K-10, were described previously (78). The K-2 and K-10 clones displayed a 10-fold decrease in radiation resistance and 5-fold reduction in their ability to form colonies in soft agar compared with control MeWo parental and neo-transfected cells (78). To determine the tumorigenicity of the KCREB transfected cells, 1 x 10^6 cells have injected s.c. over the right scapular region of BALB/c nude mice and the tumor diameter was determined 60 days after injection. As the results summarized in Table 2 show, the two KCREB-transfected clones K-2 and K-10 formed smaller tumors than the control cells, MeWo parental (MeWo-P) and neo-transfected (MeWo-Neo) cells (79,80).

Next, we analyzed the metastatic potential of KCREB-transfected cells in an experimental lung metastasis assay. To that end, 1 x 10^6 cells were injected into the lateral tail vein of BALB/c nude mice. The mice were killed 80 days after injection and the lung tumor colonies were counted. As shown in Table 2 transfection of MeWo cells with KCREB significantly reduced their potential to produce lung metastasis. The median number of metastases was lower in mice injected with KCREB-transfected cells (12 and 4 for K2 and K10, respectively). Furthermore, the lower metastatic potential of K-10 as compared with K-2 cells could be explained by the higher KCREB activity in these cells (79,80).

Table 2. Tumor growth and experimental lung metastasis of human melanoma cells transfected with KCREB in nude mice.

Cells lines	Mean tumor diameter (cm)[a]	Lung metastasis[b] median (range)
MeWo-P	1.3	27 (0-70)
MeWo-Neo	NT	22 (11-74)
K-2	0.6	12 (0-59)
K-10	0.6	4 (1-22)

[a]1 x 10^6 cells were injected s.c. and the tumor size was monitored after 60 days; [b]1 x 10^6 cells were injected into the lateral tail vein and the number of lung tumor colonies was counted after 80 days. NT – not tested.

EFFECT OF CREB/ATF-1 ON THE INVASIVE PROPERTIES OF HUMAN MELANOMA CELLS

Several molecules involved in melanoma invasion have CRE-elements in their promoter such as the type IV collagenase MMP-2 (gelatinase A) (81) and the cell adhesion molecule MCAM/MUC18 (82). MMP-2 activity directly modulates melanoma cell-adhesion and spreading to extracellular matrix (55,83) while MCAM/MUC18 expression correlates directly with the metastatic ability of melanoma cells and its ectopic expression in primary cutaneous melanoma cells rendered them tumorigenic and metastatic in nude mice (22). By the presence of CRE elements in their promoters these genes could be a target for regulation by the CREB transcription factor.

We investigated the expression of MMP-2 and MCAM/MUC18 at the mRNA level in MeWo cells following transfection with KCREB. Northern blot analysis showed that KCREB-transfected cells displayed lower levels of the 3.1 kb MMP-2 transcript and 3.3-kb MCAM/MUC18 transcript than control cells. MMP-2 and MCAM/MUC18 expression were inhibited by 2.5-fold and by 3.5-fold in K-2 and K-10 cells, respectively, as compared with Me-Wo-P and MeWo-neo cells (79,80).

To verify that the downregulation of MMP-2 expression in KCREB-transfected cells was reflected in MMP-2 activity, we analyzed supernatants from control and KCREB-transfected cells for collagenase activity by zymography. The activity of the 72-kDa collagenase (MMP-2) was significantly lower in K-2 and K-10 cells than in control MeWo-P and MeWo-Neo cells. It is possible that both the latent and the activated forms of MMP-2 were reduced in the KCREB-transfected cells. The activity of the 92-kDa collagenase (MMP-9) that does not contain CRE element in its promoter remained unchanged and served as an internal control for equal loading (79,80).

Chloramphenicol acetyltransferase (CAT) activity driven by the MMP-2 promoter was inhibited by 14- and 45-fold in KCREB-transfected K-10 and K-2 cells, respectively, as compared with control cells, suggesting that KCREB's regulation or MMP-2 expression and activity occurred at the transcriptional level (79,80).

Because we observed a downregulation of MMP-2 and MCAM/MUC18 expression in KCREB-transfected cells, we next analyzed the effect of KCREB expression on invasive properties of melanoma cells. To that end, KCREB-transfected cells were assayed for their potential to penetrate through the basement membrane, an important component in the process of tumor invasion and metastasis. Migration of cells through Matrigel-coated filters was monitored by using fibroblast-conditioned medium as a source of chemoattractant. We observed a 10-20-fold decrease in the number of the cells that invaded Matrigel-coated filters in K-2 and K-10 cells as compared with MeWo-P and MeWo-Neo cells. The results show that expression of KCREB inhibited the ability of melanoma cells to migrate and invade the basement membrane, possibly by decreasing MMP-2 and MCAM/MUC18 expression, and suggest that CREB plays an important role in the regulation of the invasive properties of melanoma cells (79,80).

CREB AND AFT-1 ACT AS A SURVIVAL FACTORS FOR HUMAN MELANOMA CELLS

Resistance to apoptosis induced by exogenous signals is an important property of tumor cells. Previous studies demonstrated that the induction of apoptosis by diverse exogenous signals depends on an elevation of cytosolic Ca^{2+} (84,85). Because CREB mediates both cAMP and Ca^{2+} transcriptional responses (86,87), we investigated the role of CREB in the resistance of melanoma cells to apoptosis induction. To induce apoptosis, we used thapsigargin (Tg), which inhibits endoplasmic reticulum-dependent Ca^{2+}-ATPase and thereby increases cytosolic Ca^{2+} (88). Furthermore, Tg has been shown to trigger apoptosis in melanoma cells (89).

The effect of Tg treatment on MeWo-P, MeWo-Neo and the two KCREB-transfected clones (K-2 and K-10) was analyzed by flow cytometry analysis after propidium iodide staining. After treament with 1 μM Tg for 48 h, KCREB-transfected cells showed a higher percentage of hypodiploid cells than the control cells. Cells with hypodiploid DNA content were increased by 2.2 and 3.3 fold in K-2 and K-10 cells, respectively, as compared with MeWo-P and MeWo-Neo cells (80,90).

Hypodiploid DNA content is indicative of apoptotic cells. To further confirm that Tg-treatment induced apoptosis in KCREB-transfected cells, we analyzed the cellular morphology of K-10 cells. Transmission electron microscopy analysis of K-10 cells after treatment with 1 μM Tg for 48 h, showed typical apoptotic morphology with plasma and nuclear membrane blebbing, chromatin condensation, and cytoplasmic vascularization. In addition, nuclear changes consistent with apoptosis, such as condensation and segregation of chromatin into compact masses aligning with the inner side of the nuclear membrane were also apparent. Control cells did not show these changes. These results demonstrated that KCREB expression in melanoma cells decreased their resistance to Tg-induced apoptosis.

To understand the mechanism by which KCREB expression increased the sensitivity to Tg-induced apoptosis in melanoma cells, we next investigated the activation of CREB and the induction of CRE-dependent transcription by Tg. Previous studies showed that CREB was activated by phosphorylation on its Ser^{133} after an increase in intracellular Ca^{2+} (87). We therefore verified first that Tg can induce CREB phosphorylation. Western blot analysis with α-p-CREB, an antibody raised against the Ser^{133}-phosphorylated peptide of CREB, that recognizes only the activated form of CREB, was performed on total extract of MeWo-P cells treated for 30 min with 1 μM Tg. After treatment with Tg, α-p-CREB detected two major proteins characterized by molecular masses of 38 kDa and 43 kDa. On the basis of their molecular weight, the 43-kDa protein could correspond to CREB and the protein of 38 kDa could correspond to ATF-1. Indeed, the transcription factor ATF-1 shares high homology with the CREB peptide sequence used to raise the α-p-CREB antibody and was previously described to be activated by an increase of intracellular Ca^{2+} (91). To verify this hypothesis, we analyzed nuclear extracts of MeWo-P cells treated for 30 min with 1 μM Tg, by Western blot with antibodies specific to CREB (α-CREB) or ATF-1 (α-ATF-1). The α-CREB antibody detected a protein that comigrated with the 43-kDa protein, and the α-ATF-1 antibody recognized a protein that comigrated with the 38-kDa protein.

Collectively, these data show that CREB and ATF-1 are involved in the resistance of MeWo cells to Tg-induced apoptosis and suggest that they act as survival factors for human melanoma cells (80,90).

TARGETING ATF-1/CREB BY SINGLE CHAIN Fv FRAGMENT (ScFv) ANTI-ATF-1

To further investigate the cellular role of ATF-1 in melanoma progression, we hereby utilized current advances in the engineering of antibodies that have made possible the cloning of small single-chain Fv (ScFv) fragments. ScFv fragments contain the antigen-binding variable domains of the light and heavy chains connected by a peptide spacer (92,93). When constructed in this manner, a single RNA transcript can be expressed and translated into an active protein that has the potential to interfere with the activity of targeted intracellular proteins. Intracellular ScFv fragments have been successfully employed to decrease the expression of ErbB-2 (94,95) the α subunit of human IL-2 receptor (96) and to restore the transcription activity of mutant p53 (97).

To investigate the role of ATF-1/CREB in tumor growth and metastasis of human melanoma, we attempted to quench their transcriptional activities by using single chain Fv fragment (ScFv) anti-ATF-1, that was derived from a monoclonal antibody (mAb) anti-ATF-1 (Fig. 3). This mAb (designated mAb 41.4) has shown to inhibit ATF-1 binding and transcriptional activation from CRE-dependent promoters *in vitro* (98). We reasoned that expression of ScFv anti-ATF-1 in melanoma cells would bind ATF-1 and inactivates the transcriptional activation of both ATF-1 and CREB (illustrated in *Figure 3*).

In this study we provide evidence that expression of ScFv anti-ATF-1 in MeWo melanoma cells inhibited their tumorigenicity and metastatic potential in nude mice. Intracellular expression of ScFv anti-ATF-1 rendered the melanoma cells susceptible to apoptosis *in vivo* compared to control untransfected cells. These studies demonstrate for the first time that intracellular ScFv anti-ATF-1 can be used to quench ATF-1 activity not only as a method to explore its function but also as a modality for cancer therapy (99, 100).

CONCLUDING REMARKS

In the chapter, we reviewed our recent studies demonstrating that the progression of human melanoma is associated with the loss of AP-2 expression on one hand and overexpression of CREB and ATF-1 transcription factors on the other. Changes in these transcription factors modulate and regulate several genes involved in the development of malignant melanoma.

Interestingly enough, some of the genes regulated by AP-2 such as MCAM/MUC18 (101), MMP-2 (79), FAS/APO-1, HER-2 (102), and Bcl-2 (103) are also regulated by CREB/ATF-1 transcription factors family (100). It is therefore feasible that the outcome of the metastatic phenotype in melanoma may well be dependent on the delicate balance between the expression of AP-2 and the CREB/ATF-1 transcription factors. Loss of AP-2 and overexpression of CREB/ATF-1 in metastatic melanoma cells may work in concert to regulate several genes contributing to the malignant phenotype.

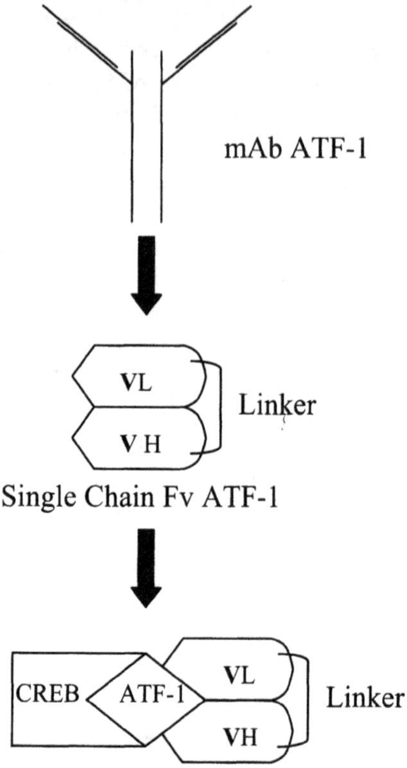

Figure 3. Schematic presentation of ScFv anti-ATF-1 generation from mAb 41.4 and its mode of action.

REFERENCES

1. Fountain JW, Bale SJ, Housman DE. Genetics of melanoma. Cancer Surv 1990; 645-71.
2. Ahmed I. Malignant melanoma: Prognostic indicators. Mayo Clin Proc 1997; 72:356-61.
3. Dooley T. Recent advances in cutaneous melanoma oncogenesis research. Oncol Res 1994; 6:1-9.
4. Clark WH, Elder DE, Guerry D, Epstein MN, Greene MH, Van Horn M. A study of tumor progression: the precursor lesions of superficial spreading and nodular melanoma. Hum Pathol 1984; 15: 1147-1165.
5. Bar-Eli M. Molecular mechanisms of melanoma metastasis. J Cell Physiol 1997; 173: 275-78.
6. Luca MR, Bar-Eli M. Molecular changes in human melanoma metastasis. Histol Histopath 1998; 13: 1225-1231.

7. Fidler IJ. Critical factors in the biology of human cancer metastasis. Cancer Res 1990; 50: 61.
8. Lassam N, Bickford S. Loss of c-KIT expression in cultured melanoma cells. Oncogene 1992; 7: 51-56.
9. Natali PG, Nicotra MR, Winkler AB, Cavaliere R, Bigotti A Ullrich A. Progression of human cutaneous melanoma is associated with loss of expression of c-KIT protooncogene receptor. Int J Cancer 1992; 52: 197-201.
10. Zakut R, Perlis R, Eliyau S, Yarden Y, Givol D, Lyman SD, Halaban R. KIT ligand (mast cell growth factor) inhibits the growth of KIT-expressing melanoma cells. Oncogene 1993; 8: 2221-2229.
11. Yarden Y, Kuang WJ, Yang-Feng T, Coussens L, Mumenitus S, Dull TJ, Chen E, Schlessinger J, Francke U, Ullrich A. A human proto-oncogene c-KIT: a new cell surface receptor tyrosine kinase for an identified ligand. EMBO J 1987; 6: 3342-3351.
12. Chabot B, Stephenson DA, Chapman VM, Besmer P, Bernstein A. The protooncogene c-KIT encoding a trans-membrane tyrosine kinase receptor maps to the mouse W locus. Nature 1988; 335: 88-89.
13. Geisler EN, Ryan MA, Housman DE. The dominant white spotting (W) locus of the mouse encodes the c-KIT protooncogene. Cell 1988; 185-192.
14. Zsebo KM, Williams DA, Geissler, EN, Broudy VC, Martin FH, Atkins HL, Hsu RY, Birkett NC, Okino KH, Murdock DC, Jacobson FW, Langley KE, Smith KA, Takeishi T, Cattanach BM, Galli SJ, Snuggs, S. Stem cell factor is encoded at the SL locus of the mouse and is the ligand for the c-KIT tyrosine kinase receptor. Cell 1990; 63: 213-224.
15. Nocka K, Majumder S, Chabot B, Ray P, Cervone M, Bernstein A, Besmer P. Expression of the c-KIT gene products in known cellular targets of W mutations in normal and W mutant mice: Evidence for impaired c-KIT kinase in mutant mice. Genes Dev 1989; 3: 816-826.
16. Brannan CI, Lyman SD, Williams DE, Eisenman J, Anderson D, Cosman D, Bedell MA, Jenkins NA, Coopeland NG. Steel-Dickie mutation encodes a c-KIT ligand lacking transmembrane and cytoplasmic domains. Proc Natl Acad Sci USA 1991; 88: 4671-4674.
17. Nishikawa S, Kusakabe M, Yoshingaga K, Ogawa M, Hayashi SI, Kunisada T, Era T, Sakakura T, Nishikawa SI. In vitro manipulation of coat color formation by a monoclonal anti-c-KIT antibody: two distinct waves of c-KIT dependency during melanocyte development. EMBO J 1991; 10: 2111-2118.
18. Giebel LB, Spritz RA. Mutation of the KIT (mast/stem cell growth factor receptor) proto-oncogene in human piebaldism. Proc Natl Acad Sci. USA 1991; 88; 8696-8699.
19. Fleischman RA, Saltman KL, Stastny V, Zneimer S. Deletion of the c-KIT protooncogene in the human development defect piebald trait. Proc Natl Acad Sci USA 1991; 88: 10885-10889.
20. Huang S, Luca M, Gutman M, McConkey DJ, Langley KE, Lyman SD, Bar-Eli M. Enforced c-KIT expression renders highly metastatic human melanoma cells susceptible to stem cell factor-induced apoptosis and inhibits their tumorigenic and metastatic potential. Oncogene 1996;13; 2339-2347.
21. Sers C, Kirsch K, Rothbacher U, Riethmuller G, Johnson JP. Genomic organization of the melanoma-associated glycoprotein MUC18: implications for the evolution of the immunoglobulin domains. Proc Natl Acad Sci USA 1993; 90: 8514-8518.

22. Xie S, Huang S, Luca M, Gutman M, Reich R, Johnson JP, Bar-Eli M. Expression of MCAM/MUC18 by human melanoma cells leads to increased tumor growth and metastasis. Cancer Res 1997; 57: 2295-2303.
23. Shih IM, Elder DE, Speicher D, Johnson JP, Herlyn M. Isolation and functional characterization of the A32 melanoma-associated antigen. Cancer Res 54; 2514-2520, 1995.
24. Johnson JP, Bar-Eli M, Jansen B, Markhuf E: Melanoma progression-associated glycoprotein MUC18/MCAM mediates homotypic cell adhesion through interaction with a heterophilic ligand. Int J Cancer 1997; 73: 769-774.
25. Lehmann JM, Holzmann B, Breitbart EW, Schmiegelow P, Riethmuller G, Johnson JP. Discrimination between benign and malignant cells of melanocytic lineage by two novel antigens, a glycoprotein with a molecular weight of 76,000. Cancer Res 1987; 47: 841-845.
26. Holzmann B, Brocker EB, Lehmann JM, Rutter DJ, Sorg C, Riethmuller G, Johnson JP. Tumor progression in human melanoma: five stages defined by their antigenic phenotypes. Int J Cancer 1987; 39: 466-471.
27. Breslow A. Thickness, cross-sectional areas and depth of invasion in the prognosis of cutaneous melanoma. Ann Surg 1970; 172: 902-908.
28. Luca M, Hunt B, Bucana CD, Johnson JP, Fidler IJ, Bar-Eli M. Direct correlation between MUC18 expression and metastatic potential of human melanoma cells. Melanoma Res 1993; 3: 35-41.
29. Bani MR, Rak J, Adachi D, Wiltshire R, Trent JM, Kerbel RS, Ben-David Y. Multiple features of advanced melanoma recapitulated in tumorigenic variants of early state (radial growth phase) human melanoma cell lines: evidence for a dominant phenotype. Cancer Res 1996; 56: 3075-3086.
30. Bar-Eli M. Role of AP-2 in tumor growth and metastasis of human melanoma. Cancer Met Rev 1999; 18: 377-385.
31. Williams T, Admon A, Luscher B, Tjian R. Cloning and expression of AP-2, a cell-type-specific transcription factor that activates inducible enhancer elements. Genes Dev 1988; 2: 1557-1569.
32. Gaynor RB, Muchardt C, Xia YR, Klisak I, Mohandas T, Sparkes RS Lusis AJ. Localization of the gene for the DNA-binding protein AP-2 to human chromosome 6p 22.3 pter. Genomics 1991;10: 1100-1102.
33. Mitchell PJ, Timmons PM, Hebert JM, Rigby PW, Tjian R. Transcription factor AP-2 is expressed in neural crest cell lineages during mouse embryogenesis. Genes Dev 1991; 5:105-119.
34. Mitchell PJ, Wang C, Tjian R. Positive and negative regulation of transcription in vitro: Enhancer-binding protein AP-2 is inhibited by SV40 antigen. Cell 1987; 50: 847-861.
35. Williams T, Tjian R. Characterization of a dimerization motif in AP-2 and its function in heterologous DNA-binding proteins. Science 1991; 251: 1067-1071.
36. Buettner R, Kannan P, Imhof A, Bauer R, Yim SO, Glackshuber R, van Dyke MW, Tainsky MA. An alternatively spliced mRNA from the AP-2 gene encodes a negative regulator of transcriptional activation of AP-2. Mol Cell Biol 1993; 13: 4174-4185.
37. Luscher B, Mitchell PJ, Williams T, Tjian R. Regulation of transcriptional factor AP-2 by the morphogen retinoic acid and by second messengers. Genes Dev 1989; 3: 1507-1517.

38. Schorle H, Meier P, Buchert M, Jaenisch R, Mitchell PJ. Transcription factor AP-2 essential for cranial closure and craniofacial development. Nature 1996; 381: 235-238.
39. Zhang J, Hagopian-Donaldson S, Serbedzija G, Elsemore J, Plehn-Dujowich D, McMahon AP, Flavell RA, Williams T. Neural tube, skeletal and body wall defects in mice lacking transcription factor AP-2. Nature 1996; 381: 238-241.
40. Yamamoto K, Tojo A, Aoki N, Shibuya A. Characterization of the promoter region of the human c-KIT protooncogene. Japan J Cancer Res 1993; 84: 1136-1144.
41. Huang S, Jean D, Luca M, Tainsky M, Bar-Eli M. Loss of AP-2 results in downregulation of c-KIT and enhancement of melanoma tumorigenicity and metastasis. EMBO J 1998; 17: 4358-4369.
42. Luca M, Xie S, Gutman M, Huang S, Bar-Eli M. Abnormalities in the CDKN2 ($p16^{INK4}$/mts-1) gene in human melanoma cells: Relevance to tumor growth and metastasis. Oncogene 1995; 11: 1399-1402.
43. Jean D, Gershenwald JE, Huang S, Luca M, Hudson MJ, Tainsky AM, Bar-Eli M. Loss of AP-2 results in upregulation of MCAM/MUC18 and an increase in tumor growth and metastasis of human melanoma cells. J Biol Chem 1988; 273: 16501-16508.
44. Gershenwald JE, Sumner W, Calderone T, Wang Z, Huang S, Bar-Eli. Dominant-negative transcription factor AP-2 augments melanoma tumor growth *in vivo*. (Submitted).
45. Liotta LA. Tumor invasion and metastases – role of the extracellular matrix: Rhoads Memorial Award Lecture. Cancer Res 1986; 46: 1-7.
46. Liotta LA, Stetler-Stevension Wg. Metalloproteinases and cancer invasion. Semin Cancer Biol 1990; 2: 99-106.
47. Blood CH, Zetter BR. Tumor interactions with the vaculature: angiogenesis and tumor metastasis. Biochim Biophys Acta 1990; 1032: 89-118.
48. Karelina TV, Goldberg GI, Eisen AZ. Matrix metalloproteinases in blood vessel development in human fetal skin and in cutaneous tumors. J Invest Dermatol 1995; 105: 411-417.
49. Frisch SM, Morisaki JH. Positive and negative transcriptional elements of the human type IV colleganase gene. Mol Cell Biol 1990; 12:6524-6532.
50. Hofmann, UB, Westphal JR, Waas ET, Zendman AJ, Cornelissen IM, Ruiter DJ, van Muijen GN. Matrix metalloproteinases in human melanoma cell lines and xenografts: increased expression of activated matrix metalloproteinase-2 (MMP-2) correlates with melanoma progression. Br J Cancer 1999; 81: 774-782.
51. Vaisanen A, Kallioinen M, Taskinen PJ, Turpeenniemi-Hujanen T. Prognostic value of MMP-2 immunoreactive protein (72 kD type IV collagenase) in primary skin melanoma. J Pathol 1998; 186: 51-58.
52. Vaisanen A, Killioinen M, von Dickhoff K, Laatikainen L, Hoyhtya M, Turpeenniemi-Hujanen T. Matrix metalloproteinase-2 (MMP-2) immunoreactive protein-a new prognostic marker in uveal melanoma? J Pathol 1999; 188: 56-62.
53. Qin H, Sun Y, Benveniste EN. The transcription factors Sp1, Sp3, and AP-2 are required for constitutive matrix metalloproteinase-2 gene expression in astroglioma cells. J Biol Chem 1999; 274: 29130-29137.
54. Somassundaram K, Jayaraman G, Williams T, Moran E, Frisch S, Thimmapaya B. Repression of a matrix metalloproteinase gene by E1A correlates with its ability to bind to cell type-specific transcription factor AP-2. Proc Natl Acad Sci USA 1996; 93: 3088-3093.

55. Luca M, Huang S, Gershenwald JE, Singh RK, Reich R, Bar-Eli M. Expression of IL-8 by human melanoma cells upregulates MMP-2 activity and increases tumor growth and metastasis. Am J Pathol 1997; 151: 1105-1113.
56. Singh RK, Gutman M, Reich R, Bar-Eli. Ultraviolet B irradiation promotes tumorigenic and metastatic properties in primary cutaneous melanoma via induction of interleukin 8. Cancer Res 1995; 55:3669-3674.
57. Cowley GP, Smith ME. Cadherin expression in melanocytic naevi and malignant melanomas. Am J Pathol 1996; 197: 183-187.
58. Jiang H, Lin J, Su ZZ, Herlyn M, Kerbel RS, Weissman BE, Welch DR, Fisher PB. The melanoma differentiation-associated gene mda-6, which encodes the cyclin-dependent kinase inhibitor p21, is differentially expressed during growth, differentiation and progression in human melanoma cells. Oncogene 1995; 10: 1855-1864.
59. Vidal MJ, Loganzo F, Jr, de Oliveira AR, Hayward NK, Albino AP. Mutations and defective expression of the WAF-1 p21 tumor-suppressor gene in malignant melanomas. Melanoma Res 1995; 5: 243-250.
60. Natali PG, Nicotra MR, Digiesi G, Cavaliere R, Bigotti A, Trizio D, Segatto O. Expression of gp 185 HER-2 in human cutaneous melanoma: Implications for experimental imunotherapeutics. Int J Cancer 1994; 56:341-346.
61. Descheemaeker KA, Syns S, Nelles L, Auwerx J, Ny T, Collen D. Interaction of AP-1-like, AP-2-like and SP-1-like proteins with two distinct sites in the upstream regulatory region of the plasminogen activator inhibitor-1 gene mediates the phorbol 12-myristate 13-acetate response. J Biol Chem 1992; 267: 15086-15091.
62. van den Oord JJ, Vandeghinste N, De Ley M, De Wolf-Peeters C. Bcl-2 expression in human melanocytes and melanocytic tumors. Am J Pathol 1994; 145: 294-300.
63. Bosher JM, Williams T, Hurst HC. The developmentally regulated transcription factor AP-2 is involved in c-*erb*B-2 overexpression in human mammary carcinoma. Proc Natl Acad Sci USA1995; 92: 744-747.
64. Zeng YX, Somasundaran K, El-Deiry WS. AP-2 inhibits cancer cell growth and activates p21/WAF-1 expression. Nature Genet 1997; 15:78-82.
65. Karjalainen JM, Kellokoski JK, Eskelinen MJ, Alhava EM, Kosma V-M. Downregulation of transcription factor AP-2 predicts poor survival in stage I cutaneous malignant melanoma. J Clin Oncol 1998; 16: 3584-3591.
66. Halaban R, Pomerantz SH, Marshall S, Lambert DT, Lerner AB. Regulation of tyrosinase in human melanocytes grown in culture. J Cell Biol 1983; 97: 480-488.
67. Halaban R, Pomerantz SH, Marshall S, Lerner AB. Tyrosinase activity and abundance in Cloudman melanoma cells. Arch Biochem Biophys 1984; 230: 383-387.
68. Halaban R, Ghosh S, Baird A. bFGF is the putative natural growth factor for human melanocytes. *In Vitro* Cell Dev Biol 1987; 26: 47-52.
69. Meyer TE, Habener JF. Cyclic adenosine 3N,5N-monophosphate response element binding protein (CREB) and related transcrption-activating deoxyribonucleic acid-binding proteins. Endocr Rev 1993; 3: 269-290.
70. Lee KA, Masson N. Transcriptional regulation by CREB and its relatives. Biochim Biophys Act 1993; 3: 221-233.
71. Lemaigre FP, Ace CI, Green MR. The cAMP response element binding protein, CREB, is a potent inhibitor of diverse transcriptional activators. Nucl Acids Res 1993; 21: 2907-2911

72. Hurst HC, Totty NF, Jones NC. Identification and functional characterization of the cellular activating transcription factor 43 (ATF-43) protein. Nucl Acids Res 1991; 19: 4601-4609.
73. Rehfuss RP, Walton KM, Loriaux MM, Goodman RH. The cAMP-regulated enhancer-binding protein ATF-1 activates transcription in response to cAMP-dependent protein kinase. J Biol Chem 1991; 266: 18431-18434.
74. Ruthberg SE, Goldstein IM, Yang YM, Stackpole CW, Ronai Z. Expression and transcriptional activity of AP-1, CRE, and URE binding proteins in B16 mouse melanoma subclones. Mol Carcinog 1994; 10: 82-87.
75. Bohm M, Moellman G, Cheng E, Alvarez-Franco M, Wagner S, Sassone-Corsi P, Halaban R. Identification of p90RSK as the probable CREB-Ser 133 kinase in human melanocytes. Cell Growth Differ 1995; 6: 291-302.
76. Walton KM, Rehfuss RP, Chriva JC, Lochner JE, Goodman RH. A dominant repressor of cyclic adenosine 3N,5N-monophosphate (cAMP)-regulated enhancer-binding protein activity inhibits the cAMP-mediated induction of the somatostain promoter *in vivo*. Mol Endocrinol 1992; 6: 647-655.
77. Woloshin PI, Walton KM, Rehfuss RP, Goodman RH, Cone RD. 3N,5N-Cyclic adenosine monophosphate-regulated enhancer binding (CREB) activity is required for normal growth and differentiated phenotype in the FRTL5 thyroid follicular cell line. Mol Endocrinol 1992; 6: 1725-1733.
78. Yang YM, Dolan LR, Ronai Z. Expression of dominant negative CREB reduces resistance to radiation of human melanoma cells. Oncogene 1996;12: 2223-2233.
79. Xie S, Price JE, Luca M, Jean D, Ronai Z, Bar-Eli M. Doninant-negative CREB inhibits tumor growth and metastasis of human melanoma cells. Oncogene 1997; 15: 2069-2075.
80. Jean D, Bar-Eli M. Regulation of tumor growth and metastasis of human melanoma by the CREB transcription factors family. Mol Cell Bioch 2000; 212: 19-28.
81. Huhtala P, Chow LT, Tryggvason K. Structure of human type IV collagenase gene. J Biol Chem 1990; 265: 11077-11082.
82. Lehmann JM, Riethmuller G, Johnson JP. MUC18, a marker of tumor progression in human melanoma, shows sequence similarity to the neural cell adhesion molecules of the immunoglobulin superfamily. Proc Natl Adad Sci USA 1989; 86: 9891-9895.
83. Ray JM, Stetler-Stevenson WG. Gelatinase A activity directly modulates melanoma cell adhesion spreading. EMBO J 1995; 14: 908-917.
84. McConkey DJ, Nicotera P, Hartzell P, Bellomo G, Wyllie AH, Orrenius S. Glucocorticoids activate a suicide process in thymocytes through an elevation of cytoslolic Ca^{2+} concentration. Arch Biochem Biophys 1996; 269: 365-370.
85. McConkey DJ, Orrenius S. The role of calcium in the regulation of apoptosis. J Leukoc Biol 1996; 59:775-59.783.
86. Gonzalez GA, Montminy MR. Cyclic AMP stimulates somatostatin gene transcription by phosphorylation of CREB at serine 133. Cell 1989; 59: 675-680.
87. Sheng M, Thompson MA, Greenberg ME. CREB: A Ca(2+)-regulated transcription factor phosphorylated by calmodulin-dependent kinases. Science 1991; 252: 1427-1430.
88. Thastrup O., Cullen PJ, Drobak BK, Hanley MR, Dawson AP. Thapsigargin, a tumor promoter, discharges intracellular Ca(2+) stores by specific inhibition of the endoplasmic reticulum Ca(2+)-ATPase. Proc Natl Acad Sci USA 1990; 87: 2466-2470.

89. Muthukkumar S, Nair P, Sells SF, Maddiwar NG, Jacob RJ Rangnekar VM. Role of EGR-1 in thapsigargin-inducible apoptosis in the melanoma cell line A375-C6. Mol Cell Biol 1995; 15: 6262-6272.
90. Jean D, Harbison M, McConkey DJ, Ronai Z, Bar-Eli M. CREB and its associated proteins act as survival factors for human melanoma cells. J Biol Chem 1998; 273: 24884-24890.
91. Liu F, Thompson MA, Wagner S, Greenberg ME, Green MR. Activating transcription factor-1 can mediate Ca(2+) and cAMP-inducible transcriptional activation. J Biol Chem 1993; 268: 6714-6720.
92. Raag R, Whitlow M. Single-chain Fvs, FASEB J 1995; 1: 73-80.
93. Winter G, Milstein C, Man-mad antibodies. Nature 1991; 349: 293-299.
94. Grim J, Deshane J, Feng M, Lieber A, Kay M, Curiel DT. erbB-2 knockout employing an intracellular single-chain antibody (sFv) accomplishes specific toxicity in erbB-2 expressing lung cancer cells. Am J Respir Cell Mol Biol 1996; 15: 348-354.
95. Graus-Porta D, Beerli R, Hyns N. Single-chain antibody-mediated intracellular retention of ErbB-2 impairs Neu differentiation factor and epidermal growth factor signaling. Mol Cell Biol 1995; 3: 1182-1191.
96. Richardson JH, Sodroski JG, Waldmann TA, Marasco WA. Phenotypic knockout of the high-affinity human interleukin 2 receptor by intracellular single-chain antibodies against the alpha subunit of the receptor. Proc Natl Acad Sci USA. 1995; 8: 3137-3141.
97. Caron de Fronmentel C, Gruel N, Venot C, Debussche L, Conseiller E, Dureuil C, Teillaud JL, Tocque B, Bracco L. Restoration of transcriptional activity of p53 mutants in tumor cells by intracellular expression of anti-p53 single chain Fv fragments. Oncogene 1999; 2: 551-557.
98. Orten DJ, Strawhecker JM, Sanderson SD, Huang D, Prytowsky MB Hinrichs SH. Differential effects of monoclonal antibodies on activating transcription factor-1 and cAMP response element binding protein interactions with DNA. J Biol Chem 1994; 51: 32254-32263
99. Jean D, Tellez C, Huang S, Davis DW, Bruns CJ, McConkey DJ, Hinrichs SH, Bar-Eli M. Inhibition of tumor growth and metastasis of human melanoma by intracellular anti-ATF-1 single chain Fv fragment. Oncogene 2000; 19: 2721-2730.
100. Jean D, Bar-Eli M. Targeting the ATF-1/CREB transcription factors by single chain Fv fragment in human melanoma: potential modality for cancer therapy. Crit Reviews Immunol 2001 (in press).
101. Karlen S, Brathen LR. Role of the initiator element in the regulation of the melanoma cell adhesion molecule gene. J Invest Dermatol 2000; 668-673.
102. Newman SP, Bates NP, Vernimmen D, Parker MG, Hurst HC. Cofactor competition between the ligand-bound oestrogen recepteror and an intron 1 enhancer leads to oestrogen repression of ERBB2 expression in breast cancer. Oncogene 2000; 19: 490-497.
103. Wilson BE, Mochon E, Boxer LM. Induction of bcl-2 expression by phosphorylated CREB proteins during B-cell activation and rescue from apoptosis. Mol Cell Biol 1996; 16: 5546-5556.

Chapter 9

HETEROCHROMATIN-ASSOCIATED PROTEIN 1, HP1$^{Hs\alpha}$, IN BREAST CANCER INVASION AND METASTASIS

Dawn A. Kirschmann, Mary J.C. Hendrix

Department of Anatomy & Cell Biology, Holden Cancer Center at the University of Iowa, College of Medicine, Iowa City, Iowa, 52242-1109, USA

Abstract

Heterochromatin has repressive effects on euchromatic gene expression. The mechanisms contributing to gene silencing are not known, however silencing has been shown to be modulated in part by heterochromatin-associated protein 1, HP1$^{Hs\alpha}$. We have identified a reduction in HP1$^{Hs\alpha}$ expression in invasive/metastatic human breast cancer cell lines and in tumor cells from distant metastases. Furthermore, HP1$^{Hs\alpha}$ transfected invasive/metastatic breast cancer cells show a reduction in the invasive phenotype indicating that HP1$^{Hs\alpha}$ is involved in modulating specific biological activities in the metastatic cascade. Here, we discuss the role of heterochromatin and HP1$^{Hs\alpha}$ expression in breast cancer invasion and metastasis with regard to gene silencing, chromatin packaging, and nuclear architecture. In addition, we introduce working models for mechanisms of HP1$^{Hs\alpha}$ function that will be addressed with future studies.

Key words: gene silencing, heterochromatin, HP1$^{Hs\alpha}$, breast cancer, invasion, metastasis

HETEROCHROMATIN

Heterochromatin and Euchromatin

In humans, the entire genome is packaged and organized in an orderly fashion such that 2 m of DNA can fit into a nucleus that is 10 μm in diameter. This packaging consists of two types of chromatin: euchromatin and heterochromatin (1, 2). Euchromatic regions of the genome comprise the majority of single copy DNA, which includes most transcribed genes. During cell division, euchromatin becomes condensed in metaphase, but is subsequently decondensed during interphase. In contrast, heterochromatin remains condensed throughout the cell cycle and replicates late in S phase. Cytologically, heterochromatin appears as

D.R. Welch (ed.), Cancer Metastasis – Related Genes, 169–189.

dark, electron dense areas located near the inner nuclear membrane and nucleolus, as well as other concentrated masses in the nucleus during interphase. Constitutive heterochromatin, generally located at pericentromeric and telomeric regions, is mostly comprised of repetitive DNA sequences that contains relatively few functionally active genes (2, 3). Facultative heterochromatin is described as euchromatic genes that have become silenced either in a specific cell lineage or during embryonic development. Inactivation of an X chromosome to prevent over-expression of genes in females is one example of facultative heterochromatin (2, 4, 5). The inactivation of genes from an entire chromosome by an alteration in chromatin packaging demonstrates the critical role of heterochromatin in the functional aspects of gene expression and the potential consequences when regulation of heterochromatinization of genes goes awry.

Heterochromatic Changes in Breast Cancer Cells

Breast cancer incidence and mortality rates remain the highest of all cancers in women in the United States (6). One of the challenges in treating breast cancer is predicting long-term prognosis. Treatment protocols are different for patients who are at "high risk" for recurrence than those at "low risk" and are based on morphologic assessment of histologic samples of primary breast tumors. However, histologic grading of tumors is not always predictive of clinical course (7, 8). Therefore, many studies have investigated biochemical, molecular, and morphological parameters for their value as prognostic markers to predict metastatic potential and recurrence of breast cancers. Included in this list are nuclear morphology (9-14) and chromosomal anomalies (15-17). In fact, the association between cell nuclei morphology (including chromatin patterns) and breast cancer patient survival has been recognized since 1925 (18).

Changes in interphase nuclear shape and size, and heterochromatin content, are significant prognostic factors for predicting long-term survival in breast cancer patients (10, 12, 19). A combination of computer-assisted image analysis of heterochromatin content and nuclear texture and shape measurements was capable of correctly predicting prognosis in 92% of breast cancer patients (10-12). In this study, poor survival was associated with an increase in nuclear area, a lower optical density of the interphase nuclei (regions with high optical density corresponded to light microscopically recognized condensed chromatin), an increase in marginal distribution of condensed chromatin, and an increase in size of chromatin clumps.

Cytogenetically, heterochromatic polymorphisms in size variation of C-bands on metaphase chromosomes from cancer patients have been observed as well (15-17, 20, 21). The C-band technique stains heterochromatic regions of metaphase chromosomes and the resulting banding patterns are then assessed for variations in size, as well as location on the chromosomal arms. These studies have demonstrated excesses and losses of C-heterochromatin on chromosomes 1, 9, and 16 in addition to frequent chromosomal rearrangements in cancer patients. The majority of breakpoints, in which chromosomal rearrangements occurred, were located in or juxtaposed to heterochromatic regions (15), suggesting that these heterochromatic variants promote changes in chromatin structure and genomic instability.

Experimentally, decreases in chromatin condensation have been observed in actively proliferating cells (22), in 17β-estradiol stimulated, but not dexamethasone treated, estrogen receptor positive non-aggressive breast carcinoma cells (23), and in *ras*- and *myc*-transformed mouse fibroblasts with a high metastatic potential (reviewed in 24). The presumably conflicting data of whether heterochromatin decreases or increases with metastatic potential demonstrates the complexity of the human genome and identifies the need for further study in this area. However, taken together, these studies demonstrate that patterns of chromatin condensation change during metastatic progression, which can lead to disturbances of ordered chromatin organization thereby resulting in aberrant gene expression and genetic instability (25, 26).

Heterochromatin and Gene Silencing

In *Drosophila*, yeast, and mammalian cells, heterochromatin has been shown to have a repressive effect on euchromatic gene expression (1, 2, 27). In *Drosophila*, investigators observed repression of euchromatic gene expression when chromosomal rearrangements placed euchromatic genes near constitutive heterochromatin, a phenomenon known as position effect variegation (PEV; 1, 28). Although the exact mechanism(s) by which PEV-mediated gene silencing occurs is unclear, it is thought to be modulated by heterochromatin-associated proteins, of which the best characterized is heterochromatin-associated protein 1, HP1 (29, 30).

Drosophila *HP1*

In *Drosophila*, HP1 is a nonhistone chromosomal protein that localizes primarily to centric regions, telomeres, and some specific euchromatic sites on polytene chromosomes (30, 31). Structurally, HP1 has two highly conserved regions; a chromodomain (chromosome organization modifier domain) at the amino terminus and a related chromo shadow domain present at the carboxy terminus (32-35). The chromo shadow domain region in HP1 contains heterochromatin and nuclear localization signals, and is involved in binding to other proteins, as well as itself (29, 35, 36). The chromodomain contains heterochromatin localization signals (34), however, no direct protein interactions with the chromodomain have been identified to date (36). Alternatively, the chromodomain may function as an RNA interacting domain since the chromodomain of a *Drosophila* histone deacetylase interacts with non-coding RNA involved in dosage compensation (37).

Developmentally, there is a marked increase in HP1 expression during the blastoderm stage in *Drosophila* which coincides with the presence of heterochromatinized DNA (38, 39) and the onset of PEV (40). Direct evidence that HP1 mediates gene silencing stems from experiments involving HP1 mutations. Mutations in the gene encoding HP1 are homozygous lethal (41). HP1 heterozygotes show a loss of silencing of transgenes at centric locations, which correlates with a more "open" (euchromatic-like) chromatin structure (42, 43). In contrast, HP1 over-expression leads to an increase in gene silencing (41, 44). Therefore, gene expression is sensitive to HP1 dosage (44, 45); however, the exact mechanism by which HP1 exerts its suppressive effect is not currently known.

Mammalian HP1-like Proteins

Subcellular Localization and Expression Patterns

Unlike *Drosophila*, multiple HP1-like proteins have been identified in mice and humans. Using the HP1 chromodomain to probe a mouse embryo cDNA library, HP1-like proteins mHP1α, M31, and M32 were identified in mice (32 46). These murine HP1-like proteins share a 51% amino acid identity to HP1 over their entire length, but have a 70% identity in the chromo domain (figure 1). Although M32 shares 86% identity to the M31 protein, these HP1-like proteins have very different localization patterns on chromosomes. M31 co-localizes to constitutive heterochromatin in interphase nuclei (47-49), and primarily to centromeric heterochromatin in metaphase chromosomes (48, 49). In contrast, M32 localizes to euchromatic regions and is excluded from constitutive heterochromatin in interphase nuclei (50). Similarly, *M31* and *M32* mRNA are differentially expressed during mouse embryogenesis, as well as in adult tissues (47, 51). Differences in localization and developmental expression suggest that M31 and M32 may function similarly, but have different chromatin targets. Alternatively, M31 and M32 may have different functions during cell development and the cell cycle. To date, no localization studies for mHP1α have been published.

Three HP1-like proteins, $HP1^{Hs\alpha}$, $HP1^{Hs\beta}$, and $HP1^{Hs\gamma}$ (arising from three distinct genes) have been identified in humans. Their protein structure is similar to HP1 in that they contain an amino terminal chromodomain and a carboxy terminal chromo shadow domain (52). In comparing the mouse and human HP1-like proteins, $HP1^{Hs\alpha}$ is 80% and 76% similar to M31 and M32, respectively. $HP1^{Hs\beta}$ is 100% identical to M31, whereas $HP1^{Hs\gamma}$ is 98% identical to M32, and only 38% identical to $HP1^{Hs\alpha}$ (53, Figure 1). Anti-centromere antibodies from patients with scleroderma, systemic lupus erythematosus, and Sjøgren's disease were used to characterize $HP1^{Hs\alpha}$ and $HP1^{Hs\beta}$ expression in human cells (48, 54-57). $HP1^{Hs\alpha}$ colocalizes primarily to centromeric regions of constitutive heterochromatin, as well as minor sites within euchromatic regions (48, 58, 59). Similarly, $HP1^{Hs\beta}$ showed discrete speckled nucleoplasmic and pericentromeric heterochromatin staining (49, 55, 56, 58, 59). The localization of $HP1^{Hs\gamma}$ remains controversial since one study demonstrates its localization almost exclusively to euchromatin (58, while other studies show localization to centromeres as well (59, 60). These differences may in part be due to a conformational change in heterochromatic $HP1^{Hs\gamma}$ that is not recognized by antibodies generated to the amino terminus of the protein (60). In a detailed study comparing all three human HP1-like proteins in HeLa cells (a cervical carcinoma cell line), Minc and colleagues not only demonstrate differences in distribution of these proteins during interphase, but also differences in localization during other stages of the cell cycle (58). Similar to M31 and M32, differences in chromosomal localization suggest that $HP1^{Hs\alpha}$, $HP1^{Hs\beta}$, and $HP1^{Hs\gamma}$ may either have completely different functions, or rather a similar function with different chromatin targets.

```
hHP1α  MGKK..TKRT ADSSS..SED EEEYVVEKVL DRRVVKGQVE YLLKWKGFSE EHNTWEPEKN 56
mHP1α  ----..---- -----..--- ---------- ---------- ---------- ---------- 56
hHP1β  ----.QN-KK VEEVL..E-E ---------- -------K-- ---------D -D------E- 57
M31    ----.QN-KK VEEVL..E-E ---------- -------K-- ---------D -D------E- 57
hHP1γ  ----.QNGKS .KKVE..EAE P--F------ -----N-K-- -F------TD AD------E- 56
M32    ----.QNGKS .KKVE..EAE P--F------ -----N-K-- -F------TD AD------E- 56
DmHP1  ----IDNPES SAKV-DAE-E ---YA---II ----R--K-- YY-----YP- TE------N- 60

hHP1α  LDCPELISEF MKKYKKMKEG ENNKPREKSE SNKRK..SNF SNSADDIKSK KKREQ....S 110
mHP1α  ---------- ---------- ---------- G----..-S- ---------- -----....- 110
hHP1β  ----D--A-- LQSQ-TAH-. .....TD--- GG---AD-DS EDKGEES-P- --K-E....- 107
M31    ----D--A-- LQSQ-TAH-. .....TD--- GG---AD-DS EDKGEES-PK --K-E....- 107
hHP1γ  -------EA- LNSQ-AG--. .....KD... GT---SL-DS E..S--S--- ---DA....A 101
M32    -------ED- LNSQ-AG--. .....KD... GT---SL-DS E..S--S--- ---DA....A 101
DmHP1  ---QD--QQY EASR-DEEKS AAS-KDRP-S SA-A-ETQGR ASSSTSTA-- R-S-EPTAP- 120

hHP1α  NDIAR..... .......... .GFERGLEPE KIIGATDSCG DLMFLMKWKD TDEADLVLAK 154
mHP1α  -----..... .......... .--------- ---------- ---------- ---------- 154
hHP1β  .EKP-..... .......... .--A------ R-------S- E--------N S------P-- 150
M31    .EKP-..... .......... .--A------ R-------S- E--------N S------P-- 150
hHP1γ  .-KP-..... .......... .--A---D-- R----I--S- E--------- S--------- 144
M32    .-KP-..... .......... .--A---D-- R-------S- E--------- S--------- 144
DmHP1  GNKSKRTTDA EQDTIPVSGS T--D----A- --L--S-NN- R-T--IQF-G V-Q-EM-PSS 180

hHP1α  EANVKCPQIV IAFYEERLTW HAYP.EDAENK EKETAKS 191
mHP1α  ---------- ---------- ----.------ ------- 191
hHP1β  --------V- -S-------- -S--S--DDK- DDKN... 185
M31    --------V- -S-------- -S--S--DDK- DDKN... 185
hHP1γ  ---M------ ---------- -SC-.--EAQ. ....... 173
M32    ---M------ ---------- -SC-.--EAQ. ....... 173
DmHP1  V--E-I-RM- -H------S- YSDN.--.... ....... 206
```

Figure 1. Comparison of the amino acid sequences of human, mouse, and Drosophila melanogaster homologs of HP1. Positions of amino acid identity with $HP1^{Hs\alpha}$ are represented by dashes and dots are gaps used to align the sequences. The chromodomain and chromo shadow domain are underlined.

Role in Gene Silencing

The gene silencing function of mammalian HP1-like proteins (mHP1) has only been demonstrated within the past two years. The first evidence demonstrating that mHP1 can repress gene expression stems from HP1-GAL4 DNA binding domain tethering experiments in which $HP1^{Hs\alpha}$- and $HP1^{Hs\gamma}$-GAL4 fusion proteins were capable of repressing reporter gene expression when bound to GAL4 sites in mammalian cells (61, 62). In addition, by using a yeast two-hybrid screen, these reports indirectly demonstrated that $HP1^{Hs\alpha}$ and $HP1^{Hs\gamma}$ were part of a larger repression complex. In fact, all three human HP1-like proteins (hHP1s) have been shown to bind to many proteins involved in chromatin structure and function including the lamin B receptor (inner nuclear membrane protein; 53, 63), SP100B (nuclear body-associated protein; 61, 62), BRG-1 (SWI/SNF gene silencing complex member; 46), TIF-1α/TIF-1β (transcriptional intermediary factors; 46, 59, 64, 65), CAF-1 (chromatin assembly factor; 65, 66), INCENP (inner centromere protein; 67), and HP1 (46, 52, 63, 68, 69). The formation of HP1 repression complexes via protein-protein interactions is thought to direct gene-specific transcriptional silencing. The self-association of hHP1s and interaction of hHP1s with other proteins is primarily through the chromo shadow domain (62, 63, 68, 69). The chromo shadow domain may also confer a species-specific function (69). In this study, no silencing activity was observed when the chromo shadow domain of swi6 (HP1-like protein in *Schizosaccharomyces pombe*) was substituted with the M31 chromo shadow domain. However the silencing activity in *S. pombe* was restored when the M31 chromo shadow domain was substituted with the swi6 chromo shadow domain. Interestingly, no species specificity was observed when the chromodomains were swapped.

The gene silencing function of mHP1s has also been verified *in vivo* with murine and *Drosophila* variegating transgene models. Using a centromeric variegating CD2 transgenic mouse, an increase in M31 expression increased the percentage of CD2 negative T cells in a dose-dependent manner, demonstrating mHP1-mediated gene silencing (70). In addition, *Drosophila* with $HP1^{Hs\alpha}$ or $HP1^{Hs\gamma}$ transgenes enhanced silencing of a FX1 array containing a variegating *white* gene (flies variegating for eye color) upon induction of hHP1 expression (60). Since hHP1s appear to have a similar localization pattern as HP1 and are capable of silencing gene expression in *Drosophila*, these data suggest that *Drosophila* variegating transgene models might be useful for elucidating the gene targeting and gene silencing mechanism(s) of hHP1 orthologs.

Although there is an accumulating wealth of data implicating *Drosophila* and mammalian HP1s in transcriptional repression of transgene expression, the mechanism by which genes are silenced remains unclear. The self-association or multimerization of HP1 proteins has been proposed to mediate the packaging of chromatin to a condensed structure inaccessible for transcriptional activation (52, 63, 64, 68, 69). Complicating this model of HP1-mediated gene silencing is the observation that some genes in *Drosophila* require a heterochromatin context for normal expression (1, 71). In addition, it is not known how or why the differential chromosomal localization of mHP1 orthologs occurs and whether particular target genes are susceptible to mHP1-mediated silencing. Nevertheless, HP1 proteins appear to play an important role in gene expression.

$HP1^{Hs\alpha}$ EXPRESSION IN BREAST CANCER

It is now widely accepted that the progression of a non-neoplastic cell to a hyperplastic cell and eventually to one that is capable of metastasis, requires the stepwise accumulation of many genetic alterations (72-75). The molecular mechanisms contributing to breast cancer metastasis are not completely understood, however generally accepted is the perception that additional genetic alterations are required from those involved in tumorigenesis. The expression of metastasis-stimulating genes and the loss of metastasis-suppressor genes are acquired by increasing activity (mutation, amplification, and/or epigenetic/epistatic regulation) and decreasing activity (mutation, loss of heterozygosity, and/or epigenetic/epistatic regulation) of these genes, respectively. Furthermore, the progression of a poorly invasive/non-metastatic cancer cell to one that is highly invasive and metastatic can arise slowly via alterations in less critical genes or more rapidly due to genetic alterations in important controlling genes.

One putative controlling gene, whose encoded protein is involved in gene silencing, and possibly chromatin packaging and genomic stability, is $HP1^{Hs\alpha}$. Alterations in $HP1^{Hs\alpha}$ expression in poorly invasive/non-metastatic breast cancer cells could lead to the differential expression of genes whose products are involved in cellular detachment, extracellular matrix degradation, cell motility, angiogenesis and vasculogenesis, and cell growth; steps that are quintessential in invasion and metastasis (76-78). Indeed, such a connection between altered $HP1^{Hs\alpha}$ expression and the breast cancer invasive/metastatic phenotype has been observed (79).

Down-Regulation of $HP1^{Hs\alpha}$ Expression in Invasive/Metastatic Breast Cancer Cell Lines

A down-regulation in $HP1^{Hs\alpha}$ expression was first identified in MDA-MB-231 cells, a breast carcinoma cell line of high invasive/metastatic potential (79). Differential display analysis was used to compare gene expression in MDA-MB-231 cells to MCF-7 cells, a poorly invasive/non-metastatic breast carcinoma cell line. $HP1^{Hs\alpha}$ was among 18 different gene products identified that were down-regulated in MDA-MB-231 cells, compared to MCF-7 cells. The reduction in $HP1^{Hs\alpha}$ mRNA expression was verified by reverse transcriptase-polymerase chain reaction using $HP1^{Hs\alpha}$-specific primers and RNA hybridization analyses. Furthermore, $HP1^{Hs\alpha}$ protein expression was reduced up to 95% as demonstrated by Western blot and immunofluorescence confocal microscopy (Figure 2, 80). In MCF-7 cells, $HP1^{Hs\alpha}$ was localized to the nucleus and nuclear membrane with a heavy punctate staining pattern reminiscent of that observed in other cell types (48, 58, 59). In contrast, a low-level diffuse nuclear staining pattern with little to no localization to the nuclear membrane was observed in the fibroblast-like MDA-MB-231 cells.

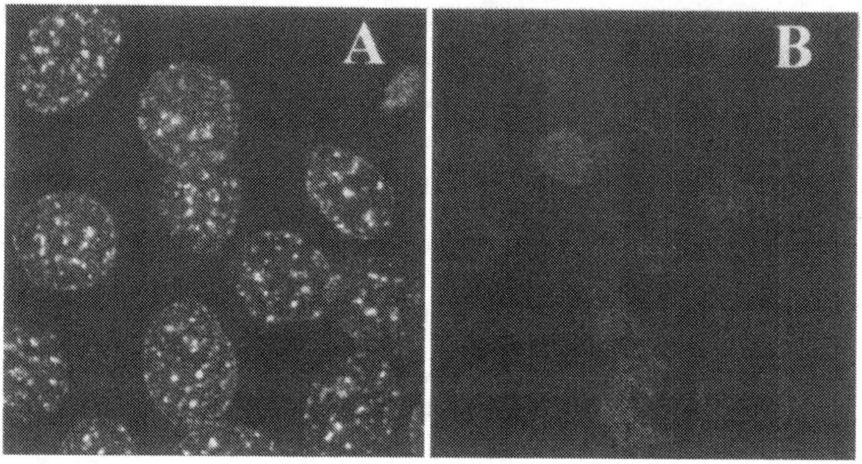

Figure 2. $HP1^{Hs\alpha}$ expression and localization in breast cancer cells. (A) MCF-7 cells (poorly invasive/non-metastatic) show $HP1^{Hs\alpha}$ localization at the nuclear envelope and heavy punctate staining throughout the nucleus. (B) MB-231 cells (highly invasive/metastatic) show little $HP1^{Hs\alpha}$ staining. Cells were methanol fixed, treated with a polyclonal anti-$HP1^{Hs\alpha}$ antibody and a rhodamine-conjugated goat anti-mouse IgG, and analyzed by confocal fluorescent microscopy.

Interestingly, the MDA-MB-231 cell line is very heterogeneous with respect to cellular morphology. With continuing passage of this cell line in culture, there is a gradual increase in large round epithelial-like cells. These epithelial-like cells are strongly positive for $HP1^{Hs\alpha}$ expression and cannot invade a collagen IV/laminin/gelatin matrix *in vitro*, further demonstrating that a down-regulation in $HP1^{Hs\alpha}$ expression is correlated with an increase in *in vitro* invasive potential (unpublished observations). These results also demonstrate that the $HP1^{Hs\alpha}$ gene is not deleted from the genome of MDA-MB-231 cells. Furthermore, the decrease in $HP1^{Hs\alpha}$ expression is not due to mutation, as the DNA sequence of the entire $HP1^{Hs\alpha}$ coding region and approximately 500 bp upstream of the putative transcription start site in MDA-MB-231 cells and genomic clones obtained from a placental library was identical to MCF-7 cells (unpublished observations). The regulation of $HP1^{Hs\alpha}$ in breast cancer cells is currently unknown. However, the lack of mutations or deletions in $HP1^{Hs\alpha}$ suggest that the regulation of $HP1^{Hs\alpha}$ transcription is through an epigenetic mechanism such as CpG methylation or histone deacetylation. Conversely, regulation of $HP1^{Hs\alpha}$ transcription in breast cancer cells maybe mediated by an epistatic mechanism through modulation of repressor and/or enhancer transcription factors that bind to the $HP1^{Hs\alpha}$ promoter. Interestingly, differential display analysis identified two zinc finger proteins (Nil-2/ZEB/AREB6 and LD5-1) whose expression was up-regulated in MDA-MB-231 cells, however it is not known whether these transcription factors are involved in $HP1^{Hs\alpha}$ regulation (79).

Identification and functional analysis of the *HP1*$^{Hs\alpha}$ promoter is currently underway.

The reduction of *HP1*$^{Hs\alpha}$ mRNA and protein expression in invasive/metastatic breast cancer cell lines is specific only for this ortholog of the hHP1-like family of gene silencing proteins. Using RT-PCR analysis with *HP1*$^{Hs\beta}$ and *HP1*$^{Hs\gamma}$-specific primers, very little difference in expression was observed between poorly invasive and invasive/metastatic breast cancer cell lines (80). These results not only implicate HP1$^{Hs\alpha}$ involvement in breast cancer invasion and metastasis, but also suggest that the different HP1-like proteins are functionally distinct.

The down-regulation in *HP1*$^{Hs\alpha}$ mRNA and protein expression was not limited to MDA-MB-231 cells, but was also down-regulated in other invasive/metastatic breast cancer cell lines (MDA-MB-435 and Hs578T), but not in T47D, a poorly invasive/non-metastatic breast cancer cell line (79, 80). A down-regulation in *HP1*$^{Hs\alpha}$ mRNA expression was also observed in an invasive, cisplatin-resistant ovarian cancer cell line (A2780 C20), compared to the poorly invasive parental cell line (A2780 PAR), by microarray gene chip analysis, suggesting a more general role for HP1$^{Hs\alpha}$ in the acquisition of an invasive/metastatic phenotype (unpublished observations). Down-regulation of HP1$^{Hs\alpha}$ expression and its correlation with an invasive/metastatic phenotype in other cancers needs to be further characterized.

Reduced HP1$^{Hs\alpha}$ Expression in Metastatic Tumor Tissues from Breast Cancer Patients

In a recent study, the expression of HP1$^{Hs\alpha}$ was evaluated in clinically relevant breast cancer tumor tissues to determine whether the down-regulation of HP1$^{Hs\alpha}$ was limited to cultured breast cancer cell lines (80). Normal mammary tissue and tissue sections from lobular and ductal carcinomas *in situ*, invasive lobular and ductal carcinoma, and metastatic tumors from distant sites (lung, skin, omentum, and lymph node) were analyzed immunohistochemically for HP1$^{Hs\alpha}$ expression. The percentage of tumor cells staining positive and the intensity of staining was used to score HP1$^{Hs\alpha}$ expression. All of the tumor tissues from non-metastatic tumors contained greater than 80% HP1$^{Hs\alpha}$ immunoreactive tumor cells with an intensity of staining of up to 3+ (with 3+ being the highest staining observed). In contrast, seven of nine tissues from distant metastases contained up to 95% less HP1$^{Hs\alpha}$ immunoreactive tumor cells (table 1). The reduction in HP1$^{Hs\alpha}$ expression in tumors from distant metastatic lesions is consistent with the data generated from cultured breast cancer cell lines with an invasive/metastatic phenotype.

Table 1. $HP1^{Hs\alpha}$ protein expression in primary and metastatic breast carcinoma tissues.

Tissue	% Cells Staining[a]	Intensity[b]
Ductal carcinoma *in situ*	>80	1-3+
Lobular carcinoma *in situ*	>90	2-3+
Invasive ductal carcinoma, Gr. 2	>80	1-3+
Invasive ductal carcinoma, Gr. 2-3	>95	2-3+
Invasive ductal carcinoma, Gr. 3	80	2-3+
Invasive lobular carcinoma, Gr. 1	>90	1-3+
Invasive lobular carcinoma, Gr. 2	>90	1-3+
Invasive lobular carcinoma, Gr. 3	>90	2-3+
Metastasis to lymph node	>90	2-3+
Metastasis to skin (scalp)	40	1-2+
Metastasis to lung	5-10	1+
Metastasis to lung	<5	1+
Metastasis to omentum	30	1-2+
Metastasis to soft tissue (hip)	20	1-2+
Metastasis to lung	<5	1+
Metastasis to lung	60	1-2+
Metastasis to lung	80	2-3+

[a] The percent of tumor cells showing $HP1^{Hs\alpha}$ immunoreactivity was estimated visually and quantitated in approximately 5% increments.
[b] Intensity of $HP1^{Hs\alpha}$ immunoreactivity was visually graded on a scale from 1+ to 3+, with 3+ representing the highest intensity.

Several reasons can be proposed to explain the lack of a reduction in $HP1^{Hs\alpha}$ expression in two of the nine distant metastatic tumors analyzed. The high $HP1^{Hs\alpha}$ expression in one lung metastasis could be indicative of a tumor that has arrested or reverted to a non-metastatic phenotype. Secondly, the high $HP1^{Hs\alpha}$ expression observed in tumor cells in a lymph node metastatic lesion may not represent a true metastasis. Through localized trauma, cancer cells (whether *in situ* or invasive) could be transferred by draining lymph to a regional lymph node where they could be trapped and grow. Thirdly, it is possible that the putative epigenetic event that causes a reduction in $HP1^{Hs\alpha}$ expression occurs very late in metastatic progression and may not have occurred at the time of observation in the majority of lymph node tumor cells. A subsequent genetic alteration leading to the reduction in $HP1^{Hs\alpha}$ expression could then induce those lymph node tumor cells to become more aggressive and allow them to reach and survive in non-lymphoid organs such as lung and bone. This could explain why all patients with lymph node metastases may not develop distant metastases and die from their disease. Fourthly, it is also conceivable that not all mechanisms of invasive and metastatic progression in breast cancer cells involve the down-regulation of $HP1^{Hs\alpha}$ expression. An expanded study to analyze $HP1^{Hs\alpha}$ expression in primary and metastatic breast cancer tumors and patient survival is currently underway.

Experimental Expression of $HP1^{Hs\alpha}$ in an Invasive/Metastatic Breast Cancer Cell Line

It was not known whether the down-regulation of $HP1^{Hs\alpha}$ expression that was observed only in invasive/metastatic breast cancer cells and not in poorly invasive/non-metastatic cells was directly influencing invasion of these cells or whether it was a consequence of other genetic alterations. Thus, to determine whether re-expression of $HP1^{Hs\alpha}$ in invasive/metastatic breast cancer cells could directly affect invasion, mammalian expression vectors containing a green fluorescent protein (*GFP*) gene alone or fused with the $HP1^{Hs\alpha}$ gene (*GFP-*$HP1^{Hs\alpha}$) were constructed and stably transfected into MDA-MB-231 cells. *GFP-*$HP1^{Hs\alpha}$ transfected MDA-MB-231 cells showed GFP fluorescence that was localized to the nucleus and was primarily excluded from the nucleoli, consistent with other reports (61, 68). In contrast, *GFP* transfected MDA-MB-231 cells had GFP fluorescence throughout the cell with the majority of fluorescence present in the cytoplasm. These results demonstrate that the GFP-$HP1^{Hs\alpha}$ fusion protein is functional, at least with respect to nuclear localization, and that GFP did not appear to interfere with chromatin binding (based on exclusion from the nucleoli).

Control GFP and GFP-$HP1^{Hs\alpha}$ expressing MDA-MB-231 were assessed for their ability to invade a collagen IV/laminin/gelatin matrix *in vitro*. When compared to untransfected or GFP expressing MDA-MB-231 cells, a 50 percent reduction in *in vitro* invasion was observed, demonstrating that re-expression of $HP1^{Hs\alpha}$ can modulate the invasive potential of MDA-MB-231 cells (figure 3). The metastatic potential of GFP-$HP1^{Hs\alpha}$ expressing MDA-MB-231 cells *in vivo* is currently being analyzed by orthotopic injection into the mammary fat pads of SCID mice. These experiments will determine whether $HP1^{Hs\alpha}$ can be classified as a breast cancer metastasis suppressor gene.

ROLE OF $HP1^{HS\alpha}$ IN THE BREAST CANCER INVASIVE/METASTATIC PHENOTYPE

Our observations are the first to provide direct evidence that down regulation of $HP1^{Hs\alpha}$ expression contributes to breast cancer cell invasion. The mechanism by which $HP1^{Hs\alpha}$ can inhibit invasion is not currently known, however several possibilities can be hypothesized based on the current knowledge of HP1 function in *Drosophila* and mammalian cells.

$HP1^{Hs\alpha}$-mediated Gene Silencing in Breast Cancer

The literature overwhelmingly demonstrates that HP1 and HP1-like proteins are involved in the transcriptional repression of gene expression. We propose that a reduction in $HP1^{Hs\alpha}$ in breast cancer cells leads to a loss of silencing of a gene or cassette of genes necessary for invasion and metastasis (Figure 4a). Such putative "target" genes could potentially be metastasis suppressor genes or genes that facilitate an epithelial to mesenchymal transformation (EMT). Cells that have undergone EMT have an aggressive invasive/metastatic phenotype (81, 82). To date, $HP1^{Hs\alpha}$-mediated silencing has been demonstrated by DNA tethering experiments or by monitoring the expression of transgenes inserted into heterochromatin. Less clear is whether $HP1^{Hs\alpha}$ can silence genes in their native chromosomal locations. No known natural gene targets of HP1 have been identified in *Drosophila* or mammalian cells. In *Drosophila*, HP1 consistently colocalizes to specific euchromatic regions of polytene chromosomes. This observation suggests that these regions contain a HP1 "targeting signal" and presumably genes in these regions are susceptible to HP1-mediated silencing.

In mammalian cells, targeting of $HP1^{Hs\alpha}$ to specific genes may occur through interactions with the KAP-1 co-repressor (59, 64, 65). KAP-1 is a universal corepressor for the KRAB (Krüppel-associated box) domain found in Cys_2His_2-type zinc finger transcription factors (84, 85). Recruitment of $HP1^{Hs\alpha}$ to specific genes via KRAB domain containing zinc finger protein-KAP-1 complexes could lead to silencing of these genes by the induction of a transcriptionally "closed" chromatin conformation (heterochromatinized). It is therefore possible that gene-specific targeting of mHP1s could be mediated through the binding of different KRAB domain containing transcription factors. Evidence to support this hypothesis has been reported by Brown and colleagues in which the Ikaros protein, a KRAB domain containing transcription factor required for normal lymphoid cell development, colocalized with transcriptionally inactive lymphoid-specific genes, M31, and centromeric DNA (86). Taken together, their data suggest that the recruitment and physical interaction of genes to centromeric heterochromatin can mediate transcriptional repression.

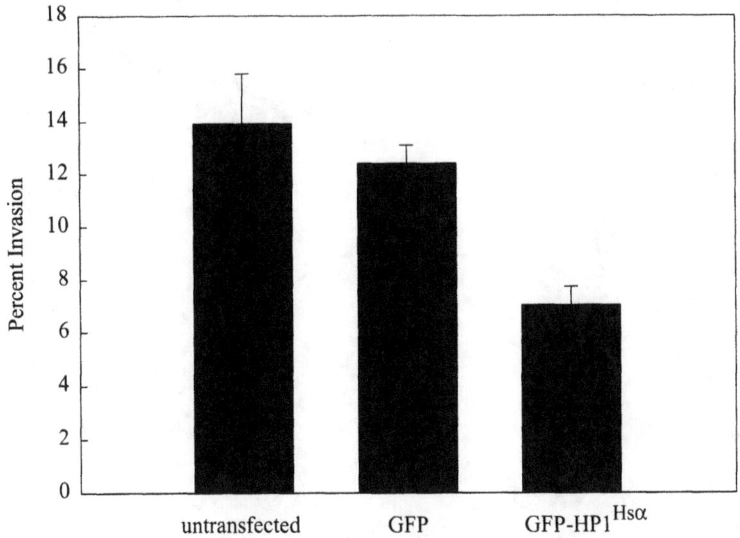

Figure 3. HP1$^{Hs\alpha}$ re-expression in MDA-MB-231 cells decreases invasive potential. MDA-MB-231 cells were stably transfected with a GFP-HP1 fusion protein mammalian expression vector and analyzed for the ability to invade a collagen IV/laminin/gelatin matrix in vitro.

Alternatively, HP1$^{Hs\alpha}$ may be required for the normal expression of a gene, or cassette of genes, located within heterochromatin. The requirement of HP1 for the transcriptional activation of genes located within heterochromatin has recently been demonstrated (71). Although Festenstein and colleagues have demonstrated that M31 can silence expression of a CD2 transgene located in centromeric heterochromatin, they have also shown that over expression of M31 can suppress silencing of a CD2 transgene that is not located in centromeres (70). These results suggest that the function of M31 varies depending on the chromosomal location of the targeted gene (70). The seemingly contrasting effects of HP1-mediated gene silencing and enhancement of gene expression appears to be dependent on the local chromosomal environment in which the gene resides or contacts. What genes and in which chromosomal context the reduction of HP1$^{Hs\alpha}$ expression in breast cancer cells modulates cellular invasive and metastatic potential is currently unknown and will be the basis for future studies.

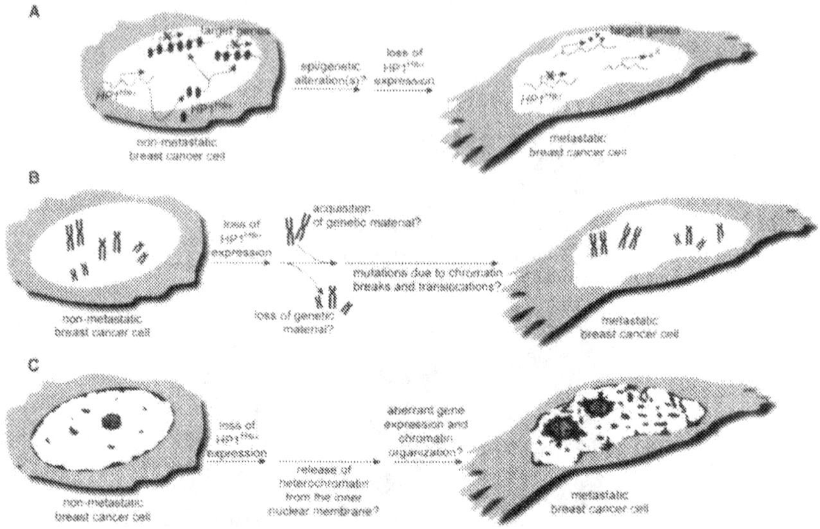

Figure 4. Working models for the role of HP1$^{Hs\alpha}$ in breast cancer invasion and metastatic phenotype. (A) HP1$^{Hs\alpha}$ mediates silencing of genes involved in invasion and metastasis; (B) HP1$^{Hs\alpha}$ mediates genomic stability; and (C) HP1$^{Hs\alpha}$ mediates genomic organization and gene expression. These models are not mutually exclusive.

HP1$^{Hs\alpha}$-mediated genomic stability in breast cancer

A second possible function for the reduction in HP1$^{Hs\alpha}$ expression in modulating breast cancer invasive/metastatic potential may be in altering genomic stability. One thing that cancers have in common is an unstable genome leading to chromosomal aneuploidy, inversions, translocations, and DNA amplifications and deletions (87). Many cancer cells that demonstrate size heteromorphisms in chromosomal C-bands (equated with heterochromatin) are also associated with numerical and structural chromosomal changes (17). These studies observed that chromosomal breakage frequently occurs in or near heterochromatic regions that could lead to mutations, amplifications, and deletions during replication and homologous recombination. We propose that a reduction in HP1$^{Hs\alpha}$ expression may lead to variations in heterochromatin content, which in turn could contribute to chromosome abnormalities (Figure 4b). While no experimental evidence linking HP1$^{Hs\alpha}$ to chromosomal abnormalities in humans has been reported to date, in *S. pombe*, mutations in the *swi6* gene resulted in an increased rate of chromosome loss (88). This loss was attributed to a decreased ability of centromeres to migrate to the poles of the spindle during anaphase (89). The *swi6* gene in yeast encodes an HP1-like chromodomain protein (46% identical to HP1) that colocalizes to centromeres and telomeres (89). The *swi6* gene product is required for transcriptional repression of silent mating-type loci and centromeres (88, 90). These data, coupled with the observations that neoplastic cells frequently have abnormal karyotypes, suggest that $HP1^{Hs\alpha}$ may be involved in genomic stability.

$HP1^{Hs\alpha}$-mediated Genomic Organization in Breast Cancer

Accompanying genetic instability in cancers are alterations in nuclear morphology (reviewed in 91), which are thought to be mediated in part by changes in the composition of nuclear matrix proteins (reviewed in 26). The nuclear matrix is defined as the nuclear structure that remains following salt extraction of nuclease treated nuclei. The nuclear matrix is composed of the internal nuclear matrix, residual nucleoli, and the nuclear pore-lamina complex of the nuclear membrane (92). Growing evidence suggests that the nuclear matrix not only defines nuclear shape, but also plays an active role in chromatin organization, replication, and gene expression (93, 94). Recently, the lamin B receptor (LBR), an integral inner nuclear membrane protein, has been shown to bind to $HP1^{Hs\alpha}$ (53, 63). The interaction of LBR with $HP1^{Hs\alpha}$ maybe a putative mechanism by which a large fraction of heterochromatin is localized to the periphery of the nucleus (63, 95). We propose that a direct destabilization of heterochromatic regions from the inner nuclear membrane via a reduction in $HP1^{Hs\alpha}$ expression, and thereby a loss in appropriate genomic organization, may be one mechanism by which gene expression is altered in invasive/metastatic breast cancer cells (Figure 4c). Indirect evidence to support this hypothesis stems from observations that 17β-estradiol treatment of MCF-7 breast cancer cells increased the irregularity of the inner nucleoplasmic border which corresponded with a decrease in condensed heterochromatin in these cells (23). However, it is not currently known whether $HP1^{Hs\alpha}$ expression is modulated by17β-estradiol treatment.

The proposed models for the involvement of $HP1^{Hs\alpha}$ in breast cancer invasion and metastasis are not mutually exclusive and could all, to some degree, contribute to the invasive/metastatic phenotype. Furthermore, these models demonstrate the enormous potential that modulation of $HP1^{Hs\alpha}$ expression can have on inducing subsequent genetic alterations in breast cancer cells. Since breast cancer cell lines that have undergone EMT also have reduced levels of $HP1^{Hs\alpha}$, it is not known whether down-regulation of $HP1^{Hs\alpha}$ is a putative "master switch" that causes EMT or rather is a consequence of EMT. Nevertheless, a down-regulation in $HP1^{Hs\alpha}$ expression by breast cancer cells contributes to the acquisition of an aggressive invasive/metastatic phenotype.

ACKNOWLEDGEMENTS

The authors would like to express their gratitude to Dr. Lori Wallrath for critically reviewing this manuscript. This work was supported by the University of Iowa Carver Collaborative Award and NIH/NCI CA59702.

REFERENCES

1. Weiler KS, Wakimoto BT. Heterochromatin and gene expression in *Drosophila.* Ann. Rev. Gen. 29: 577-605, 1995.
2. Hennig W. Heterochromatin. Chromosoma. 108: 1-9, 1999.
3. Klein CB, Costa M. DNA methylation, heterochromatin, and epigenetic carcinogens. Mutation Res. 386: 163-180, 1997.
4. Riggs AD, Pfeifer GP. X-chromosome inactivation and cell memory. Trends Genet. 8: 169-174, 1992.
5. Lyon MF. X-chromosome inactivation: a repeat hypothesis. Cytogenet. Cell. Genet. 80: 133-137, 1998.
6. Brown PH, Lippman SM. Chemoprevention in breast cancer. Breast Can. Res. Treat. 62: 1-17, 2000.
7. Baak JP, Van Dop H, Kurver PH, Hermans J. The value of morphometry to classic prognisticators in breast cancer. Cancer. 56: 374-.82, 1985.
8. Stenkvist B, Bengtsson E, Dahlqvist B, Eklund G, Eriksson O, Jarkrans T, Nordin B. Predicting breast cancer recurrence. Cancer. 50: 2884-2893, 1982.
9. Komitowski DD, Hart MM, Janson CP. Chromatin organization and breast cancer prognosis. Cancer. 72: 1239-1246, 1993.
10. Aubele M, Auer G, Falkmer U, Voss A, Rodenacker K, Jütting U, Höfler H. Identification of a low-risk group of stage I breast cancer patients by cytometrically assessed DNA and nuclear texture parameters. J. Pathol. 177: 377-384, 1995.
11. Komitowski D, Janson C. Quantitative features of chromatin structure in the prognosis of breast cancer. Cancer. 65: 2725-2730, 1990.
12. Umbricht C, Oberholzer M, Gschwind R, Christen H, Torhorst J. Prognostic significance (relapse, non-relapse) of nuclear shape parameters in lymph node negative breast cancer. Anal. Cell. Pathol. 1: 11-23, 1989.
13. Weyn B, van de Wouwer G, van Daele A, Scheunders P, van Dyck D, van Marck E, Jacob W. Automated breast tumor diagnosis and grading based on wavelet chromatin texture description. Cytometry. 33: 32-40, 1998.
14. Larsimont D, Kiss R, d'Olne D, de Launoit Y, Mattheiem W, Paridaens R, Pasteels, JL, Gompel C. Correlation between nuclear cytomophometric parameters and estrogen receptor levels in breast cancer. Cancer. 63: 2162-2168, 1989.
15. Dutrillaux B, Gerbault-Seureau M, Zafrani B. Characterization of chromosomal anomalies in human breast cancer: a comparison of 30 paradiploid cases with few chromosome changes. Cancer Genet. Cytogenet. 49: 203-217, 1990.
16. Adhvaryu SG, Rawal UM. C-band heterochromatin variants in individuals with neoplastic disorders: carcinoma of breast and ovary. Neoplasma. 38: 379-384, 1991.
17. Atkin NB, Brito-Babapulle V. Heterochromatin polymorphism and human cancer. Cancer Genet. Cytogenet. 3: 261-272, 1981.
18. Greenough RB. Varying degrees of malignancy in cancer of the breast. Cancer Res. 9: 453-463, 1925.

19. Doudkine A, Macaulay M, Poulin N, Palcic B. Nuclear texture measurements in image cytometry. Pathologica. 87: 286-299, 1995.
20. Berger R, Bernheim A, Kristoffersson U, Mitelman F, Olsson H. C-band heteromorphism in breast cancer patients. Cancer Genet. Cytogenet. 18: 37-42, 1985.
21. Sivakumaran, TA, Ghose S, Kumar H, Singha U, Kucheria K. Absence of pericentromeric heteorchromatin (9qh-) in a patient with bilateral retinoblastoma. Acta Genet. Med. Gemellol. 46: 193-198, 1977.
22. Kendall FM, Wu CT, Giaretti W, Nicolini CA. Multiparameter geometric and densitometric analysis of the G0-G1 transition of WI-38 cells. J. Histochem. Cytochem. 7: 724-729, 1977.
23. Losa GA, Graber R, Baumann G, Nonnenmacher TF. Steroid hormones modify nuclear heterochromatin structure and plasma membrane enzyme of MCF-7 cells. A combined fractal, electron microscopical and enzymatic analysis. Eur. J. Histochem. 42: 21-29, 1998.
24. Davie JR, Samuel SK, Spencer VA, Holth LT, Chadee DN, Peltier CP, Sun JM, Chen HY, Wright JA. Organization of chromatin in cancer cells: role of signalling pathways. Biochem. Cell Biol. 77: 265-275, 1999.
25. Nickerson JA. Nuclear dreams: the malignant alteration of nuclear architecture. J. Cell. Biochem. 70: 172-180, 1998.
26. Pienta KJ, Ward WS. An unstable nuclear matrix may contribute to genetic instability. Med. Hypotheses. 42: 45-52, 1994.
27. Boyer O, Zhao JC, Cohen JL, Depetris D, Yagello M, Lejeune L, Burel S, Mattei MG, Klatzmann D. Position-dependent variegation of a CD4 minigene with targeted expression to mature CD4+ T cells. J. Immunol. 159: 3383-3390, 1997.
28. Muller HJ. Types of visible variations induced by X-rays in *Drosophila*. J. Genetics. 22: 299-335, 1930.
29. Wallrath LL. Unfolding the mysteries of heterochromatin. Curr. Opin. Gen. Dev. 8: 147-153, 1998.
30. Eissenberg JC, Elgin SCR. The HP1 protein family: getting a grip on chromatin. Curr. Opin. Gen. Dev. 10: 204-210, 2000.
31. James TC, Elgin SCR. Identification of a nonhistone chromosomal protein associated with heterochromatin in *Drosophila melanogaster* and its gene. Mol. Cell. Biol. 6: 3862-3872, 1986.
32. Singh PB, Miller JR, Pearce J, Kothary R, Burton RD, Paro R, James TC, Gaunt SJ. A sequence motif found in a *Drosophila* heterochromatin protein is conserved in animals and plants. Nucl. Acids Res. 19: 789-794, 1991.
33. Aasland R, Stewart AF. The chromo shadow domain, a second chromo domain in heterochromatin-binding protein 1, HP1. Nucl. Acids Res. 23: 3168-3173, 1995.
34. Platero JS, Hartnett T, Eissenberg JC. Functional analysis of the chromo domain of HP1. EMBO J. 14: 3977-3986, 1995.
35. Powers JA, Eissenberg JC. Overlapping domains of the heterochromatin-associated protein HP1 mediate nuclear localization and heterochromatin binding. J. Cell Biol. 120: 291-299, 1993.
36. Cowieson NP, Partidge JF, Allshire RC, McLaughlin PJ. Dimerisation of a chromo shadow domain and distinctions from the chromodomain as revealed by structural analysis. Curr. Biol. 10: 517-525, 2000.
37. Akhtar A, Zink D, Becker PB. Chromodomains are protein-RNA interaction modules. Nature. 407: 405-409.

38. Kellum R, Raff JW, Alberts BM. Heterochromatin protein 1 distribution during development and during the cell cycle in *Drosophila* embryos. J. Cell Sci. 108: 1407-1418, 1995.
39. James TC, Eissenberg JC, Craig C, Dietrich V, Hobson A, Elgin SCR. Distribtuion patterns of HP1, a heterochromatin-associated nonhistone chromosomal protein of Drosophila. Eur. J. Cell Biol. 50: 170-180, 1989.
40. Lu BY, Ma J, Eissenberg JC. Developmental regulation of heterochromatin-mediated gene silencing in *Drosophila*. Development. 125: 2223-2234, 1998.
41. Eissenberg JC, James TC, Foster-Hartnett DM, Hartnett T, Ngan, V, Elgin SCR. Mutation in a heterochromtian-specific chromosomal protein is associated with suppression of position-effect variegation in *Drosophila melanogaster*. Proc. Natl. Acad. Sci. USA. 87: 9923-9927, 1990.
42. Cryderman DE, Cuaycong MH, Elgin SCR, Wallrath LL. Characterization of sequences associated with position-effect variegation at pericentric sites in *Drosophila* heterochromatin. Chromosoma. 107: 277-285, 1998.
43. Eissenberg JC, Hartnett T. A heat shock-activated cDNA rescues the recessive lethality of mutations in the heterochromatin-associated protein HP1 of *Drosophila melanogaster*. Mol. Gen. Genet. 240: 333-338, 1993.
44. Eissenberg JC, Morris GD, Reuter G, Hartnett T. The heterochromatin-associated protein HP-1 is an essential protein in *Drosophila* with dosage-dependent effects on position-effect variegation. Genetics. 131: 345-352, 1992.
45. Locke J, Kotarski MA, Tartof KD. Dosage-dependent modifiers of position effect variegation in *Drosophila* and a mass action model that explains their effect. Genetics. 120: 181-198, 1988.
46. Le Douarin B, Nielsen AL, Garnier JM, Ichinose H, Jeanmougin F, Losson R, Chambon P. A possible involvement of TIF1α and TIF1β in the epigenetic control of transcription by nuclear receptors. EMBO J. 15: 6701-6715, 1996.
47. Peterson K, Wang G, Horsley D, Richardson JC, Sapienza C, Latham KE, Singh PB. The *M31* gene has a complex developmentally regulated expression profile and may encode alternative protein products that possess diverse subcellular localisaton patterns. J. Exp. Zoology. 280: 288-303, 1998.
48. Nicol L, Jeppesen P. Human autoimmune sera recognize a conserved 26 kD protein associated with mammalian heterochromatin that is homologous to heterochromatin protein 1 of *Drosophila*. Chromosome Res. 2: 245-253, 1994.
49. Wreggett KA, Hill F, James PS, Hutchings A, Butcher GW, Singh PB. A mammalian homologue of *Drosophila* heterochromatin protein 1 (HP1) is a component of constitutive heterochromatin. Cytogenet. Cell. Genet. 66: 99-103, 1994.
50. Horsley D, Hutchings A, Butcher GW, Singh PB. M32, a murine homologue of *Drosophila* heterochromatin protein 1 (HP1), localizes to euchromatin within interphase nuclei and is largely excluded from constitutive heterochromatin. Cytogenet. Cell. Genet. 73: 308-311, 1996.
51. Miyado K, Sato M, Kimura M. Differential expression of mRNAs for M31 and M32, murine homologues of *Drosophila* heterochromatin protein 1 (HP1), during murine embryogenesis. Biochem. Mol. Biol. Intl. 44: 1051-1058, 1998.
52. Brasher SV, Smith BO, Fogh RH, Nietlispach D, Thiru A, Nielsen PR, Broadhurst RW, Ball LJ, Murzina NV, Laue ED. The structure of mouse HP1 suggests a unique mode of single peptide recognition by the shadow chromo domain dimmer. EMBO J. 19: 1587-1597, 2000.

53. Ye Q, Worman HJ. Interaction between an integral protein of the nuclear envelope inner membrane and human chromodomain proteins homologous to *Drosophila* HP1. J. Biol. Chem. 271: 14653-14656, 1996.
54. Saunder WS, Chue C, Goebl M, Craig C, Clark RF, Powers JA, Eissenberg JC, Elgin SC, Rothfield NF, Earnshaw WC. Molecular cloning of a human homologue of *Drosophila* heterochromatin protein HP1 using anti-centromere autoantibodies with anti-chromo specificity. J. Cell Sci. 104: 573-582, 1993.
55. Furuta K, Chan EKL, Kiyosawa K, Reimer G, Luderschmidt C, Tan EM. Heterochromatin protein $HP1^{Hs\beta}$ (p25β) and its localization with centromeres in mitosis. Chromosoma. 106: 11-19, 1997.
56. Furuta K, Hildebrandt B, Matsuoka S, Kiyosawa K, Reimer G, Luderschmidt C, Chan EKL, Tan EM. Immunological characterization of heterochromatin protein p25β autoantibodies and relationship with centromere autoantibodies and pulmonary fibrosis in systemic scleroderma. J. Mol. Med. 76: 54-60, 1998.
57. Muro Y, Yamada T, Iwai T, Sugimoto K. Epitope analysis of chromo antigen and clinical features in a subset of patients with anti-centromere antibodies. Mol. Biol. Reports. 23: 147-151, 1996.
58. Minc E, Allory Y, Worman HJ, Courvalin JC, Buendia B. Localization and phosphorylation of HP1 proteins during the cell cycle in mammalian cells. Chromosoma. 108: 220-234, 1999.
59. Nielsen AL, Ortiz JA, You J, Oulad-Abdelghani M, Khechumian R, Gansmuller A, Chambon P, Losson R. Interaction with members of the heterochromatin protein 1 (HP1) family and histone deacetylation are differentially involved in transcriptional silencing by members of the TIF1 family. EMBO J. 18: 6385-6395, 1999.
60. Ma J, Hwang KK, Worman HJ, Courvalin JC, Eissenberg JC. Expression and functional analysis of three isoforms of human heterochromatin-associated protein HP1 in *Drosophila*. Chromosoma. In press.
61. Seeler JS, Marchio A, Sitterlin D, Transy C, Dejean A. Interaction of SP100 with HP1 proteins: a link between the promyelocytic leukemia-associated nuclear bodies and the chromatin compartment. Proc. Natl. Acad. Sci. USA. 95: 7316-7321, 1998.
62. Lehming N, Le Saux A, Schüller J, Ptashne M. Chromatin components as part of a putative transcriptional repressing complex. Proc. Natl. Acad. Sci. USA. 95: 7322-7326, 1998.
63. Ye Q, Callebaut I, Pezhman, A, Courvalin JC, Worman HJ. Domain-specific interactions of human HP1-type chromodomain proteins and inner nuclear membrane protein LBR. J. Biol. Chem. 272: 14983-14989, 1997.
64. Ryan RF, Schultz DC, Ayyanathan K, Singh PB, Friedman JR, Fredericks WJ, Rauscher FJ, III. KAP-1 corepressor protein interacts and colocalizes with heterochromatic and euchromatic HP1 proteins: a potential role for Krüppel-associated box-zinc finger proteins in heterochromatin-mediated gene silencing. Mol. Cell. Biol. 19: 4366-4378, 1999.
65. Lechner MS, Begg GE, Speicher DW, Rauscher FJ, III. Molecular determinants for targeting heterochromatin protein 1-mediated gene silencing: direct chromoshadow domain-KAP-1 corepressor interaction is essential. Mol. Cell. Biol. 20: 6449-6465, 2000.
66. Murzina N, Verreault A, Laue E, Stillman B. Heterochromatin dynamics in mouse cells: interaction between chromatin assembly factor 1 and HP1 proteins. Mol. Cell. 4: 529-540, 1999.

67. Ainsztein AM, Kandels-Lewis SE, Mackay AM, Earnshaw WC. INCENP centromere and spindle targeting: identification of essential conserved motifs and involvement of heterochromatin protein HP1. J. Cell Biol. 143: 1763-1774, 1998.
68. Yamada T, Fukuda R, Himeno M, Sugimoto K. Functional domain structure of human heterochromatin protein $HP1^{Hs\alpha}$: involvement of internal DNA-binding and C-terminal self-association domains in the formation of discrete dots in interphase nuclei. J. Biochem. 125: 832-837, 1999.
69. Wang G, Ma A, Chow CM, Horsley D, Brown NR, Cowell IG, Singh PB. Conservation of heterochromatin protein 1 function. Mol. Cell. Biol. 20: 6970-6983, 2000.
70. Festenstein R, Sharghi-Namini S, Fox M, Roderick K, Tolaini M, Norton T, Saveliev A, Kioussis D, Singh P. Heterochromatin protein 1 modifies mammalian PEV in a dose- and chromosomal-context-dependent manner. Nat. Gen. 23: 457-461, 1999.
71. Lu BY, Emtage PCR, Duyf BJ, Hilliker AJ, Eissenberg JC. Heterochromatin protein 1 is required for the normal expression of two heterochromatin genes in Drosophila. Genetics. 155: 699-708, 2000.
72. Moustafa AS, Nicolson GL. Breast cancer metastasis-associated genes: prognostic significance and therapeutic implications. Oncology Res. 9: 505-525, 1997.
73. Yokota J. Tumor progression and metastasis. Carcinogenesis. 21: 497-503, 2000.
74. Bechmann MW, Niederacher D, Schnürch HG, Gusterson BA, Bender HG. Multistep carcinogenesis of breast cancer and tumour heterogeneity. J. Mol. Med. 75: 429-439, 1997.
75. Hanahan D, Weinberg RA. The hallmarks of cancer. Cell. 100: 57-70, 2000.
76. Stracke ML, Liotta LA. Multi-step cascade of tumor cell metastasis. In vivo. 6: 309-316, 1992.
77. Woodhouse EC, Chuaqui RF, Liotta LA. General mechanisms of metastasis. Cancer. 80: 1529-1537, 1997.
78. Hendrix MJC, Seftor EA, Kirschmann DA, Seftor REB. Molecular biology of breast cancer metastasis: molecular expression of vascular markers by aggressive breast cancer cells. Breast Can. Res. 2: 417-422, 2000.
79. Kirschmann DA, Seftor EA, Nieva DRC, Mariano EA, Hendrix MJC. Differentially expressed genes associated with the metastatic phenotype in breast cancer. Breast Can. Res. Treat. 55: 127-136, 1999.
80. Kirschmann, DA, Lininger RA, Gardner LMG, Seftor EA, Odero VA, Ainsztein AM, Earnshaw WC, Wallrath LL, Hendrix MJC. Down-regulation of $HP1^{Hs\alpha}$ expression is associated with the metastatic phenotype in breast cancer. Can. Res. 60: 3359-3363, 2000.
81. Thompson EW, Paik S, Brunner N, Sommers CL, Zugmaier G, Clarke R, Shima TB, Torri J, Donahue S, Lippman ME, Martin GR, Dickson RB. Assocition of inceased basement membrane invsivness with absence of estrogen receptor and expressio nof vimentin in human breast cancer cell lines. J. Cell. Physiol. 150: 534-544, 1992.
82. Boyer B, Thiery JP. Epithelial-mesenchyme interconversion as example of epithelial plasticity. APMIS. 101: 257-268, 1993.
83. Fanti L, Dorer DR, Berloco M, Henikoff S, Pimpinelli S. Heterochromatin protein 1 binds transgene arrays. Chromosoma. 107: 286-292, 1998.
84. Margolin JF, Friedman JR, Meyer WK, Vissing H, Thiesen HJ, Rauscher FJ, III. Kruppel-associated boxes are potent transcriptional repression domains. Proc. Natl. Acad. Sci. USA. 91: 4509-4513, 1994.

85. Bellefroid EJ, Poncelet DA, Lecocq PJ, Revelant O, Martial JA. The evolutionarily conserved Kruppel-associated box domain defines a subfamily of eukaryotic multifingered proteins. Proc. Natl. Acad. Sci. USA. 88: 3608-3612, 1991.
86. Brown KE, Guest SS, Smale ST, Hahm K, Merkenschlager M, Fisher AG. Association of transcriptionally silent genes with Ikaros complexes at centromeric heterochromatin. Cell. 91: 845-854, 1997.
87. Dos Santos NR, Van Kessel AG. Chromosomal abnormalities: detection and implications for cancer development. Anticancer Res. 19: 4697-4714, 1999.
88. Allshire RC, Nimmo ER, Ekwall K, Javerzat JP, Cranston G. Mutations derepressing silent centromeric domains in fission yeast disrupt chromosome segregation. Genes Dev. 9: 218-233, 1995.
89. Ekwall K, Javerzat JP, Lorentz A, Schmidt H, Cranston G, Allshire R. The chromodomain protein swi6: a key component at fission yeast centromeres. Science. 269: 1429-1431, 1995.
90. Lorentz A, Ostermann K, Fleck O, Schmidt H. Switching gene swi6, involved in repression of silent mating-type loci in fission yeast, encodes a homologue of chromatin-associated proteins from *Drosophila* and mammals. Gene. 143: 139-143, 1994.
91. Bosman FT. The nuclear matrix in pathology. Virchows Arch. 435: 391-399, 1999.
92. Berezney R. The nuclear matrix: a heuristic model for investigating genomic organization and function in the cell nucleus. J. Cell. Biochem. 47: 109-123, 1991.
93. Berezney R, Mortillaro MJ, Ma H, Wei X, Samarabandu J. "The nuclear matrix: a structural milieu for genomic function". In *Nuclear Matrix: Structural and Functional Organization*. Berezney R, Jeon KW, Eds. San Diego, CA. Associated Press, 1995.
94. Pienta KJ, Getzenberg RH, Coffey DS. Cell structure and DNA organization. Crit. Rev. Eukaryotic Gene Express. 1: 355-385, 1991.
95. Goldberg, M, Harel A, Gruenbaum Y. The nuclear lamina: molecular organization and interaction with chromatin. Crit. Rev. Eukaryotic Gene Express. 9: 285-293, 1999.

Chapter 10

INHIBITION OF INVASION AND METASTASIS DURING SPECIFIC AMINO ACID RESTRICTION ASSOCIATED WITH METASTASIS SUPPRESSOR AND OTHER GENE CHANGES

Gary G. Meadows, Xiaokang Ge, Hui Zhang, Daniel R. Oros, Ya-Min Fu

Cancer Prevention and Research Center, Department of Pharmaceutical Sciences, College of Pharmacy, Washington State University, Pullman, WA 99164-6510

Abstract

Dietary tyrosine (Tyr) and phenylalanine (Phe) restriction dramatically decreases metastasis of a number of implanted murine tumors including melanoma, hepatocarcinoma, lung carcinoma, and leukemia without inducing host toxicity. Herein, we show that Tyr/Phe deprivation *in vitro* rapidly decreases the intracellular free Tyr concentration in A375 melanoma cells. The decreased availability of Tyr and Phe to melanoma, prostate, and breast cancer cells *in vitro* significantly inhibits their invasive ability. MKK4/SEK1 protein expression and/or phosphorylation increased in A375 melanoma and MDA-MB-231 breast cancer cells suggesting an important role for this metastasis suppressor gene in control of invasion during Tyr/Phe deprivation. Additionally, 14 other genes are differentially modulated by Tyr/Phe deprivation. These results suggest that specific intracellular free amino acids regulate the metastatic potential of tumor cells by modulating gene expression and function.

Keywords: metastasis suppressor gene, invasion, mitogen-activated protein kinase kinase 4 (MKK4)/stress-activated protein/Erk kinase 1 (SEK1), amino acid, tyrosine, phenylalanine, methionine, melanoma, breast cancer, and prostate cancer.

D.R. Welch (ed.), Cancer Metastasis – Related Genes, 191–208.

1. INTRODUCTION

It has become increasingly evident that diet and dietary factors can impact not only the incidence of cancer, but also modulate its spread. We have previously documented that restriction of tyrosine (Tyr) and phenylalanine (Phe) in the diet specifically modulates cell cycle protein expression to inhibit tumor growth and inhibit metastasis of pigmented B16BL6 melanoma in mice without toxicity to the host (1,2). There is very little information about how nutrients control the expression of mammalian genes, and there is even less known about the role of specific amino acids in regulation of gene expression in tumor cells. General amino acid starvation can increase, as well as decrease gene expression in rat hepatoma cells (3). Also in rat hepatoma cells, the specific limitation of Phe, methionine (Met), leucine (Leu), and tryptophan leads to changes in abundance of mRNA for 19 genes (4). Herein we present evidence that Tyr and Phe deprivation *in vitro* leads to a decrease in intracellular free Tyr concentration. This in turn is associated with inhibition of invasion, enhancement of the metastasis suppressor gene, mitogen-activated protein kinase kinase 4 (MKK-4)/stress-activated protein/Erk kinase 1 (SEK1), and alteration in the expression of other genes.

While most of our research has concentrated on the effects of Tyr/Phe restriction on melanoma, dietary restriction of these amino acids also inhibits growth and metastasis of a number of other cancers including L1210 leukemia (5), RT7-4bs hepatocarcinoma (6), and 3LL Lewis lung carcinoma (6). The growth inhibitory and anti-metastatic effects of dietary Tyr/Phe restriction are not due to body weight loss, decreased food intake, or general starvation (6-11) indicating a more specific mechanism associated with Tyr/Phe restriction.

Dietary restriction of Tyr/Phe, in addition to suppression of metastasis, also enhances the chemotherapeutic effect of L-dopa methylester against B16 melanoma *in vivo* and permits the anti-melanoma activity of vitamin C (7, 11, 12). Furthermore, it prevents the emergence of drug resistance during chemotherapy (12), and abolishes drug resistance of adriamycin-resistant B16BL6 melanoma cells and P388 leukemia cells to adriamycin *in vitro* (13). Tyr/Phe deprivation *in vitro* synergistically augments the chemotherapeutic effect of melphlan against A375 melanoma while at the same time protecting normal cells from the toxic effects of this drug (unpublished observations). Additionally, restriction of these amino acids enhanced the cytotoxicity of chimeric heparin-binding transforming growth factor-alpha and *Pseudomonas* exotoxin A against B16BL6 melanoma (14). These data indicate that restriction of Tyr/Phe could be a useful addition to conventional chemotherapeutic treatment of melanoma and possibly other cancers.

Metastasis is a multifactorial process, and tumor cells must complete a number of steps before establishing themselves at secondary sites. Elstad *et al.* (15) showed that B16BL6 melanoma cells isolated from mice that were fed a diet restricted in Tyr/Phe or cells cultured in media with reduced levels of these amino acids, were greatly impaired in their ability to establish pulmonary tumor nodules in experimental metastasis assays (1). B16BL6 melanoma cells with the suppressed metastatic phenotype also exhibit reduced invasion through Matrigel™ and growth factor reduced Matrigel™ (16, 17). The fact that the tumor cells do not immediately revert back to a metastatic phenotype when repleted with Tyr/Phe suggests that gene expression is altered.

In section 3, we show that the anti-invasive effects of Tyr/Phe deprivation *in vitro* extended to other human tumor systems. We present further evidence in section 4 that the metastasis suppressor gene, MKK4/SEK1, is expressed in melanoma as well as breast cancer cells, and that the protein expression and/or phosphorylation of the MKK4/SEK1 protein is enhanced by Tyr/Phe deprivation. Moreover, a number of other genes relevant to tumor invasion and metastasis are differentially expressed in response to Tyr/Phe deprivation and these are presented in section 5.

2. EFFECT OF TYR/PHE DEPRIVATION ON INTRACELLULAR TYR AND PHE LEVELS IN A375 MELANOMA

As discussed above dietary restriction of Tyr/Phe inhibits metastasis and the expression in the metastatic phenotype of B16BL6 melanoma cells. The exact trigger leading to the reduced metastatic potential is unknown. However it is well known that cells sense extracellular signals and as a result can modify metabolism and/or gene expression. A classical example of this is the response of the pancreas to glucose, insulin, and glucagon levels in the blood. Thus, we examined the effect of Tyr/Phe deprivation on the intracellular levels of these amino acids in melanoma and normal cells. The data in Figure 1 show that for

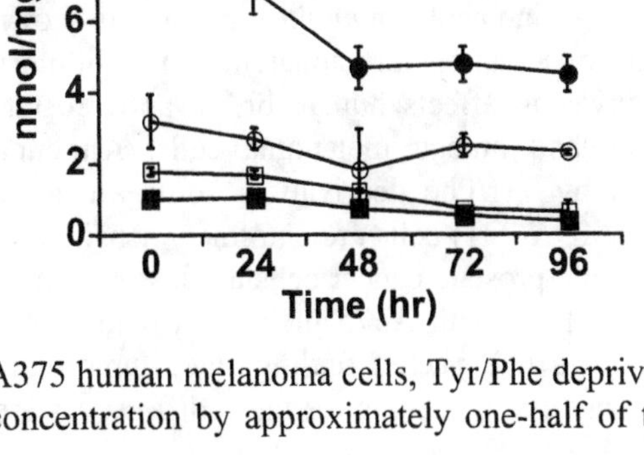

Figure 1. Intracellular free Tyr and Phe concentrations in A375 human melanoma and human infant foreskin fibroblasts (HIFF) cultured in Tyr/Phe-deprived media. Amino acids were measured by HPLC using o-phthaldialdehyde derivatization with fluorescence detection (18). Values are means of triplicate data points and bars are standard error.

A375 human melanoma cells, Tyr/Phe deprivation significantly decreases the Tyr concentration by approximately one-half of the starting concentration over time

($p < 0.05$), beginning 24 hr after deprivation. It is important to note that complete depletion of Tyr/Phe is not necessary to achieve an antimetastatic effect *in vivo*. In fact, dietary restriction of Tyr/Phe in mice lowers plasma Tyr by 33% and Phe by 21% (11). This suggests that melanoma cells are especially sensitive to alterations in the levels of these amino acids.

The Phe concentration did not significantly change over time ($p > 0.05$). This finding is consistent with the results in Table 1 indicating that deprivation of this amino acid alone does not have a major effect on *in vitro* invasion. In human infant foreskin fibroblasts (HIFF), a normal cell line, neither Tyr nor Phe concentrations significantly change over time ($p > 0.05$). In these cells the steady state levels of Tyr and Phe are lower than in the A375 melanoma cells. These data suggest that Tyr and Phe levels are tightly regulated in normal cells and support the hypothesis that the cellular and molecular changes associated with inhibition of melanoma invasion and metastasis are a consequence of decreased intracellular Tyr concentration. Moreover, melanoma cells become arrested in the G0/G1 phase of the cell cycle within 24 hr after Tyr/Phe limitation (2), further indicating the finely tuned response of these cells to changes in intracellular amino acid concentrations.

3. ROLE OF SPECIFIC AMINO ACID DEPRIVATION ON TUMOR INVASION

The metastatic phenotype of B16BL6 melanoma cell is suppressed and the altered phenotype is stable after culturing these cells in Tyr/Phe restricted media *in vitro* (1). Moreover, culturing B16BL6 melanoma cells in media low in Met, glutamine (Gln), Leu, or arginine does not induce the change in phenotype (1). Phe, Met, and Leu are essential amino acids and Gln and Tyr are nonessential amino acids. Unlike Tyr and Phe restriction, Gln and Leu restriction *in vitro* does not alter the metastatic phenotype of B16BL6 melanoma cells (1). Gln is a major source of nitrogen and energy in the body. The corresponding alpha keto acid, glutamate, is involved in the synthesis of a wide variety of other amino acids. Therefore, deprivation of this amino acid would be expected to have a major impact on the growth of melanoma cells. Leu was selected because it is an essential amino acid and also because it is a neutral amino acid like Tyr and Phe.

The effect of deprivation of these amino acids on *in vitro* invasion is shown in Table 1 and a number of important observations are apparent. The inhibitory activity of Tyr/Phe deprivation on invasion affects human breast and prostate cancer cell lines, as well as murine and human melanoma cells. Although invasion of all cell lines is inhibited by Tyr/Phe deprivation, some are more sensitive than others. MDA-MB-231 (MDA-231) cells are extremely sensitive to the deprivation of these amino acids while prostate cancer cells are less sensitive. The effect varies in the melanoma cell lines with A375 melanoma being more sensitive than SK-Mel-28. This is not an unexpected finding since tumors are highly heterogeneous. However, the underlying reasons for these differences are not known. Interestingly, in A375 and B16BL6 melanoma, Tyr deprivation has a greater effect on *in vitro* invasion than Phe deprivation. This is consistent with the results in Figure 1 indicating that intracellular Tyr, but not Phe concentration is altered. Gln deprivation has no effect, while Leu deprivation enhances invasion of these cell lines. These results point to the selectivity of amino acid deprivation in control of invasion. This conclusion is further supported by the fact that serum starvation does not inhibit *in vitro* invasion (data not shown) or *in vivo* metastatic potential of B16BL6 melanoma (1). The increased invasion associated with Leu deprivation, and the modest effect of Phe deprivation on invasion, demonstrate that the effects observed with combined Tyr/Phe deprivation are not a consequence of essential amino acid deprivation.

Whereas invasion of melanoma cell lines is not influenced by Gln deprivation, DU-145 and PC-3 prostate cells are differentially sensitive to deprivation of this amino acid. Gln deprivation does not alter invasion of PC-3 cells, but does inhibit invasion of DU-145 cells. These results further underscore the different regulatory effects of amino acids on tumor cells.

All of the cell lines tested were sensitive to Met deprivation. The enhanced sensitivity of the PC-3 prostate cancer cell line may reflect the Met-dependency of these cells (19). The fact that Met deprivation inhibits invasion of melanoma is not consistent with the relative ineffectiveness of dietary Met restriction to impact metastasis of B16BL6 melanoma *in* vivo (7). The reason for this discrepancy is not known. It is not related to the inability of Met restriction to reduce plasma Met levels since the levels of this amino acid are reduced by 50% in mice fed a Met-restricted diet, while Tyr levels are only reduced by 33% (7).

Table 1. Relative influence of specific amino acid deprivation on tumor invasion.

Cell Line	Day[b]	Relative Inhibition of Invasion[a]					
		Gln	Leu	Met	Tyr	Phe	Tyr/Phe
Breast Cancer							
MDA-231	1						↓↓↓↓
Melanoma							
A375	1						↓↓↓
	2						↓↓↓↓
	3	O	↑	↓↓	↓↓↓	↓	↓↓↓↓
B16BL6	3	O	↑	↓↓	↓↓↓	↓	↓↓↓↓
SK-MEL-28	1			↓↓			↓
	2			↓↓↓			↓↓
	3			↓↓↓			↓↓↓
Prostate							
DU-145	1	↓↓		↓↓			↓↓
	2	↓↓↓		↓↓↓			↓↓↓
	3	↓↓↓		↓↓↓			↓↓↓
PC-3	1	O		↓↓↓			↓↓
	2	O		↓↓↓			↓↓↓
	3	O		↓↓↓↓			↓↓↓

[a]↑, denotes an increase in invasion; O, denotes no effect on invasion; ↓ to ↓↓↓↓, denotes 20% to 80% inhibition of invasion.

[b]number of days that cells were cultured in amino acid-deprived media ± 10% FBS.

4. INFLUENCE OF TYR/PHE DEPRIVATION ON METASTASIS-SUPPRESSOR GENE EXPRESSION

To date, about five genes have been identified that conform to the criteria of metastasis-suppressor genes. The properties and characteristics of these genes, which include BRMS1, KAI1, Kiss1, MKK4, and nm23 (NME1), were recently discussed in an excellent review by Yoshida et al. (20). We examined the effect of Tyr/Phe deprivation on two of these genes, nm23 and MKK4.

Nm23 was initially identified in the murine K1735 melanoma and six human homologues have also been identified. Although the function of this gene is still unknown, the nm23 protein is highly homologous to the nucleoside diphosphate kinases of *Dictyostelium discoideum* and of human cells (21). It is proposed that these enzymes control invasion through their novel cell signaling mechanisms (20).

We examined the mRNA expression of nm23 in the highly invasive and metastatic B16BL6 murine melanoma cells that were isolated from mice fed a normal diet and a diet restricted in Tyr/Phe. Dietary Tyr/Phe restriction decreased tumor heterogeneity and selected for tumor variants with decreased metastatic potential (7, 15). We found that nm23 mRNA expression was not altered by dietary Tyr/Phe restriction in B16BL6 melanoma (21).

The mitogen-activated protein kinase kinase 4 (MKK4), a member of mitogen-activated protein kinase (MAPK) kinase family, was first cloned in the mouse and termed as stress-activated protein kinase kinase 1 (SEK1) in 1994 (22). It was subsequently cloned in humans and named MKK4 in 1995 (23). MKK4/SEK1 has been identified as an intermediate in signal transduction pathways between upstream MEKK1 and downstream stress-activated protein kinase/c-Jun NH2-terminal kinase (SAPK/JNK1) and p38 MAPK, but not ERK1 (23). MKK4 is activated by dual phosphorylation on serine and threonine amino acid residues by MAPK and MEKK1 (24). MKK4 activates JNKs by phosphorylation on threonine and Tyr residues of JNKs (25). Interestingly, activation of JNK, but not the activation of p38 is abolished in MKK4 deficient MKK4 (-/-) ES cells (26).

It is suggested that MKK4/SEK1 is an essential gene for maintaining cell viability and embryonic development since mice with a homozygous null allele for MKK4/SEK1 are not viable (26). Cells with disrupted MKK4/SEK1 genes exhibit defective AP-1-dependent transcription activity (26). There is evidence that metastasis suppressor genes are functionally inactivated as tumor cells acquire metastatic capacity resulting in a conversion of the tumor from a non-metastatic to a metastatic phenotype (27). Mutations and deletions of the MKK4/SEK1 gene are reported in some lung, colon, testicular, pancreatic, biliary, and breast carcinoma tissues and in some cancer cell lines (28, 29). Rat AT6.1 prostate cancer cells, which do not express the endogenous SEK1/MKK4 gene, fail to respond to stress stimuli and are not able to form macroscopic metastases in the lung (30). Mice carrying AT6.1 cells stably transfected with human MKK4/SEK1 cDNA have a significant reduction in the number of macroscopic lung metastases in comparison to mice injected with AT6.1 cells transfected with an empty vector (31). These two studies suggest that MKK4/SEK1 functions as a metastasis suppresser gene.

We investigated whether MKK4/SEK1 is also involved in invasion of human breast cancer and melanoma cell lines using a standard *in vitro* Matrigel invasion system. We then examined the total expression of MKK4/SEK1 protein and the expression of the phosphorylated protein. Figure 2A shows that the human estrogen receptor positive MCF-7 breast cancer cells are highly non-invasive as compared to estrogen receptor negative MDA-231 breast cancer cells and A375 melanoma cells. Among the three cell lines examined, MCF-7 cells express the highest level of phosphorylated MKK4/SEK1 protein (Figure 2B).

The phosphorylated MKK4/SEK1 protein is not detectable in MDA-231 cells. MCF-7 cells express 25 times more MKK4/SEK1 protein than MDA-231 breast cancer cells, and 22 times more than A375 melanoma cells. These data suggest that both the level of MKK4/SEK1 protein, as well as the level of phosphorylation, are important to the invasive ability of these cells.

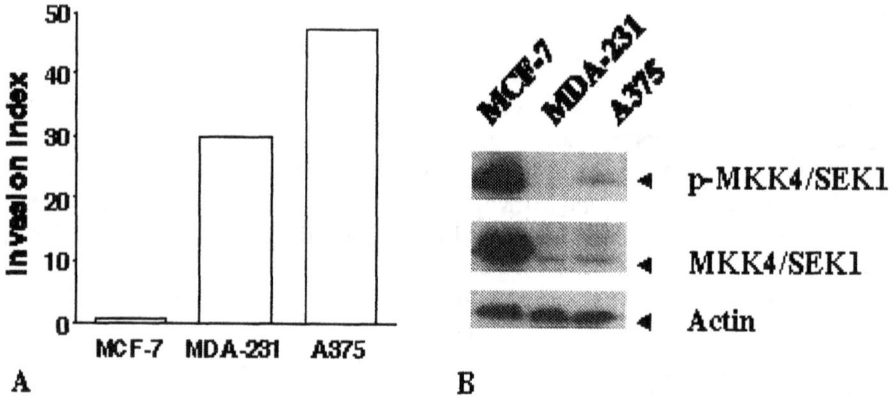

Figure 2. Comparison between MKK4/SEK1 expression and invasion of tumor cell lines. A. Invasion of MCF-7, MDA-231, and A375 cells through growth factor reduced Matrigel™. Each cell line was suspended in chamber in Minimum Essential Medium (MEM) supplemented with 0.5% bovine serum albumin, and 5 X 10^4 cells were added to the upper compartment of an invasion insert. After an 8-hr incubation at 37 ℃ in 5% CO_2, the cells on the upper side of the insert filter (8 μM size) were removed using cotton swabs. The filters were fixed with methanol and stained with Diff-Quick. Cells that fixed to the underside of the filter were counted under a microscope at 200 × magnification. Five random fields per well were counted. Results are expressed as an invasion index in which the number of MCF-7 cells invading through the Matrigel to the underside of the filter was designated as 1. The invasion index for MDA-231 and A375 cells is defined as the degree of increase in invasion relative to invasion of MCF-7 cells. B. Western blot analysis indicating the amount of phosphorylated (upper bands) and non-phosphorylated (middle bands) MKK4/SEK1 protein from MCF-7, MDA-231, and A375 cells. Cells were harvested in a lysis buffer containing 150 mM NaCl, 50 mM Tris, pH 7.4, 5 mM EDTA, 1% NP40, 1% SDS, 50 mM DTT, and a protease inhibitor mixture. After centrifugation at 13,000 × g for 5 min at 4 ℃, 25 μg of protein from each sample was subjected to 10% SDS-PAGE gel electrophoresis, transferred onto nitrocellulose membrane, and blotted with rabbit anti-phosphorylated MKK4/SEK1 (New England Biolabs, Beverly, MA). The filter was then reprobed with rabbit anti-MKK4/SEK1 (Sigma, St. Louis, MO). The amount of actin (lower bands) from each sample was detected and used as loading control.

Invasion of MDA-231 breast cancer and A375 melanoma cell lines cultured in complete medium and in Tyr/Phe-deprived medium were compared in the same invasion assay. Deprivation of Tyr/Phe significantly decreases invasion of both cell lines, as shown in Table 1.

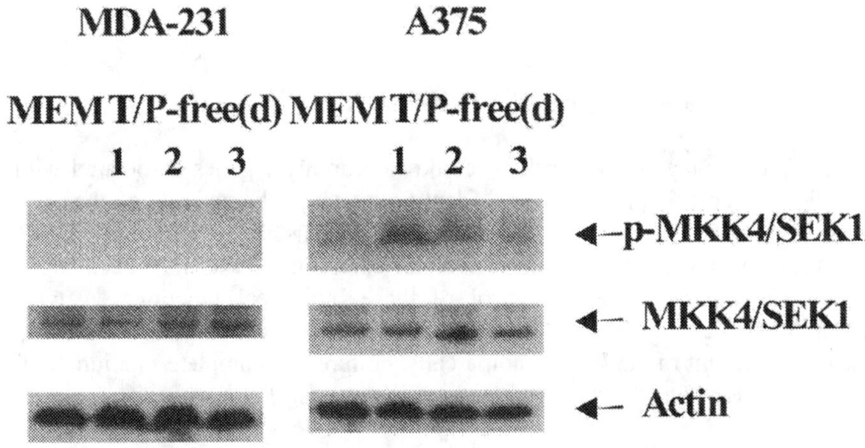

Figure 3. Western blot analysis showing the amounts of phosphorylated (upper bands) and non-phosphorylated (middle bands) MKK4/SEK1 protein from MDA-231 and A375 cells cultured in Tyr/Phe-deprived medium. Cells were cultured in MEM or MEM deprived of Tyr/Phe from 1-3 days in the presence of 10% FBS. Cells were harvested in a lysis buffer containing 150 mM NaCl, 50 mM Tris, pH 7.4, 5 mM EDTA, 1% NP40, 1% SDS, 50 mM DTT, and a protease inhibitor mixture. After centrifugation at 13,000 × g for 5 min at 4 °C, 25 μg of protein from each sample was subjected to 10% SDS-PAGE gel electrophoresis, transferred onto nitrocellulose membrane, and blotted with rabbit anti-phosphorylated MKK4/SEK1 (New England Biolabs). The filter was then reprobed with rabbit anti-MKK4/SEK1 (Sigma). The amount of actin (lower bands) from each sample was detected and used as a loading control.

Since Tyr/Phe deprivation is a nutritional stress to cells, we hypothesized that deprivation of these two amino acids would modulate MKK4/SEK1 expression. Figure 3 shows that deprivation of Tyr/Phe has no effect on the amount of phosphorylated MKK4/SEK1 protein in MDA-231 breast cancer cells. However, the expression of MKK4/SEK1 protein is increased at 2 and 3 days of culture in Tyr/Phe-deprived medium. In A375 melanoma cells, deprivation of Tyr/Phe increased the amount of phosphorylated MKK4/SEK1 protein. The phosphorylated protein increases to the highest level (3 times that of control) after 24 hr, and then returns to the baseline level after 72 hr.

Tyr/Phe deprivation does not affect MKK4/SEK1 protein expression in A375 cells. The data indicate that both expression of MKK4/SEK1 protein and expression of its phosphorylated form can be regulated by amino acid deprivation and that the responses vary depending on the type of cancer. Although these data show an inverse relationship between MKK4/SEK1 expression/phosphorylation and tumor invasion, further *in vivo* studies are needed to corroborate this association.

5. DIFFERENTIAL MODULATION OF GENE EXPRESSION BY TYR/PHE DEPRIVATION

Differential display is a powerful method to search for genes associated with a specific biological phenomenon. Identifying and characterizing the genes associated with the anti-invasive and anti-metastatic effects of Tyr/Phe deprivation in melanoma cells is crucial for examining molecular mechanism(s) associated with amino acid control of these biological functions. A typical example of the degree of difference that we observed between A375M, a metastatic variant of A375 melanoma cells, cultured in complete medium and in cells cultured in Tyr/Phe-deprived medium is shown in Figure 4.

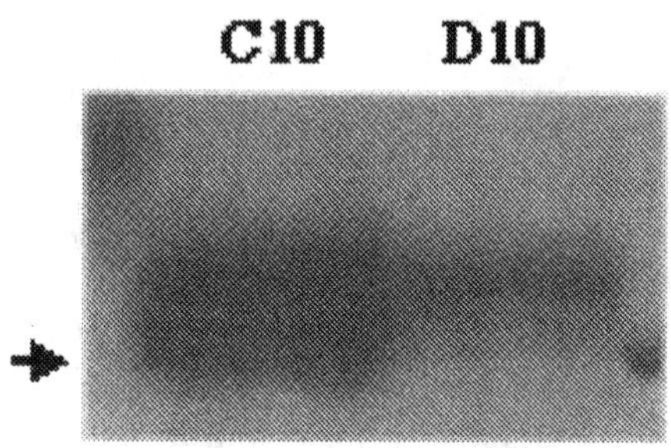

Figure 4. Expression of the FUS/CHOP fusion gene as determined by differential display of A375M melanoma. Cells were cultured in complete medium (C10) or in Tyr/Phe-deprived medium (D10) for 48 hr. The cells were harvested and total RNA was extracted with Trizol reagent (GIBCO-BRL, Grand Island, NY), according to the manufacturer's instruction. Three primers, (dT)12G, (dT)12C, and (dT)12A, were used to synthesize the first strand of cDNA with M-MLV reverse transcriptase (Promega, Madison, WI). The cDNA was used as a template to run PCR with arbitrary primers (Operan, Alameda, CA). [α-^{33}P]-d ATP was used in the PCR reaction and the PCR products were fractionated on a urea-denatured 4.5% PAGE sequencing gel. The gel was dried in a Bio-Rad gel drier at 80 °C for 2 hr, exposed to X-ray film, and subsequently developed. The FUS/CHOP fusion gene band, indicated by the arrow in Figure 4, was cloned into the pGEM-T easy vector system (Promega). Specific sequences were

identified with the PRISM Ready Reaction Dye Deoxy Terminator Cycle Sequencing Kit (Applied Biosystems, Foster City, CA). Rapid amplification of the cDNA 5' ends was used to recover the full cDNA sequence. Both cDNA and deduced amino acid sequences were used to search the NIH GeneBank database to determine the identity of the cloned genes.

Table 2 contains a list of the differentially expressed genes associated with the regulation of melanoma during Tyr/Phe deprivation. At present, fourteen differentially expressed genes have been identified. The expression of four genes increased while the expression of ten genes decreased in cells cultured in Tyr/Phe-deprived medium. Although the list is far from inclusive, it is apparent that the identified genes are important to mRNA processing, cell cycle, mobility, cytoskeleton organization, and signal transduction of melanoma cells. All are important to the process of invasion and metastasis, however, little is known regarding the role of these genes on invasion and metastasis. We are currently investigating the function of several of these genes in melanoma invasion.

6. SUMMARY AND CONCLUSIONS

Free amino acids are important regulators of invasion and metastasis. In an earlier report we showed that the metastatic phenotype of B16BL6 melanoma cells is specifically modulated *in vitro* by Tyr/Phe restriction, but not by restriction of other amino acids (1). Table 1 shows that the *in vitro* invasive ability of human breast, melanoma, and prostate cancer cell lines also vary in their response to specific amino acid deprivation. Whether the effects of amino acid deprivation are direct or indirect are not completely understood. However, the fact that the invasive phenotype can be modulated in the absence of the host response clearly indicates that the altered phenotype is related to the availability of amino acids to the tumor cells. The data in Figure 1 clearly show that intracellular free Tyr levels are reduced in melanoma cells in response to external Tyr/Phe deprivation. Intracellular Phe levels were resistant to change. This was reflected in the fact that Tyr deprivation alone has a greater impact on invasion than Phe deprivation. Thus, we believe that the intracellular concentration of particular amino acids serves as a signal to regulate a variety of cellular events including growth, invasion, and metastasis.

We now show that MKK4/SEK1 metastasis suppressor protein expression and phosphorylation is positively influenced by Tyr/Phe deprivation (Figure 3). From Figure 4 it is also apparent that deprivation of these amino acids can result in both increased and decreased gene expression. An example of the negative influence of Tyr/Phe restriction is the decrease in expression and phosphorylation of focal adhesion kinase protein (47). Very little is known regarding the modulation of gene expression and function by amino acids and this is a very fertile area of research. Changes in gene expression appear not to be associated with the general or total inhibition of protein synthesis resulting from amino acid limitation (48). In fact, Tyr/Phe limitation of B16BL6 melanoma cells for 4 days does not inhibit protein synthesis (14), however, limitation has a major impact on a number of processes that modulate tumor growth, invasion, and metastasis (1, 2, 17).

Table 2. Genes regulated by Tyr and Phe deprivation in A375M melanoma cells identified by differential display

Gene	Expression[a] (Control)	Expression[a] (Tyr/Phe Free)	Function	References
Human HAPC071 protein	↑	↓	Unknown	
Human mitochondrial 16S RNA	↓	↑	Involved in protein synthesis in the mitochondria.	
Semaphorin cytoplasmic domain-associated protein 3A	↑	↓	Semaphorin plays an important role in cell guidance. The function of semaphorin cytoplasmic domain-associated protein 3A is unclear.	(32-34)
Human Tropomodulin 3	↑	↓	Tropomodulin is a pointed end capping protein for actin filaments. It binds specifically to the N terminus of tropomyosin and blocks the elongation and depolymerization of tropomyosin-coated actin filaments.	(35, 36)
Human Archain-1 (ARCN1)	↓	↑	ARCH1 is also known as coatomer protein delta-COP. It is involved in vesicle structure and trafficking.	(37-39)

Table 2 (cont.)

Gene	Expression[a] (Control)	(Tyr/Phe Free)	Function	References
Human Mig-6/ Rat gene 33	↓	↑	This gene is involved in cell cycle progression and terminal differentiation in response to chronic stress.	(40, 41)
Human *ras*-related GTP-binding protein gene	↓	↑	Unknown	
Heterogeneous nuclear ribonucleoprotein complex K (hnRNP K)	↑	↓	HnRNP K participates in the processing of pre-mRNAs and in the export of mRNAs from the nucleus.	(42-44)
FUS gene, Fusion of FUS/ CHOP protein in malignant liposarcoma	↑	↓	Fusion of FUS/CHOP results from the chromosomal translocation of t(12;16)(q13;p11). It has been used as a clinical diagnostic marker in malignant liposarcoma.	(42 ,45, 46)
Unknown gene	↑	↓	Unknown	
Unknown gene	↑	↓	Unknown	
Unknown gene	↓	↑	Unknown	
HPPK.332_H_18	↑	↓	Unknown	

[a]↑, denotes increased expression; ↓, denotes decreased expression.

One of the best known genes that is regulated by amino acids is the CHOP (C/EBP homologous protein)/gadd153 gene (49). This gene encodes for a transcription factor that regulates perturbations of the endoplasmic reticulum in response to stress (48). Interaction of the amino acid response elements (AARE) in the promoter region of the gene with activated transcription factor 2 (ATF2) is essential for transcriptional activation of this gene (49, 50). It is likely that AARE are present in other genes important to the control of metastasis. In fact, two known genes regulated *in vitro* by amino acids that are relevant to tumor invasion and metastasis are matrix metalloproteinase-1 (MMP-1) and tissue inhibitor of metalloproteinase-1 (TIMP-1) (51). Supraphysiological concentrations (1 mg/ml) of Tyr increase the steady-state level of MMP-1 mRNA by 2-fold in human skin fibroblasts, and decrease the steady-state level of TIMP-1 by one-half relative to fibroblasts cultured under regular conditions. In searching the NIH GeneBank we found that the promoter regions of the MMP-1 and TIMP-1 genes contain the AARE consensus sequences. Further research is needed to determine functionality of these putative AARE sequences in the MMP-1 and TIMP-1 genes. Identification of the regulatory elements associated with amino acid responsive genes could lead to the development of novel approaches to control the key signaling molecules that enable tumors to grow, invade, and metastasize.

ACKNOWLEDGMENTS

This work was supported by grant number CA77604 from the National Cancer Institute and Department of Defense Prostate Cancer Research Program grant number DAMD17-00-1-0107 to Dr. Meadows and by a minority supplement sponsored by the Continuing Umbrella of Research Experiences (CURE) program to Dr. Oros.

REFERENCES

1. Elstad, C. A. and Meadows, G. G. Modulation of B16-BL6 murine melanoma metastatic phenotype by tyrosine and phenylalanine restriction in the absence of host selection pressures. Anticancer Res, 13: 523-528, 1993.
2. Fu, Y.-M., Yu, Z.-X., Ferrans, V. J., and Meadows, G. G. Tyrosine and phenylalanine restriction induces G0/G1 cell cycle arrest in murine melanoma in vitro and in vivo. Nutr Cancer, 29: 104-113, 1997.
3. Shay, N. F., Nick, H. S., and Kilberg, M. S. Molecular cloning of an amino acid-regulated mRNA (amino acid starvation-induced) in rat hepatoma cells. J Biol Chem, 265: 17844-18748, 1990.
4. Marten, N. W., Burke, E. J., Hayden, J. M., and Straus, D. S. Effect of amino acid limitation on the expression of 19 genes in rat hepatoma cells. FASEB J, 8: 538-544, 1994.
5. Pine, M. J. Improved host defense against L1210 leukemia by deprivation of dietary phenylalanine. Nutr Cancer, 3: 94-102, 1981.
6. Abdallah, R. M., Starkey, J. R., and Meadows, G. G. Dietary restriction of tyrosine and phenylalanine: inhibition of metastasis of three rodent tumors. J Natl Cancer

Inst, 78: 759-766, 1987.
7. Elstad, C. A., Meadows, G. G., and Abdallah, R. M. Specificity of the suppression of metastatic phenotype by tyrosine and phenylalanine restriction. Clin Exp Met, 8: 393-416, 1990.
8. Meadows, G. G. and Oeser, D. E. Response of B16 melanoma-bearing mice to varying dietary levels of phenylalanine and tyrosine. Nutr Rep Int, 28: 1073-1082, 1983.
9. Meadows, G. G., DiGiovanni, J., Minor, L., and Elmer, G. W. Some biological properties and an in vivo evaluation of tyrosine phenol-lyase on growth of B-16 melanoma. Cancer Res, 36: 167-171, 1976.
10. Meadows, G. G., Pierson, H. F., Abdallah, R. M., and Desai, P. R. Dietary influence of tyrosine and phenylalanine on the response of B16 melanoma to carbidopa-levodopa methyl ester chemotherapy. Cancer Res, 42: 3056-3063, 1982.
11. Pierson, H. F. and Meadows, G. G. Sodium ascorbate enhancement of carbidopa-levodopa methyl ester antitumor activity against pigmented B16 melanoma. Cancer Res, 43: 2047-2051, 1983.
12. Meadows, G. G., Abdallah, R. M., and Starkey, J. R. Interaction between specific dietary factors and experimental chemotherapy of metastatic melanoma. Cancer Chemother Pharmacol, 16: 229-236, 1986.
13. Elstad, C. A., Thrall, B. D., and Meadows, G. G. Tyrosine and phenylalanine restriction sensitizes adriamycin-resistant P388 leukemia cells to adriamycin. Nutr Cancer, 25: 47-60, 1996.
14. Fu, Y.-M., Li, Y.-Q., and Meadows, G. G. Influences of tyrosine and phenylalanine limitation on cytotoxicity of chimeric TGF-α on B16BL6 murine melanoma in vitro. Nutr Cancer, 31: 1-7, 1998.
15. Elstad, C. A. and Meadows, G. G. Phenotypic stability of B16-BL6 melanoma exposed to low levels of tyrosine and phenylalanine. Anticancer Res, 10: 1313-1318, 1990.
16. Uhlenkott, C. E., Huijzer, J. C., Cardeiro, D. J., Elstad, C. A., and Meadows, G. G. Attachment, invasion, chemotaxis, and proteinase expression of B16-BL6 melanoma cells exhibiting a low metastatic phenotype after exposure to dietary restriction of tyrosine and phenylalanine. Clin Expl Met, 14: 125-137, 1996.
17. Pelayo, B. A., Fu, Y.-M., and Meadows, G. G. Inhibition of B16BL6 melanoma invasion by tyrosine and phenylalanine deprivation is associated with decreased secretion of plasminogen activators and increased plasminogen inhibitors. Clin Expl Met, 17: 841-848, 1999.
18. Jones, B. N., Paabo, S., and Stein, S. Amino acid analysis and enzymatic sequence determination of peptides by an improved o-phthaldialdehyde precolumn labeling procedure. J Liquid Chromatogr, 4: 565-586, 1981.
19. Poirson-Bichat, F., Gonfalone, G., Bras-Gonçalves, R. A., Dutrillaux, B., and Poupon, M. F. Growth of methionine-dependent human prostate cancer (PC-3) is inhibited by ethionine combined with methionine starvation. Br J Cancer, 75: 1605-1612, 1997.
20. Yoshida, B. A., Sokoloff, M. M., Welch, D. R., and Rinker-Schaeffer, C. W. Metastasis-suppressor genes: a review and perspective on an emerging field. J Natl Cancer Inst, 92: 1717-30, 2000.
21. Huijzer, J. C., McFarland, M., Niles, R. M., and Meadows, G. G. Phorbol 12-myristate 13-acetate enhances nm23 gene expression in murine melanocytes but not in syngeneic B16-BL6 melanoma variants. J Cell Phys, 166: 487-494, 1996.

22. Sánchez, I., Hughes, R. T., Mayer, B. J., Yee, K., Woodgett, J. R., Avruch, J., Kyriakis, J. M., and Zon, L. I. Role of SAPK/ERK kinase-1 in the stress-activated pathway regulating transcription factor c-Jun. Nature, 372: 794-800, 1994.
23. Dérijard, B., Raingeaud, J., Barrett, T., Wu, I.-H., Han, J., Ulevitch, R. J., and Davis, R. J. Independent human MAP kinase signal transduction pathways defined by MEK and MKK isoforms. Science, 267: 682-685, 1995.
24. Yan, M., Dai, T. D. J. C., Kyriakis, J. M., Zon, L. I., Woodgett, J. R., and Templeton, D. J. Activation of stress-activated protein kinase by MEKK1 phosphorylation of its activator SEK1. Nature, 372: 798-800. 1994.
25. Dérijard, B., Hibi, M., Wu, I.-H., Barrett, T., Su, B., Deng, T., Karin, M., and Davis, R. J. JNK1: a protein kinase stimulated by UV light and Ha-Ras that binds and phosphorylates the c-Jun activation domain. Cell, 76: 1025-1037, 1994.
26. Yang, D., Tournier, C., Wysk, M., Lu, H.-T., Xu, J., Davis, R. J., and Flavell, R. A. Targeted disruption of the MKK4 gene causes embryonic death, inhibition of c-Jun NH_2-terminal kinase activation, and defects in AP-1 transcriptional activity. Proc. Natl. Acad. Sci. USA, 94: 3004-3009, 1997.
27. Dear, T. N. and Kefford, R. F. Molecular oncogenetics of metastasis. Mol Aspects Med, 11: 243-324, 1990.
28. Teng, D. H.-F., Perry III, W. L., Hogan, J. K., Baumgard, M., Bell, R., Berry, S., Davis, T., Frank, D., Frye, C., Hattier, T., Hu, R., Jammulapati, S., Janecki, T., Leavitt, A., Mitchell, J. T., Pero, R., Sexton, D., Schroeder, M., Su, P.-H., Swedlund, B., Kyriakis, J. M., Avruch, J., Bartel, P., Wong, A. K. C., Oliphant, A., Thomas, A., Skolnick, M. H., and Tavtigian, S. V. Human mitogen-activated protein kinase kinase 4 as a candidate tumor suppressor. Cancer Res, 57: 4177-4182, 1997.
29. Su, G. H., Hilgers, W., Shekher, M. C., Tang, D. J., Yeo, C. J., Hruban, R. H., and Kern, S. E. Alterations in pancreatic, biliary, and breast carcinomas support MKK4 as a genetically targeted tumor suppressor gene. Cancer Res, 58: 2339-2342, 1998.
30. Chekmareva, M. A., Kadkhodaian, M. M., Hollowell, C. M. P., Kim, H., Yoshida, B. A., Luu, H. H., Stadler, W. M., and Rinker-Schaeffer, C. W. Chromosome 17-mediated dormancy of AT6.1 prostate cancer micrometastases. Cancer Res, 58: 4963-4969, 1998.
31. Yoshida, B. A., Dubauskas, Z., Chekmareva, M. A., Christiano, T. R., Stadler, W. M., and Rinker-Schaeffer, C. W. Mitogen-activated protein kinase kinase 4/stress-activated protein/Erk kinase 1 (MKK4/SEK1), a prostate cancer metastasis suppressor gene encoded by human chromosome 17. Cancer Res, 59: 5483-5487, 1999.
32. Brambilla, E., Constantin, B., Drabkin, H., and Roche, J. Semaphorin SEMA3F localization in malignant human lung and cell lines: A suggested role in cell adhesion and cell migration. Am J Pathol, 156: 939-950, 2000.
33. Martin-Satue, M. and Blanco, J. Identification of semaphorin E gene expression in metastatic human lung adenocarcinoma cells by mRNA differential display. J Surg Oncol, 72 : 18-23, 1999.
34. Vikis, H. G., Li, W., He, Z., and Guan, K.-L. The semaphorin receptor plexin-B1 specifically interacts with active Rac in a ligand-dependent manner. Proc Natl Acad Sci USA, 97: 12457-12462, 2000.
35. Chu, X., Thompson, D., Yee, L. J., and Sung, L. A. Genomic organization of mouse and human erythrocyte tropomodulin genes encoding the pointed end capping protein for the actin filaments. Gene, 256: 271-281, 2000.
36. Cooper, J. A. and Schafer, D. A. Control of actin assembly and disassembly at

filament ends. Curr Opin Cell Biol, 12: 97-103, 2000.

37. Kordula, T., Bugno, M., Rydel, R. E., and Travis, J. Mechanism of interleukin-1 and tumor necrosis factor α-dependent regulation of the α_1-antichymostrypsin gene in human astrocytes. J Neurosci, 20 : 7510-7516, 2000.
38. Hwang, S. R., Steineckert, B., Kohn, A., Palkovits, M., and Hook, V. Y. Molecular studies define the primary structure of α_1-antichymotrypsin (ACT) protease inhibitor in Alzheimer's disease brains. Comparison of ACT in hippocampus and liver. J Biol Chem, 274 : 1821-1827, 1999.
39. Ikari, Y., Mulvihill, E., and Schwartz, S. M. Alpha1-Proteinase inhibitor, alpha 1-antichymotrypsin and alpha 2-marcoglobulin are the anti-apoptotic factors of vascular smooth muscle cells. J Biol Chem , in press: 2001.
40. Wick, M., Burger, C., Funk, M., and Muller, R. Identification of a novel mitogen-inducible gene (mig-6): regulation during G1 progression and differentiation. Exp Cell Res, 219: 527-535, 1995.
41. Makkinje, A., Quinn, D. A., Chen, A., Cadilla, C. L., Force, T., Bonventre, J. V., and Kyriakis, J. M. Gene 33/Mig-6, a transcriptionally inducible adapter protein that binds GTP-Cdc42 and activates SAPK/JNK. A potential marker transcript for chronic pathologic conditions, such as diabetic nephropathy. Possible role in the response to persistent stress. J Biol Chem, 275: 17838-17847, 2000.
42. Matunis, M. J., Michael, W. M., and Dreyfuss, G. Characterization and primary structure of the poly(C)-binding heterogeneous nuclear ribonucleoprotein complex K protein. Mol Cell Biol, 12: 164-171, 1992.
43. Tomonaga, T. and Levens, D. Heterogeneous nuclear ribonucleoprotein K is a DNA-binding transactivator. J Biol Chem, 270: 4875-4881, 1995.
44. Miau, L.-H., Chang, C.-J., Shen, B.-J., Tsai, W.-H., and Lee, S.-C. Identification of heterogeneous nuclear ribonucleoprotein K (hnRNP K) as a repressor of C/EBPß-mediated gene activation. J Biol Chem, 273: 10784-10791, 1998.
45. Rabbitts, T. H., Forster, A., Larson, R., and Nathan, P. Fusion of the dominant negative transcription regulator CHOP with a novel gene FUS by translocation t(12;16) in malignant liposarcoma. Nat Genet, 4: 175-180, 1993.
46. Perez-Losada, J., Pintado, B., Gutierrez-Adan, A., Flores, T., Banares-Gonzalez, B., del Campo, J. C., Martin-Martin, J. F., Battaner, E., and Sanchez-Garcia, I. The chimeric FUS/TLS-CHOP fusion protein specifically induces liposarcomas in transgenic mice. Oncogene, 19: 2413-2422, 2000.
47. Fu, Y.-M., Yu, Z.-X., Pelayo, B. A., Ferrans, V. J., and Meadows, G. G. Focal adhesion kinase-dependent apoptosis of melanoma induced by tyrosine and phenylalanine deficiency. Cancer Res, 59 : 758-765, 1999.
48. Jousse, C., Bruhat, A., Ferrara, M., and Fafournoux, P. Evidence for multiple signaling pathways in the regulation of gene expression by amino acids in human cell lines. J Nutr, 130: 1555-1560, 2000.
49. Bruhat, A., Jousse C., Carraro, V., Reimold, A. M., Ferrara, M., and Fafournoux, P. Amino acids control mammalian gene transcription: activating transcription factor 2 is essential for the amino acid responsiveness of the CHOP promoter. Mol Cell Biol, 20: 7192-7204, 2000.
50. Fafounoux, P., Bruhat, A., and Jousse C. Amino acid regulation of gene expression. Biochem J, 351: 1-12, 2000.
51. Varga, J., Li, L., Mauviel, A., Jeffrey, J., and Jimenez, S. A. L-tryptophan in supraphysiologic concentrations stimulates collagenase gene expression in human skin fibroblasts. Lab Invest, 70: 183-191, 1994.

Chapter 11

ROLE OF BRMS1 IN BREAST CARCINOMA METASTASIS

Rajeev S. Samant, Lalita R. Shevde, Danny R Welch*

Jake Gittlen Cancer Research Institute, The Pennsylvania State University College of Medicine, Hershey, Pennsylvania, USA

INTRODUCTION AND BACKGROUND

Breast cancer is a leading cause of deaths among women worldwide. Mortality from breast carcinoma in United States exceeded 40,000 in 1999. The American Cancer Society estimates that in 2001 about 192,200 new cases of invasive breast cancer (Stages I-IV) will be diagnosed among women in the United States. The breast cancer incidence rate, a measure of the number of new breast cancers per 100,000 women, increased by about 4% per year during the 1980s. During the 1990s, the incidence leveled off to 110.6 cases per 100,000 women. In 2001, there will be about 40,600 deaths from breast cancer in the United States (40,200 among women, and 400 among men). Breast cancer is the second leading cause of cancer death in women, exceeded only by lung cancer. When breast carcinoma cells are confined to breast tissue, surgical resection is easy, thus long term survival rates are high. The cure rate and the quality of life drops significantly once the tumor cells leave the primary site and colonize distant tissues (stage IV, metastatic disease) (1, 2) . Thus an effective treatment and/or prevention of the metastatic disease is of a great importance.

A recent cataloging of differential gene/protein expression and chromosomal abnormalities occurring progression of breast carcinoma (2, 3) revealed that some karyotypic alterations typically occur (at loci 1p,1q, 3p, 6q, 7q, 11p and 11q) in the later stages of breast cancer. Among the most common (40-65% of cases) changes in both familial and sporadic breast carcinomas are alterations of chromosome 11q particularly surrounding the region near 11q13. We hypothesized that chromosome 11q encodes metastasis suppressor gene(s). To test this hypothesis we (in collaboration with Drs. Bernard Weissman and Karen Phillips, University of North Carolina at Chapel Hill), introduced a neomycin tagged normal human chromosome 11(neo11) into highly metastatic MDA-MB-435 (435) cell line by microcell mediated chromosome transfer. Testing the chromosome 11 hybrid cell lines in athymic mice for metastasis from an orthotopic (mammary fat pad) site showed that

D.R. Welch (ed.), Cancer Metastasis – Related Genes, 209–217.

chromosome 11 significantly suppressed the metastatic ability of 435 cell line without affecting the tumorigenicity (4).

DISCOVERY OF BRMS1

To identify the gene(s) responsible for this phenotypic change, the gene expression profile in metastasis competent (435) was compared to that of metastasis suppressed (neo11/ 435) variants using differential display (5). Out of 64 differentially expressed mRNAs, six candidate transcripts with at least 5-fold greater expression in neo11/435 were identified and differential expression confirmed by RNA blotting. Three of the six were novel transcripts with no significant homology in the combined molecular database. Human tissue expression profile of the three candidates showed high level in kidney, with a unique transcript sizes of 1, 1.2 and 1.5 kb respectively. A full-length clone corresponding to the 1.5 kb transcript was obtained by screening a human kidney cDNA library. This novel cDNA was termed as **Br**east **M**etastasis Suppressor **1** (***BRMS1***) (6). *BRMS1* cDNA encompassed an open reading frame of 738 bp, with a predicted polypeptide of 246 amino acids. (M_r. 28,500). Sequence homology searches at various databases revealed potential nuclear localization sequence, N-terminal glutamic acid rich region, an imperfect leucine zipper etc. Though apparently suggestive of a transcription factor, *BRMS1* did not appear to be a part of any known major protein families.

BRMS1 cDNA was used to screen bacterial artificial chromosome libraries at Genome Systems Inc (St. Louis, MO) to obtain a genomic BAC clone of *BRMS1*. The genomic sequence analysis revealed that *BRMS1* is organized into 10 exons spanning ~8.5 kb (7, 8). The first exon is outside the coding region. Most of the intron-exon junctions correspond to the junction of predicted protein motifs. *BRMS1* shows classical intron-exon junctions obeying the GU-AG splicing rule (9). To date no evidence of alternate splicing is detected by RT-PCR or northern blot analysis.

Fluorescence *in situ* hybridization analysis mapped *BRMS1* gene at human chromosome 11q13.1-q13.2, a region described to be involved in later stages of breast carcinoma (10). In our efforts to explore the region upstream to *BRMS1* for identifying the regulatory sequences, we have sequenced approximately 2 kb upstream of *BRMS1*. We found the gene encoding i-b-1,3-N-acetylglucosaminyl transferase (*iGnT*) located in this region with direction of transcription same as BRMS1. Also, RIN1, Ras inhibitor 1 coding gene is located at a close proximity, downstream of BRMS1. *iGnT* is involved in synthesis of poly-N-acetyl-lactosamine which is essential for the formation of the i antigen, which are the first human alloantigens shown to display developmental changes. *iGnT* may also have a role to play in transformed phenotype (11). Though it will be very interesting to obtain insight about the status of *iGnT* and *RIN1* in metastasis, thus far, there are no obvious implications of close localization of these genes.

A 500 bp region upstream of *BRMS1* (majority of the intergenic region) shows presence of TATA independent promoter elements with interesting *cis*- elements such as c-Myb, AP2, CREB etc. and is able to turn on a luciferase reporter gene. Involvement of the various *cis*-elements in regulation of *BRMS1* expression and their possible effect on metastatic cells is an interesting study (7).

Thus, we have developed an insight about the *BRMS1* genomic organization, which will be useful in understanding the biology of this gene as well as help in generation of necessary reagents for performing clinically important studies with *BRMS1*.

BRMS1 SUPPRESSES METASTASIS OF BREAST CARCINOMA

To assess the effect of *BRMS1* on breast carcinoma metastasis, *BRMS1* (under control of a constitutive promoter) was transfected into two independently derived metastatic human breast carcinoma cell lines, MDA-MB-435 and MDA-MB-231. Clones expressing low, medium or high level of *BRMS1* were isolated and tested for tumorigenicity and metastasis in athymic mice. MDA-MB-435 shows metastasis to lungs and regional lymph nodes following a primary tumor formation at the mammary fat pad. This assay most closely mimics the situation in patients (i.e. all steps of the metastatic cascade must be completed). MDA-MB-231 on the other hand forms a primary tumor, but does not metastasize from this site. It does however form lung metastasis when injected intravenously into the lateral tail vain. In both the cell lines, BRMS1 expression suppressed metastasis significantly (from the mammary fat pad and following intravenous injection, respectively). In MDA-MB-231 the suppression was seen to be in an expression dependent manner. As expected, *BRMS1* did not influence the growth of primary tumor at orthotopic site (6).

BRMS1 SUPPRESSES METASTASIS OF MELANOMA

Research of Robertson *et al.* suggested the presence of a tumor suppressor(s) on the long arm of human chromosome 11 in human melanomas (12). We tested the hypothesis that *BRMS1* is a tumor or metastasis suppressor of human melanoma cell line MelJuSo, by stably transfecting *BRMS1* in the pcDNA3 vector (CMV promoter) and testing their abilities to form tumors and metastasize in athymic mice. *BRMS1* did not alter the tumorigenicity but significantly reduced metastasis to lungs (60% suppression as compared to vector only transfectant control), suggesting that it does not function as a tumor suppressor but does function as a melanoma metastasis suppressor as well. We have seen similar suppression of metastasis of C8161 by BRMS1 (13).

MOUSE HOMOLOG OF BRMS1 AND ITS FUNCTIONAL CHARACTERIZATION

BRMS1 cDNA sequence showed very striking homology to some of the mouse ESTs in GenBank. In order to identify the murine homolog of BRMS1 (*brms1*), a multi-tissue blot was screened using human BRMS1 cDNA as a probe, murine testis showed highest levels of *brms1*. Screening of murine testis cDNA library led to the identification of a 1.4 kb cDNA (14). The sequence reveals 85% homology to the human *BRMS1* at DNA level. the predicted protein sequence for the murine homolog is 95% identical, suggesting that it is strongly conserved across these species. Northern blot analysis of cell lines isolated from *brms1* was then tested for suppression of metastasis of mouse mammary carcinoma cell line 66cl4 (15) in syngeneic BALB/c mice. Similar to results with human breast carcinoma cells and BRMS1, transfection with *brms1* did not inhibit 66cl4 primary tumor formation, but significantly suppressed its metastatic capability. This suggests that the murine homolog has similar functionality as *BRMS1* (16).

The cloned cDNA was used to screen a BAC library to obtain *BRMS1* genomic DNA (Genome Systems). Three BAC clones 226/I4, 226/H4, and 239/N7 were confirmed to encode the entire *BRMS1* gene. Detailed analysis of BAC clone 226/I4 shows that the gene spans 8.5 kb, and like the human gene, is organized into 10 exons and 9 introns, also, exon 1 is not translated in both genes (17). The size of the exons is conserved between human and mouse; however there are some differences in the sizes of the introns. Notably, intron 1 in human is 3.3 kb, while murine intron 1 is 4.4 kb. Both the genes follow the typical rules of splicing. The *BRMS1* promoter shares about a 70% homology to its human counterpart, retaining many, but not all of the putative promoter elements. Both promoters also appear to be TATA independent (18). Also of interest is that, like the human genome, the murine ortholog of the *iGnT* gene, (i-Beta-1, 3-N-acetyl-glucosaminyl transferase) is found upstream of *BRMS1*. Sequencing downstream of *brms1* revealed murine Ras inhibitor protein (*RIN1*) (19-21) gene located in close proximity (~500 bp). This has been also confirmed by mapping studies (Kent W. Hunter, Karl W. Broman, Thomas Le Voyer, Luanne Lukes, Diana Cozma, Michael T. Debies, Jessica Rouse, and Danny R. Welch, manuscript in preparation) in strains of mice with differential susceptibility to mammary tumor metastasis.

Though the proximity of these genes is intriguing further studies are needed to establish any kind of *cis* effects on *brms1*.

HOW DOES BRMS1 SUPPRESS METASTASIS?

To unravel involvement of *BRMS1* in the phenomenon of metastasis suppression many parallel approaches were employed. Examination of various tissue RNA for *BRMS1* showed a. unique transcript of approximately 1.5 kb in almost all tissues. *BRMS1* expression was high in peripheral blood lymphocytes, placenta, kidney, pancreas, spleen, testis and low in brain and lung. No tissue specific spliced variants were detected (6). The expression of unique *BRMS1* mRNA was also detected at a low level form breast tissue samples obtained from reduction mammoplasty (data not shown). Since the normal tissue expression pattern was ubiquitous, expression of *BRMS1* was determined in various breast cell lines (with varying degrees of aggressive behavior when grown in athymic mice) and compared with MDA-MB-435 and neo11/435; the latter one being its high expressing control. MDA-MB-435, SUM185, MCF7 and SV40T-transformed MCF10A express *BRMS1* at a low level. MDA-MB-231, LCC15 and T47D^{Co} express moderate levels of *BRMS1*. MKL4 (FGF 4-transfected MCF7 cells) expresses a relatively high level of BRMS1. These results clearly show quantitative differences in expression, but relationship to behavior is more complex. Except for MCF10AT, all of the other cell lines were derived from metastases, but in our laboratory, only MDA-MB-435, MDA-MB-231 and LCC15 metastasize in athymic mice. Although *BRMS1* mRNA levels *per se* do not predict metastatic ability in athymic mice, there was no evidence for gross rearrangement, deletion or mutation. To rule out these possibilities, more extensive studies will be required and are underway.

The mechanism of metastasis suppression by *BRMS1* is currently being investigated with respect to various steps in the metastatic cascade. *In vitro* assays assessing individual steps predict a complex role for this molecule. It apparently blocks metastasis downstream of local invasion since invasive cords are present in histologic sections of primary tumor. Likewise matrix metalloproteinase-2 and 9 activities (zymography) are practically unaltered. Blockage at steps subsequent to intravasation is also argued by data with 231 cells injected intravenously. There is a modest (30-70%) reduction in motility as measured by *in vitro* wound healing assays. Motility as measured by the ability to migrate through polyethylene filters was significantly lower for *BRMS1* transfectants compared to untransfected or vector only transfected 231. The highest expressors of *BRMS1* were significantly suppressed for motility (75-85% suppression); whereas the suppression was not significant for lowest expressor clone. It is also observed that the mixed population of clones expressing non tagged *BRMS1* retains this reduced motility. Similar results were observed in *BRMS1* expressing clones of MDA-MB-435.

BRMS1 encodes a novel protein of 246 amino acids. Analysis of the amino acid sequence of *BRMS1* reveals several putative phosphorylation sites for cAMP/cGMP, protein kinase C and casein kinase II. To determine whether *BRMS1* protein indeed undergoes any kind of phosphorylation, 901 epitope (epitope derived from SV40T antigen, amino acids 684-698; (22, 23)) tagged *BRMS1*-231 clone (high expressor) was grown in phosphate free media and then was supplemented with radio-labeled (P^{32}) ortho-phosphate. *BRMS1* was immunoprecipitated using anti-901 antibody and was resolved on a SDS-PAGE and the gel was electro-blotted onto a PVDF membrane. Autoradiography did not reveal any phosphate incorporation in *BRMS1* a simultaneous western blot analysis using chemiluminescent detection was performed as a control to establish the success of immunoprecipitation.

BRMS1 protein sequence also shows putative two nuclear localization sequences at amino acids 198-205 and 239-245. Hence to know the sub-cellular localization of *BRMS1*, we compared equal amounts nuclear protein extract from 901 tagged *BRMS1* expressing MDA-MB-231 with cytosolic protein from the same cell line. We observed much higher levels of BRMS1 in the nucleus. We also used immunofluorescence using the anti-901 antibodies to look at the cellular localization of tagged *BRMS1* in 231. It was evident from the co-localization with DAPI staining that *BRMS1* was predominantly nuclear. We studied expression of BRMS1 protein deleted for the nuclear localization signals; using nuclear fractionation as well as the immunofluorescence. This deletion renders it distributed uniformly between cytosolic as well as nuclear compartments. Studies are underway to determine whether such a truncated version of BRMS1 protein will still retain the metastasis suppression properties.

BRMS1 also shows presence of a putative coiled coli, leucine zipper and an N-terminal glutamic acid rich region. Considering that it is localized predominantly to nucleus, it is very tempting to speculate that it may be acting as a transcription factor. We have yet not directly addressed this issue but it was only intuitive to look for the levels of known metastatic supper genes such as *NM23, Kai1* and *KiSS1*. A western blot analysis of the MDA-MB 231-*BRMS1* clones with variable gradation of *BRMS1* expression did not show any direct or inverse correlation of NM23 or Kai1protein levels with the expression of *BRMS1*. Similar lack of correlation was noted for KiSS1 with Northern blot analysis (data not shown). Since heparanase enzyme synthesis is implicated to be directly correlated to metastatic potentials, we did an RT-PCR analysis of heparanase expression using MDA-MB-231 and it's *BRMS1* expressing clones. The parental 231 shows a very high expression of heparanase which remains unaltered in the *BRMS1* expressing clones. Although only results using 231 are shown, similar results were obtained with 435.

DISCUSSION

Though *BRMS1* seems to be ubiquitously expressed in various normal tissues as well as breast carcinoma cell lines with variable aggression, when transfected into 435 and 231 cells, there was significant decrease in both, the incidence of metastasis and the number of lung metastasis per mouse. Although the tumor development in *BRMS1* transfected 435 cells was slightly delayed as compared to the controls, the tumors still formed and grew at comparable rates. Even when tumor bearing animals were studied for a longer duration (in order to compensate for the slower growth of the locally growing tumor), metastasis was suppressed. *BRMS1* mRNA was still detectable within the primary tumor (data not shown). Taken together, these data fulfil the functional definition that *BRMS1* is a bonafide metastasis suppressor (3) i.e., metastasis is suppressed whereas tumorigenicity is not. It is also very interesting to note that at least in the case of 231, there is a direct correlation of *BRMS1* expression to the suppression.

The mechanism by which *BRMS1* suppresses metastasis is not unraveled yet. We have investigated its effect on various steps in the metastatic cascade. *BRMS1* possibly functions downstream of the local invasion step as invasion cords are observed at the edge of the locally growing tumor. We do not see changes in adhesion, proteinase activity or *in vitro* growth rates. The decrease in motility and the results of wound healing assays suggest a clear effect on the cell motility. The trypsin sensitivity observed in *BRMS1* clones of 435 and the motility and consistent slight (though insignificant) decrease in adhesion suggest a cell surface related effect. It correlates with our independent finding that *BRMS1* is involved in a potentially novel mechanism of action; re-establishment of homotypic GAP junctional communication (GJIC). Gap junctions are channels that allow passage of small molecules (<1 kDa). It is observed that GJIC is diminished or absent in many neoplastic cell lines and primary tumor. Moreover loss of GJIC tends to correlate inversely with progression in neoplastic mammary tissue (24). When doubly labeled (diI and calcein) cells (435, 231 or their respective *BRMS1* transfectants) were dropped onto a confluent monolayer of unlabeled acceptor cells, parental cells (435 and 231) did not exhibit capacity to transfer calcine. However *BRMS1* transfected cell lines showed restored capacity to transfer the dye. Furthermore, the expression profile of connexins (Cxs), the protein subunits of gap junctions, changes. Specifically, the expression of BRMS1 in MDA-MB-435 cells increases Cx43 expression and reduces Cx32 expression, resulting in a gap junction phenotype more similar to normal breast tissue. Taken together, these results suggest that gap junctional communication and the Cx expression profile may be important contributors to the metastatic potential of breast cancer cells (25). Albeit necessary to further analyze the *BRMS1* gene in other breast carcinoma during various stages of progression, as it is possible that the defects in the breast carcinoma could be due to down regulation and /or mutation of *BRMS1*. Such experiments underway and will require investigations using collection of matched samples from primary tumor and metastasis. Understanding *BRMS1* function may help prevent metastasis and improve breast cancer survival.

ACKNOWLEDGMENTS

This research was supported grants to DRW from the U.S. Public Health Service (CA-87728, CA-62168) and from the U.S. Army Medical Research and Materiel Command DAMD-96-1-6152. RSS is supported by a postdoctoral fellowship (DAMD17-01-1-0362) from the U.S. Army Medical Research and Materiel Command. LRS is supported by a postdoctoral fellowship (PDF-2000-218) from the Susan G. Komen Breast Cancer Research Fund.

In addition to the funding sources listed above, we have received generous support from the National Foundation for Cancer Research and the Jake Gittlen Memorial Golf Tournament.

We are grateful to the comments and suggestions from all members fo the laboratory and the Jake Gittlen Cancer Research Institute.

REFERENCES

1. Woodhouse EC, Chuaqui RF, Liotta LA. General mechanisms of metastasis. Cancer 1997; 80(8): 1529-37.
2. Welch DR, Wei LL. Genetic and epigenetic regulation of human breast cancer progression and metastasis. Endocrine-related Cancer 1998; 5(3): 155-97.
3. Welch DR, Rinker-Schaeffer CW. What defines a useful marker of metastasis in human cancer? J.Natl.Cancer Inst. 1999; 91(16): 1351-53.
4. Phillips KK, Welch DR, Miele ME, Lee J-H, Wei LL, Weissman BE. Suppression of MDA-MB-435 breast carcinoma cell metastasis following the introduction of human chromosome 11. Cancer Res. 1996; 56(10): 1222-26.
5. Liang P, Pardee AB. Differential display of eukaryotic messenger RNA by means of the polymerase chain reaction. Science (Wash.D.C.) 1992; 257(5072): 967-71.
6. Seraj MJ, Samant RS, Verderame MF, Welch DR. Functional evidence for a novel human breast carcinoma metastasis suppressor, *BRMS1*, encoded at chromosome 11q13. Cancer Res. 2000; 60(11): 2764-69.
7. Samant, R. S., Debies, M. T., Seraj, M. J., Verderame, M. F., and Welch, D. R. Genomic organization and chromosomal localization of the breast metastasis suppressor gene [BRMS1]. Proceedings of the American Association for Cancer Research 41, 1461. 2000. Ref Type: Abstract
8. Samant RS, Debies MT, Welch DR. Human *BRMS1* complete genomic sequence. GenBank Accession Number 2001; AF281036.
9. Crick F. Split genes and RNA splicing. Science (Wash.D.C.) 1979; 204(4390): 264-71.
10. Zhuang Z, Merino MJ, Chuaqui R, Liotta LA, Emmert-Buck MR. Identical allelic loss on chromosome 11q13 in microdissected *in situ* and invasive human breast cancer. Cancer Res. 1995; 55(3): 467-71.
11. Sasaki K, Kurata-Miura K, Ujita M et al. Expression cloning of cDNA encoding a human beta-1,3-N- acetylglucosaminyltransferase that is essential for poly-N-acetyllactosamine synthesis. Proc.Natl.Acad.Sci.U.S.A 1997; 94(26): 14294-99.
12. Robertson G, Coleman A, Lugo TG. A malignant melanoma tumor suppressor on human chromosome 11. Cancer Res. 1996; 56(19): 4487-92.
13. Shevde, L. A., Samant, R. S., and Welch, D. R. Suppression of human melanoma metastasis by breast metastasis suppressor [BRMS1]. Proceedings of the American Association for Cancer Research 42, 646. 2001. Ref Type: Abstract
14. Samant RS, Debies MT, Welch DR. *Mus musculus* breast metastasis suppressor 1-like (*brms1*) protein mRNA. GenBank Accession Number 2001; AF233580.
15. Lelekakis M, Moseley JM, Martin TJ et al. A novel orthotopic model of breast cancer metastasis to bone. Clin.Exptl.Metastasis 1999; 17(2): 163-70.
16. Samant, R. S., Debies, M. T., Shevde, L. A., and Welch, D. R. Identification and characterization of mouse homolog (brms1) of the breast cancer metastasis suppressor BRMS1. Proceedings of the American Association for Cancer Research 42, 2808. 2001. Ref Type: Abstract
17. Debies MT, Samant RS, Welch DR. Murine *brms1* complete genomic sequence. GenBank Accession Number 2001; AF368292.
18. Debies, M. T., Samant, R. S., and Welch, D. R. Genomic characterization of brms1, the murine homolog of the human breast metastasis suppressor gene, BRMS1. Proceedings of the American Association for Cancer Research 42, 2762. 2001. Ref Type: Abstract
19. Han L, Colicelli J. A human protein selected for interference with Ras function interacts

directly with Ras and competes with Raf1. Molec.Cell.Biol. 1995; 15(3): 1318-23.

20. Afar DE, Han L, McLaughlin J et al. Regulation of the oncogenic activity of BCR-ABL by a tightly bound substrate protein RIN1. Immunity. 1997; 6(6): 773-82.
21. Han L, Wong D, Dhaka A et al. Protein binding and signaling properties of RIN1 suggest a unique effector function. Proc.Natl.Acad.Sci.U.S.A 1997; 94(10): 4954-59.
22. Fu TM, Bonneau RH, Epler M et al. Induction and persistence of a cytotoxic T lymphocyte (CTL) response against a herpes simplex virus-specific CTL epitope expressed in a cellular protein. Virology 1996; 222(1): 269-74.
23. Kierstead TD, Tevethia MJ. Association of p53 binding and immortalization of primary C57BL/6 mouse embryo fibroblasts by using simian virus 40 T-antigen mutants bearing internal overlapping deletion mutations. J.Virol. 1993; 67(4): 1817-29.
24. Nicolson GL, Dulski KM, Trosko JE. Loss of intercellular junction communication correlates with metastatic potential in mammary adenocarcinoma cells. Proc.Natl.Acad.Sci.(USA) 1988; 85(2): 473-76.
25. Saunders MM, Seraj MJ, Li ZY et al. Breast cancer metastatic potential correlates with a breakdown in homospecific and heterospecific gap junctional intercellular communication. Cancer Res. 2001; 61(5): 1765-67.

Chapter 12

THE ROLE OF KISS1 IN MELANOMA METASTASIS SUPPRESSION

John F. Harms, Danny R. Welch

Jake Gittlen Cancer Research Institute. The Pennsylvania State University College of Medicine, Hershey, PA.

Abstract

Among neoplasms, the severity and relevance of metastasis is perhaps most striking in the case of melanoma where surgical resection would conduce cure if not for the subsequent complications of distant metastatic foci. Compounding this notion are statistics revealing that the number of cases of malignant melanoma have doubled each of the last four decades (1) and autopsies of patients presenting with melanoma reveal lung invasion in approximately ninety percent of cases (2). Thus, a clear understanding of the mechanisms governing melanoma metastasis and the role of metastasis suppressor genes in controlling this cascade is essential. The KiSS1 metastasis suppressor gene, identified and functional in melanoma models, provides us with just such an opportunity. As we endeavor to elucidate the role of KiSS1, an understanding of melanoma metastasis suppression mediated by chromosome 6, which provided the platform for the discovery of KiSS1, similarly proffers important clues to understanding the mechanism of KiSS1 metastasis suppression.

COMPLETE METASTASIS SUPPRESSION BY CHROMOSOME 6

Of the relatively few models of metastasis suppression in the literature, arguably the most penetrant, is the impact of human chromosome 6 within the highly metastatic human melanoma cell line C8161. In this model, metastasis suppression is complete. An amelanotic human melanoma cell line derived from an abdominal wall metastasis, C8161, exhibits reproducible metastatic potential from orthotopic (subcutaneous or intradermal) and intravenous injections in athymic mice, generating an average of 100 lung metastases per mouse following tail vein injection (3). Introduction of an intact neomycin tagged human chromosome 6 into C8161 by microcell-mediated chromosome ransfer (MMCT) produced a series of hybrid clones (designated neo6/C8161) which, with the exception of one clone, exhibited unchanged *in vitro* growth rates (4). *In vivo*, despite a slight latency in tumor formation and retarded growth rates, the incidence of tumor formation in the neo6/C8161 hybrids was unchanged. However, the cells were no longer metastatic. neo6/C8161 cells introduced either

D.R. Welch (ed.), Cancer Metastasis – Related Genes, 219–229.

by i. v. or s.c./i.d. injection failed to produce any macroscopic metastases and when mice in spontaneous metastasis assays were retained up to 30 weeks following tumor excision to compensate for the reduced growth rate of the hybrids, still no metastases were evident. While Welch *et al.* demonstrated this suppression with three distinct neo6/C8161 clones in the original communication (4), Miele *et al.* recently expanded the evidence showing complete suppression with four additional clones in both spontaneous and experimental assays (5,6).

The impact of chromosome 6 is not limited solely to C8161. Introduction of chromosome 6 into the metastatic melanoma cell line, MelJuSo, significantly inhibited the average number of lung metastases obtained in both the spontaneous and experimental metastasis assays (5). Tumorigenicity and *in vivo* growth rates of the neo6/MelJuSo clones were indistinguishable from the metastatic parental line and exhibited similar vascular invasion.

Nevertheless, the inherent genetic deficiencies of melanoma cell lines and the influences of chromosome 6 cannot be over-generalized. Tumorigenicity, not merely metastasis, was suppressed upon chromosome 6 introduction into the tumorigenic (but non~metastatic) human melanoma cell lines UACC-903 and UACC-091 (7).

DISCOVERY OF KISSI

The clear metastasis suppression mediated by chromosome 6 is consistent with the hypothesis that a metastasis-suppressor locus resides on this chromosome. Based on this idea, Lee and Welch utilized subtractive hybridization and differential display to identify differentially expressed genes, upregulated in neo6/C8161 and neo6/MelJuSo hybrids (8). In all, seven differentially expressed genes were isolated, including sequences bearing homology with rat nucleophosmin B23, transcription factor AP-2A, high-mobility group protein HMG-I(Y) and a partially sequenced fragment, I6A7. The three remaining novel clones exhibited no significant homology to DNA sequences known at that time. Among these, two clones – designated KiSS1 and KiSS16 – displayed a qualitative difference in expression between the metastatic parent and non-metastatic hybrids and were ideal subjects for further analysis.

Expression of the ~1.0 kb KiSS1 transcript was undetectable in the metastatic parental C8161 pool or subclones by northern analysis, but was present in all chromosome 6 hybrid clones (9). This qualitative difference sharply contrasted to the other differentially expressed clones which were often detectable in the metastatic cell lines but exhibited quantitative upregulation between 1.9- and 36-fold in either the C8161 or MelJuSo hybrids (8).

KiSSl EXPRESSION AND REGULATION

KiSSl Expression in Melanoma and Normal Tissues

The unambiguous upregulation of KiSS 1 in the chromosome 6 hybrid cell lines is bolstered by the expression profile of KiSS1 in a panel of unrelated melanoma cell lines representing the various stages of melanoma progression. First, Northern analyses of poly(A^+)-RNA isolated from several metastatic human melanoma lines unrelated to C816l (A375P, A375M, MeWo, MeWo-3S5. MeWo-70W, M24met and OM431) demonstrated a lack of KiSS1 expression (9). Second, semi-quantitative RT-PCR analysis of a panel of cell lines representing normal melanocytes (FM1085), radial growth phase melanoma (WM35), early vertical growth phase (WM793), vertical growth phase (WM115) and metastatic melanoma (WM239A), suggested that KiSS1 expression is correlated with melanoma progression (8). KiSS1 was highly expressed in the lines derived from melanocytes and a lesion in radial growth phase, but was lost with further progression toward metastasis. Recent unpublished data from the laboratory of Minoru Takata (Kanazawa University) using *in situ* hybridization of melanoma specimens shows high expression in benign dysplatic nevi and radial growth phase melanomas, but markedly decreased (or absent) expression in invasive melanomas (personal communication). These data support the *in vitro* findings found using the panel of cell lines (FM1085, WM35, WM793, WM115, WM239A) provided by Dr. Meenhard Herlyn (Wistar Institute).

The mRNA expression of KiSS1 in a variety of human tissues assayed by northern blot, including heart, brain, placenta, liver, lung, skeletal muscle, kidney and pancreas, revealed considerable KiSS1 expression in placenta.

Significantly weaker, but detectable, expression was evident in pancreas and kidney where smaller transcript sizes (0.8kb and 0.9kb, respectively) imply that alternate splicing may be at work in different tissues (9). RNA dot blots also revealed expression in brain tissues, findings recently confirmed elsewhere (10).

Chromosome 6 and KiSS1: A Regulatory Relationship

Any discussion of KiSSl expression, however, must address the relationship of KiSSl with the presence of chromosome 6. The identification of KiSS1 following subtractive hybridization between C8161 ± chromosome 6 led to the hypothesis that KiSSl is encoded on chromosome 6. However, fluorescence *in situ* hybridization and radiation hybrid mapping have mapped he KiSSl gene to chromosome lq32 (9, 11). This finding has an important implication – wild-type, functional KiSSl is present in the metastatic cells. Moreover, the defect leading to metastasis would appear to be upstream. This arrangement obliges the hypothesis that a locus on chromosome 6 contributes a *trans*-acting regulatory product required for KiSSl expression.

Identifying the *trans*-Regulatory Locus on Chromosome 6

The identity of the hypothesized regulatory locus on chromosome 6 is still unknown. Nevertheless, the location of the gene locus has been successfully narrowed (6). Early in the characterization of chromosome 6 metastasis suppression, a separate MMCT donor was also utilized that possessed a human chromosome 6 harboring a deletion on its q-arm (abbreviated neo6qdel). This incomplete chromosome failed to suppress metastasis following introduction into parental C8161 cells. Correspondingly, KiSS1 expression was not detectable in neo6qdel/C8161 hybrids (6,9). Miele *et al.* have subsequently characterized this donor chromosome by PCR karyotyping, identifying a deletion of 6q16.3-q23, and have documented its failure to suppress in four neo6qdel/C8161 clones (6). Thus, the regulatory locus on chromosome 6, which itself satisfies the definition of a metastasis-suppressor locus, is hypothesized to reside within this approximately 40cM region.

This interaction between chromosome 6 and the KiSS1 genomic locus on chromosome 1 also hinges on the *cis* regulatory regions surrounding KiSS1. The general genomic structure of KiSS1 includes 4 exons, of which only two comprise coding sequences (11). The last exon, possessing the translation stop codon, also encodes the polyadenylation signal. Efforts to identify upstream consensus sequences which may mediate transcriptional regulation are ongoing (RS Samant and DR Welch, in preparation).

METASTASIS SUPPRESSION BY KiSS1

The specific absence of KiSS1 from metastatic cell lines; its loss correlated with increased melanoma staging; and its conspicuous restoration in non-metastatic hybrids; each provided powerful circumstantial evidence for the role of KiSS1 as a metastasis suppressor gene. To assess whether KiSS1 alone was capable of suppressing C8161 metastasis, KiSS1 cDNA was subcloned into the pcDNA3 expression vector, under the constitutive transcriptional control of the cytomegalovirus immediate-early promoter.

Transfection of KiSS1 into the highly metastatic C8161 cell line produced a number of subclones with varying expression of KiSS1, as determined by northern blotting (8). When clones representing these differing expression levels were each tested in athymic mice, a dose-dependent suppression of metastasis was revealed. In both the experimental and spontaneous metastasis assays, the presence of KiSS1 significantly reduced the number of metastases in lung (Figure 1). In addition, regional lymph node metastases were also suppressed in the KiSS1 clones. Importantly, the classification of KiSSl as a metastasis suppressor gene was verified since primary tumor growth occurred in all KiSS1-transfected clones (9). Moreover, growth *rates* were not suppressed.

KiSS1 suppression of metastasis was further demonstrated in the most highly metastatic subclone of C8161, C8161cl.9 (8). Suppression of C8161cl.9

was nearly 100% in the experimental assay and over 93% in the spontaneous assay. Of the two highest KiSS1-expressors assessed, the greatest number of metastases identified in a mouse was only two.

As in the case of chromosome 6 suppression, KiSS1 expression also mediated metastasis suppression in human melanoma cell line, MelJuSo (8). Metastasis of MelJuSo-KiSS1 clones, was suppressed 77-98% and 51-99% in the experimental and spontaneous assays, respectively, compared to parental MelJuSo and vector-only controls.

Despite parallels between chromosome 6 and KiSS1 metastasis suppression, a key distinction is noted. Chromosome 6 suppression is complete. KiSS1 suppression, while significant, is incomplete (5,9). With mean numbers of experimental lung metastases # 1, the two highest expressing clones derived from parental C8161 presented with #4 metastases per mouse. The lowest expressing KiSS1 transfectants, however, peaked in certain mice to levels comparable to parental C8161 cells (> 200 metastases per mouse). Nevertheless, the mean number of metastases in these mice was still significantly less than the control cell lines ($P < 0.05$) (8).

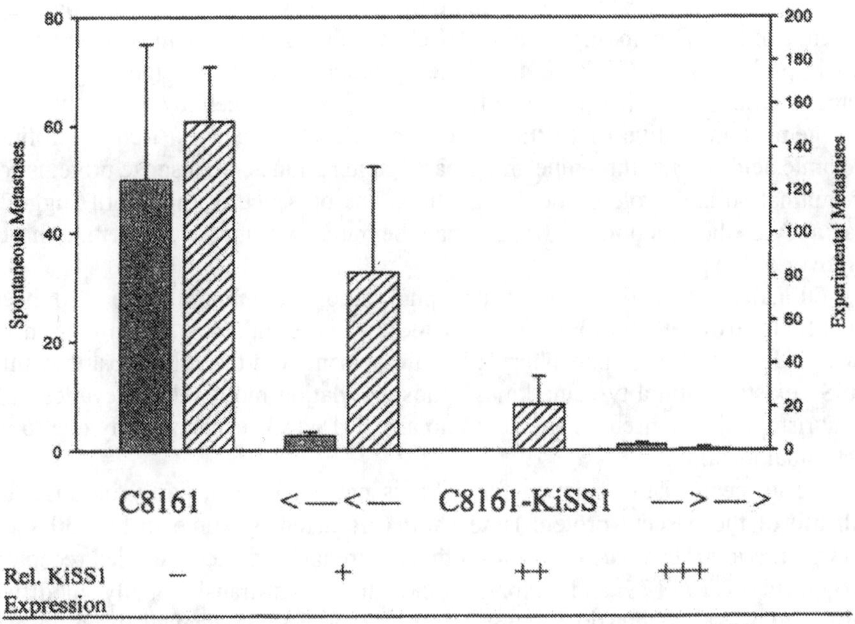

Figure 1. Mean number ofmetastases per mouse (± SEM). KiSS1 metastasis suppression is dose dependent. Mean number of metastases per mouse (± SEM) of parental C8161 cells and three C8161-KiSS1 clones in the spontaneous (solid) and experimental (hashed) metastasis assays. Relative expression of KiSS1 in each clone was determined by Northern blotting. Adapted from Lee et al.

Why doesn't KiSS1 completely suppress metastasis? First, there may be more than one metastasis suppressor gene *trans*-activated by introduction of chromosome 6. Second, there could be more than one metastasis suppressor (or regulator of metastasis suppression) encoded on chromosome 6. Third, expression of KiSS1 may be lost in some of the metastases. Cytomegalovirus promoters sometimes become inactivated in eukaryotic systems (12). Indeed, loss of KiSS1 expression has been seen in late-passage pcDNA3-KiSS1 transfectants that have restored metastatic potentials (DR Welch, SF Goldberg; unpublished observations).

THE KISS1 PROTEIN

The ~1.0 kb KiSS1 transcript encodes an open reading frame of 435 nucleotides. The predicted protein is predominantly hydrophilic protein with a predicted molecular weight of 15.4 kDa (145 a.a.). Analysis of the KiSS1 amino acid sequence in such engines as PROSITE (13), SignalP V1.1 (14), PSORT (15) and PESTfind (16) exposes a number of potentially important domains. First, a signal sequence and cleavage site at the N-terminus of the protein (aa. 1-19) is predicted with up to 100% probability. According to PSORT (*k*-NN prediction), the presence of this domain implies a 33.3% chance the KiSS1 protein is secreted and equal probabilities (22.2%) that the protein resides in the cytoplasm, mitochondria or nucleus. Secondly, amino acid residues 67-80 exhibit an arrangement indicative of a PEST sequence. PEST sequences, rich in proline, glutamic acid, serine, threonine and aspartic acid residues, predispose proteins for ubiquitination and proteosome degradation. The presence of this motif suggests that a cytosolic fraction of KiSS1 may be rapidly degraded, begetting hasty turnover of the protein.

Finally, several consensus phosphorylation site motifs have also been identified; however, the role of any of these sites is unknown. Both consensus sites for PKC and PKA phosphorylation are present, and the sole tyrosine within KiSS1 exists within a tyrosine kinase phosphorylation motif. The relevance of a N-myristoylation consensus site at amino acids 118-123 seems unlikely due to its c-terminal location.

The mechanism of action for KiSS is not yet known, partly because the half-life of the nascent protein is so short (estimated by some to be <30 sec). However, two groups recently showed that an orphan G-protein coupled receptor, designated hOT7T17S and Axor12, binds to a post-translationally modified variant of KiSS1, Metastin (Figure 2) (10, 17). KiSS1 is cleaved to be 54 a.a. potypeptide which is subsequently amidated. Receptor binding is significantly affected by the amidation. Preliminary assays suggest that activation of the receptor can alter signaling and focal adhesion kinases (17).

However, this does not agree entirely with other data showing that chromosome 6-melanoma hybrids and KiSS1 transfectants display essentially no differences in adhesion to a variety of substrates or endothelial cells. Yan *et al.* recently showed that KiSS1 specifically reduced expression of MMP9 (Note: not

MMP2) in HT1080 cells (18), suggesting an alternative mechanism of action. Most data strongly suggest that KiSS1 is acting at a late step in the metastatic cascade (19); however, definitive proof is still required.

KISS 1 MECHANISM OF SUPPRESSION

Due to the abundant expression of KiSS1 in the neo6/C8161 hybrids, analyses of chromosome 6 metastasis suppression also provide insight into possible roles of KiSS1. Qualitatively, the neo6/C8161 hybrids appear to possess increased cytoplasmic extensions and dendritic processes and grow to lower saturation densities in culture. However, comparisons of neo6/C8161 cells with parental C8161 show no difference for either *in vivo* or *in vitro* invasion and only a slight difference in pseudopod extension into a collagen type IV gradient, a possible reflection on motility (20). The capacity to grow in both soft and hard agar was reduced for both neo6 hybrids and KiSS1 transtectants (9) and similarly, growth on suboptimal (bacterial plate) surfaces has revealed only a slight advantage for the parental line (21). Thus gross impacts upon invasion, motility and adherence are not immediately apparent.

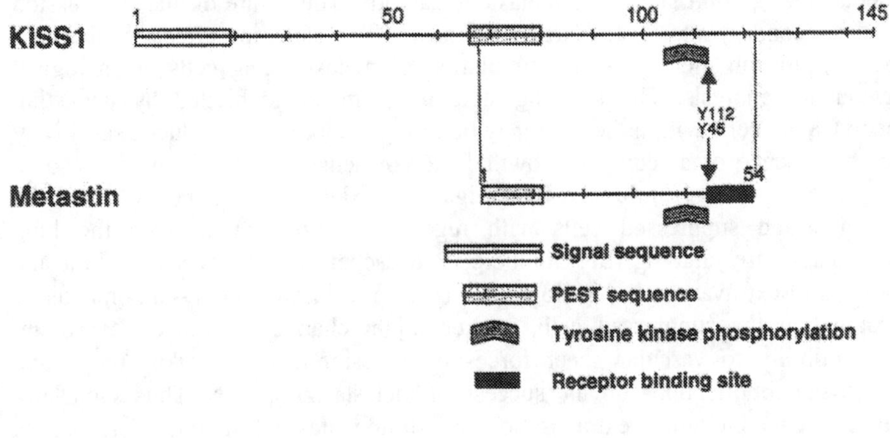

Figure 2. Schematic diagram of the KiSS1 protein and its postanslationally modified polypeptide, Metastin. Major predicted protein domains are shown.

Two observations suggest that the mechanism by which chromosome 6 and/or KiSS1 are suppressing metastasis is not due to restoring sensitivity to immune cell killing. Injection of metastasis-suppressed cells into severe combined immunodeficient mice (SCID; T- and B-cell deficient) or athymic-beige (natural killer cell- and T-cell-deficient) argues that restoration of immunogenicity to these cell types is not responsible for metastasis suppression.

Chromosome 6, the Lung Micro-Environment and Dormancy

Promising clues toward the function of KiSS1 have been proffered by tracking the fate of cells tagged with enhanced green fluorescent protein (GFP). Expression of GFP within C8161 and neo6/C8161 cells has offered considerably greater resolution for the identification and enumeration of metastases, micrometastases and single cells *in vivo*. Tail vein administration of the labeled cells into athymic mice exposed a comparable number of cells lodging within the lung and similar rates of clearance. As expected, the C8161 cells demonstrated obvious proliferation within one week and macroscopic metastases by one month; however, the neo6/C8161 cells remained predominantly as single cells. While retained fluorescence suggested continued metabolic activity of the cells, the continued viability of the cells was substantiated by recovery of the few single cells which persisted at one and two months through culturing of lung explants under dual antibiotic selection. The cells preserved their capacities to both grow in culture and proliferate at the orthotopic site when subsequently introduced intradermally (19).

The evidence that the KiSS1-expressing neo6/C8161 cells are capable of persisting in the lung microenvironment but fall short of proliferation corroborates previously mentioned evidence suggesting that suppression is mediated at a point late in the metastatic cascade. The nature of this suppression at the secondary site is still largely unknown. The paracrine effects of the lung parenchyma on the mitogenic stimulation of metastasizing cells are a logical avenue of analysis. The seeming state of dormancy exhibited by persisting neo6/C8161 cells within the lung may be brought about by an induced sensitivity to the absence of a requisite growth factor or sensitivity to the presence of a negative regulatory influence. Investigations examining the behavior of the parental and suppressed cells with regard to extravasation from the lung vasculature are ongoing (SF Goldberg, R Muschel and DR Welch). While the process of extravasation is analogous in reverse to intravasation – a competency retained in the suppressed cells – even slight changes in the extravasation, susceptibility to vascular sheer forces or adhesion-driven anoikis, may have significant implications on the success of metastasizing cells. Thus such late-stage events including extravasation and mitogenic and inhibitory responses cannot be ruled out as when determining the point of metastasis suppression mediated by KiSS1 and/or chromosome 6.

KISS1 effects on non-melanoma cell lines

Evidence obtained through work with KiSS1 in cell lines other than melanoma has also contributed to our understanding of KiSS1 metastasis suppression. KiSS1 transfected into the metastatic human breast cancer cell line, MDA-MB-435, resulted in a 95% reduction in metastatic potential (22). In addition, clonogenicity of KiSS1 transfectants was markedly reduced in both soft and hard agar and the cells displayed a significant greater rate and propensity to spread or flatten on type-IV collagen when compared to parenta MDA-MB-435 cells (22). These observations, achieved in cells of distinctly different embryonic derivation, both corroborate results of the melanoma models and imply the suppressive mechanisms mediated by KiSS1 are neither unique nor limited to melanoma.

Finally, investigations in a fibrosarcoma cell line, HT-1080, have yielded intriguing results which, while they depict a possible mechanism of KiSS1 suppression, reveal details that differ from the melanoma models. Yan *et al.* have reported that HT-1080 cells transfected with KiSS1 display a significant (65%) reduction in *in vitro* invasion through Matrigel^R coated porous filters (18). This reduction in invasiveness was further attributed to a reduction in MMP-9 type IV collagenase activity stemming from decreased (75%) transcription of the MMP-9 gene. This reduction was credited in part to diminished p65 and p50 NF-6B proteins in the nucleus as an apparent result of increased cytosolic retention achieved through increased I6B∀ protein in KiSS1 transfectants (18). Inability to detect KiSS1 by RT-PCR in five MMP-9 secreting oral squamous cell carcinoma derived lines and confirmed expression of KiSS1 in the MDA-MB-231 breast carcinoma, which scarcely expresses MMP-9, would support this notion. However, neo6/C8161 cells, which exhibit complete metastasis suppression and concomitant expression of KiSS1, do not display differences in type-IV collagenases or TIMP mRNA levels (20). Furthermore, as previously discussed, neither neo6/C8161 nor neo6/MelJuSo cells exhibited significant differences in invasiveness in comparison to their corresponding parental lines.

Thus the clues that currently shed light on the KiSS1 mechanism of metastasis suppression serve predominantly to expose what KiSS1 does not impact, while evidence specifying its particular function appears confounded from our limited vantage. Nevertheless, as indications to the mechanism and function of KiSS1 in the suppression of metastasis emerge, we gain important clues to the complexities of the metastatic cascade and the contributive processes and permissive states that facilitate and enable metastasis. Knowledge of these components will serve well in the development of diagnostic assays and therapeutic modalities commended to detect, prevent or treat metastatic disease.

ACKNOWLEDGMENTS

This research was supported by grants to DRW from the National Institutes of Health (CA-62168, CA-87728), the National Foundation for Cancer Research and the Jake Gittlen Memorial Golf Tournament.

We are also indebted to members of the laboratory for helpful discussions and recommendations.

REFERENCES

1. Clark, W. H. Human cutaneous malignant melanoma as a model for cancer. Cancer Metastasis Rev., 10: 83-88, 1991.
2. Bailly, M., Zebda, N., and Dore, J. F. Human melanoma metastasis related to specific adhesion with lung cells rather than direct growth stimulation. Anticancer Res., 14: 1791-1800, 1994.
3. Welch, D. R., Bisi, J. E., Milier, B. E., Conaway, D., Seftor, E. A., Yohem, K. H., Gilmore, L. B., Seftor, R. E. B., Nakajima, M., and Hendrix, M. J. C. Characterization of a highly invasive and spontaneously metastatic human malignant melanoma cell line. Intl. J. Cancer, 47: 227-237, 1991.
4. Welch, D. R., Chen, P., Miele, M. E., McGary, C. T., Bower, J. M., Weissman, B. E., and Stanbridge, E. J. Microcell-mediated transfer of chromosome 6 into metastatic human C8161 melanoma cells suppresses metastasis but does not inhibit tumorigenicity. Oncogene, 9: 255-262, 1994.
5. Miele, M. E., Robertson, G., Lee, J. H., Coleman, A., McGary, C. T., Fisher, P. B., Lugo, T. G., and Welch, D. R. Metastasis suppressed, but tumorigenicity and local invasiveness unaffected, in the human melanoma cell line MelJuSo after introduction of human chromosomes 1 or 6. Molec. Carcinog., 15: 284-299, 1996.
6. Miele, M. E., Jewett, M. D., Goldberg, S. F., Hyatt, D. L., Morelli, C., Gualandi, F., Rimessi, P., Hicks, D. J., Weissman, B. E., Barbanti-Brodano, G., and Welch, D. R. A human melanoma metastasis-suppressor locus maps to 6q16.3-q23. Intl. J. Cancer, 86: 524-528, 2000.
7. Trent, J. M., Stanbridge, E. J., McBride, H. L., Meese, E. U., Casey, G., Araujo, D. E., Witkowski, C. M., and Nagle, R. B. Tumorigenicity in human melanoma cell lines controlled by introduction of human chromosome 6. Science (Wash. D.C.), 247: 568-571, 1990.
8. Lee, J.-H. and Welch, D. R. Identification of highly expressed genes in metastasis-suppressed chromosome 6/human malignant melanoma hybrid cells using subtractive hybridization and differential display. Intl. J. Cancer, 71: 1035-1044, 1997.
9. Lee, J.-H.. Miele, M. E., Hicks, D. J., Phillips, K. K., Trent, J. M., Weissman, B. E., and Welch, D. R. KiSS-1, a novel human malignant melanoma metastasis-suppressor gene. J. Natl. Cancer Inst., 88: 1731-1737, 1996.
10. Muir, A. I., Chamberlain, L., Elshourbagy, N. A., Michailovich, D., Moore, D. J., Calamari, A., Szekeres, P. G., Sarau, H. M., Chambers, J. K., Murdock, P., Steplewski, K., Shabon, U., Miller, J. E., Middleton, S. E., Darker, J. G., Larminie, C. G. C., Wilson, S., Bergsma, D. J., Emson, P., Faull, R., Philpott, K. L., and Harrison, D. C. AXOR12: A novel human G protein-coupled receptor, activated by the peptide KiSS-1. J. Biol. Chem., M102743200, 2001.

11. West, A., Vojta, P. J., Welch, D. R., and Weissman, B. E. Chromosome localization and genomic structure of the KiSS-1 metastasis suppressor gene (KISS1). Genomics, 54: 145-148, 1998.
12. Gopalkrishnan, R. V., Christiansen, K. A., Goldstein, N. I., DePinho, R. A., and Fisher, P. B. Use of the human EF-1alpha promoter for expression can significantly increase success in establishing stable cell lines with consistent expression: a study using the tetracycline-inducible system in human cancer cells. Nucleic Acids Res., 27: 4775-4782, 1999.
13. PROSITE. http://, www.expasy.hcuge.ch/index.html: 2001.
14. SignalP v1.1. http://, www.cbs.dtu.dk/services/SignalP/: 2001.
15. PSORT. http://,psort.nibb.ac.jp: 2001.
16. PESTfind. http://, www.at.embnet.org/embnet/tools/bio/PESTfind: 2001.
17. Ohtaki, T., Shintani, Y, Honda, S., Matsumoto, H., Hori, A., Kanehashi, K., Torao, Y., Kumano, S., Takatsu, Y., Matsuda, Y., lshibashi, Y., Watanabe, T., Asada, M., Yamada, T., Suenaga. M., Kitada, C., Usuki, S., Kurokawa, T., Onda, H., Nishimura, 0., and Fujino, M. Metastasis suppressor gene KiSS1 encodes peptide ligand of a G-protein-coupled receptor. Nature (London), 411: 613-617, 2001.
18. Yan, C. H., Wang, H., and Boyd, D. D. KiSS-1 represses 92-kDa type IV collagenase expression by down-regulating NF-kappaB binding to the promoter as a consequence of IkappaB∀-induced block of p65/p50 nuclear translocation. J. Biol. Chem., 276: 1164-1172, 2001.
19. Goldberg, S. F., Harms, J. F., Quon, K., and Welch, D. R. Metastasis-suppressed C8161 melanoma cells arrest in lung but fail to proliferate. Clin. Exptl. Metastasis, 17: 601-607, 1999.
20. You, J., Miele, M. E., Dong, C., and Welch, D. R. Suppression of human melanoma metastasis by introduction of chromosome 6 may be partially due to inhibition of motility, but not to inhibition of invasion. Biochem. Biophys. Res. Comm., 208: 476-484, 1995.
21. Rieber, M. S., Welch, D. R. and Rieber, M. Suppression of C8161 melanoma metastatic ability by chromosome 6 induces differentiation-associated tyrosinase and decreases proliferation on adhesion-restrictive substrates mediated by overexpression of p21WAF1 and down-regulation of bcl-2 and cyclin D3. Biochem. Biophys. Res. Comm., 281: 159-165, 2001.
22. Lee, J.-H. and Welch, D. R. Suppression of metastasis in human breast carcinoma MDA-MB-435 cells after transfection with the metastasis suppressor gene, KiSS-1. Cancer Res., 57: 2384-2387, 1997.

Chapter 13

OSTEOPONTIN: A RAS-REGULATED GENE THAT CONTRIBUTES TO TUMOR METASTASIS

Ann F. Chambers, Alan B. Tuck

London Regional Cancer Centre, London, Ontario N6A 4L6 Canada

Abstract

Osteopontin (OPN) is a secreted, integrin-binding protein, which has been implicated in a variety of >stress-responsive= pathologies. In this review, we focus primarily on the possible roles for OPN in cancer. We will not attempt to provide a comprehensive review of all aspects of OPN. We instead will summarize findings from our laboratory which implicate OPN in tumor progression and metastasis, and which suggest that OPN levels in both tumors and blood plasma from cancer patients may have clinical value as a marker for metastasis and poor patient prognosis.

Key words: osteopontin, cancer, ras-regulated gene, integrin signaling, metastasis-promoting gene expression, prognostic markers

INTRODUCTION TO OSTEOPONTIN (OPN)

Osteopontin (OPN) is a secreted, integrin-binding, glycosylated phosphoprotein which is expressed in a limited number of normal tissues and whose expression can also be induced in a variety of pathologies and stress-response situations in the body. In spite of several decades of research on OPN, from many laboratories world-wide, much remains to be learned about the biological role of OPN in the normal or disease contexts in which it is found. In this review, we will consider some of the history of the discovery and initial characterization of this protein, and will then focus primarily on summarizing research from our own laboratory, especially on the role of OPN in cancer. Many questions remain about OPN in cancer, including its possible functions, its regulation and its potential role as a tumor marker, and we will highlight some of these questions and research which is beginning to address them. We will discuss evidence which links OPN to cancer metastasis, along with functional evidence that OPN is not merely associated with cancer but plays an important biological role in cancer progression. We also will discuss the clinical

D.R. Welch (ed.), Cancer Metastasis – Related Genes, 231–246.

potential of OPN measurements, in both tumor tissue and patients' blood, as a marker for tumor progression and prognosis in cancer patients. We will not attempt to provide a complete review of the many facets of OPN, in normal physiology and development, or in the various disease states where it has been implicated, and refer the reader to several reviews that cover these topics more extensively (e.g. Butler, 1989; Senger et al., 1989; Patarca et al., 1993; Denhardt and Guo, 1993; Weber and Cantor, 1996; Oates et al., 1997; Rittling and Denhardt, 1999; Sodek et al., 2000, as well as reviews contained in Denhardt et al., 1995).

In 1979, Senger and colleagues described a secreted protein produced preferentially by a variety of transformed cell lines (Senger et al., 1979, 1980; review, Senger et al., 1989). In the 1980's, several groups were studying phosphoproteins isolated from bone and other mineralized tissue (e.g. Franzen and Heinegard, 1985; Fisher et al., 1997; Prince et al., 1987), one of which appeared to be similar in size and properties to the tumor cell-secreted protein. This protein was cloned and sequenced from rat osteosarcoma cells, was shown to contain an arginine-glycine-aspartic acid (RGD) amino acid sequence, and the name osteopontin proposed, by Oldberg et al. (1986). A cDNA clone, designated as >2ar=, which was inducible in murine JB6 epidermal cells in response to a tumor promoter, was cloned by Smith and Denhardt (1987). The bone-derived and tumor-associated proteins were shown to represent the same protein, osteopontin, by Craig et al. (1988). In this study, increased expression of osteopontin (OPN) was demonstrated in ras-transformed cells. Prince (1989) provided secondary structure predictions for the protein, and discussed sequences such as the integrin-binding RGD sequence, the run of nine aspartic acid residues, and predicted calcium-binding and heparin-binding sequences, all of which provided clues to possible functions for the protein.

OPN IS A RAS ONCOGENE-INDUCED GENE

In early studies designed to determine the molecular regulation of metastasis, transfection of an activated ras oncogene into NIH 3T3 fibroblasts or other cell types was shown to confer metastatic ability on the cells (e.g. Bondy et al., 1985, Hill et al., 1988; reviewed in Chambers and Tuck, 1988, Greenberg et al., 1989, Muschel and McKenna, 1989). These studies proved to be an informative approach to understanding molecular contributors to metastasis, and led to the identification of a variety of genes regulated by ras and other oncogenes, including proteases, cell adhesion molecules, etc. (review, Chambers and Tuck, 1993). Many of these gene products have subsequently been shown to contribute functionally to metastatic ability. OPN was shown to be one of the genes significantly up-regulated by ras, and (as will be discussed below) has also been shown to play a functional role in metastasis.

OPN was identified as a ras-induced gene whose expression was significantly up-regulated in metastatic, ras-transformed NIH 3T3 cells (Craig et al., 1988, 1990; Tuck et al., 1991). (At that point in time, OPN was also designated as Secreted Phosphoprotein-1/SPP-1, as well as murine >2ar=). Accompanying the increase in OPN production by ras-transformed NIH 3T3 cells was an increased ability to adhere to OPN-coated substrates (Chambers et al., 1993). Antisense OPN transfection studies, described below, subsequently established that OPN was not only induced by ras, but also contributed to the induction of metastatic ability in the cells. The structure of the murine OPN gene and its regulatory regions were resolved (Miyazaki et al., 1990; Craig and Denhardt, 1991; Behrend et al., 1993; review, Denhardt and Guo, 1993). Details of the mechanism by which ras expression leads to increased OPN expression were then deduced by Guo et al. (1995). In that report, a novel ras-activated enhancer (RAE) sequence was identified in the murine OPN promoter. This sequence (GGAGGCAGG in the mouse) is significantly conserved, as GGANNNAGG, in the OPN promoters of other species as well. Guo et al. (1995) demonstrated that nuclei isolated from various metastatic cells, both ras-transformed and others, contained proteins able to form complexes with the RAE sequence, whereas these RAE-binding proteins were less abundant in non-metastatic cells. One of these RAE-binding proteins, designated as MATF ("Metastasis Associated Transcription Factor"), appears to be a likely candidate for a nuclear protein up-regulated in ras-transformed and other metastatic cells and responsible for some of the increased expression of OPN in metastatic cells. While this factor awaits further characterization, it is intriguing to speculate that this protein may coordinately regulate a series of genes, including OPN, whose expression is important for metastatic ability. Understanding how OPN is regulated in metastatic cells thus may provide clues to regulation of the metastatic phenotype.

OPN CAN CONTRIBUTE FUNCTIONALLY TO THE MALIGNANT BEHAVIOR OF CANCER CELLS

Several studies had shown that OPN was *associated* with the transformed phenotype, beginning with its earliest description as a protein whose secretion was increased in various transformed cell lines, as mentioned above (Senger et al., 1979, 1980; review 1989). However, a *functional* role for OPN in cancer was demonstrated by antisense transfection studies. Antisense OPN cDNA approaches in several models showed that various measures of malignancy and metastatic ability were reduced when OPN RNA or protein were specifically reduced (Behrend et al., 1994, Gardner et al., 1994, Su et al., 1995). Consistent with these studies, catalytic ribozymes targeted to different regions of the OPN mRNA similarly were shown to reduce metastatic properties of cells (Feng et al., 1995). Transfection of OPN cDNA was shown to increase metastatic ability in a rat mammary epithelial cell line (Oates et al., 1996). Recent studies using ras-transformed cells from OPN null mice have suggested that OPN may functionally enhance the effects of ras (Wu et al., 2000), providing further evidence that OPN can play a functional role in malignancy.

Growing evidence suggests mechanisms by which OPN may contribute functionally to metastatic ability and other malignant properties. OPN is a secreted protein which can bind to cell surface integrins, especially $\alpha v\beta 3$, and other integrins (including $\alpha v\beta 1$ and $\alpha v\beta 5$), as well as to other non-integrin cell surface molecules, including CD44 (Weber et al., 1996; Katagiri et al., 1999). Integrin binding has been shown to occur through the GRGDS amino acid sequence in the protein, which is strictly conserved in all species studied. Furthermore, recombinant OPN that lacks the RGD sequence, due to site-directed mutagenesis, does not support cell attachment for a variety of cell lines (Xuan et al., 1995). Thus it is reasonable to suppose that one of the functions of OPN is to signal, via integrin binding and activation, to cells carrying the appropriate receptors. Many studies have suggested what some of these signals may be, in various contexts in which OPN plays a role, and some of these studies will be mentioned here.

One proposed role for OPN in malignancy may be to protect cancer cells from the deleterious effects of host-produced nitric oxide (NO). OPN has been shown to be able to inhibit induction of nitric oxide synthase (iNOS) in kidney tubule cells, macrophages and vascular tissue (Hwang et al., 1994; Rollo et al., 1996; Scott et al., 1998). These effects have been shown to be mediated via the GRGDS sequence of OPN, implicating integrin signaling. Because OPN is induced in various reactive physiological situations in which NO is transiently induced (e.g., immune functions, vascular remodeling), it is reasonable to propose that OPN may be part of a regulatory mechanism that can limit the potentially toxic effects of NO. Similarly, it is possible that cancer cells that over-express OPN could subvert this mechanism, resulting in protection of the cancer cells from host defenses. This hypothesis is outlined in more detail by Denhardt and Chambers (1994).

OPN also may affect malignancy and metastatic ability more directly, via receptor-mediated effects on signal transduction in cancer cells. OPN has been shown to have direct effects on cell behavior and changes in gene expression associated with malignancy. In a series of studies, we have examined the effects of OPN on the malignant behavior of several human breast cancer cell lines. Metastatic MDA-MB-435 cells secrete OPN protein in culture (Bautista et al., 1994; Xuan et al., 1994). These cells also can bind to OPN-coated substrates and migrate toward OPN in vitro (Xuan et al., 1994, 1995). The behavior of MDA-MB-435 cells in response to OPN has been compared with a series of human mammary epithelial cell lines representing a progression series from the same patient (the 21T series, initially characterized by Band et al., 1990). 21PT cells are non-tumorigenic in nude mice assays, 21NT cells are weakly tumorigenic but non-metastatic, and 21MT-1 cells are weakly metastatic. OPN induced migration of the 21T series cells, as it did for MDA-MB-435 cells (Tuck et al., 1999). While MDA-MB-435 cells express high levels of OPN mRNA and protein, the 21T series cells as a group produce considerably less. However, the levels produced within the 21T series cells were highest in the most malignant of the series.

The effects of OPN on the non- and poorly tumorigenic 21T cells (21PT and 21NT, respectively) were tested in two ways: by exogenous addition of recombinant OPN protein, and by transfection of the cells to over-express endogenous OPN. Exogenous OPN was shown to increase both migration and invasion through Matrigel, in both 21PT and 21NT cells. Endogenous over-expression of OPN by both cell lines produced cells that had increased invasive ability. Thus OPN was shown to be able to induce directed migration and invasive properties in breast cancer cells (Tuck et al., 1999).

The induction of various protease activities, either by exogenous addition or endogenous over-expression of OPN in these cells, was also examined in that study (Tuck et al., 1999). By both strategies, OPN was shown to increase both mRNA and activity for urokinase-type plasminogen activator (uPA). This study thus demonstrated that OPN could alter expression and activity of a protease that has been implicated in various aspects of metastatic ability, accompanied by behavioral changes in the cells to make them more aggressive.

A subsequent study with the same series of cell lines (MDA-MB-435, and 21PT and 21NT, either with exogenous OPN or transfected to over-express endogenous OPN) showed a link between integrin-mediated effects of OPN and a growth factor pathway, that of the hepatocyte growth factor (HGF) and its cell surface receptor, Met (Tuck et al., 2000). HGF induced migration of all of the cell lines, and OPN plus HGF produced synergistic effects on migration ability. The increased migration in response to OPN was associated with an early increase in Met kinase activity, followed by a later induction of Met mRNA and protein expression. OPN effects on the malignant behavior of breast cancer cells thus involve activation of the HGF growth factor/Met receptor pathway, as well as direct OPN signaling via integrins.

The nature of the integrin signaling in these cells was also examined in this study (Tuck et al., 2000). The cell lines of high and low malignancy were shown to use different integrins to respond to OPN. The migration of MDA-MB-435 cells to OPN could be blocked by antibodies to the $\alpha v\beta 3$ integrin, while the migration of 21PT and 21NT was dependent on $\alpha v\beta 1$ and -$\beta 5$, and was $\alpha v\beta 3$-independent. This finding is intriguing, in light of studies that have linked increased expression of $\alpha v\beta 3$ with increasing malignancy in breast and other cancers (e.g., Liapis et al., 1996; van der Pluijm et al., 1997; Wong et al., 1998), as well as studies linking the

$\alpha v\beta 3$ integrin (e.g., Brooks et al.,1996; Koistinen et al.,1999) or OPN itself (Teti et al., 1998) with matrix metalloproteinase activation. Together, the above-described studies (Tuck et al., 1999, 2000) demonstrate that OPN can functionally influence the malignant behavior of cancer cells, and provide information regarding the interactions between specific integrins and growth factor receptor pathways in processing an OPN-mediated signal. OPN produced by cancer cells (or perhaps by tumor infiltrating host cells) thus has the potential to induce malignant behavior and changes in gene expression that can promote tumor progression and metastasis.

OPN might also affect tumors indirectly, by affecting angiogenesis, the development of new blood vessels which is required for growth of both primary tumors and metastases, as well as for providing a circulatory route for the escape of cells from a primary tumor. Both OPN and the $\alpha v\beta 3$ integrin can be expressed by endothelial cells during vascular remodeling, and OPN can be chemoattractive to endothelial cells (e.g. Liaw et al.,1994, 1995; Prols, 1998). In vitro, OPN has been shown to promote endothelial cell survival, mediated by endothelial cell signaling and regulation of gene expression via nuclear factor kappa B (NF-κB) activity (Scatena et al., 1998; Malyankar et al., 2000). A role for OPN in angiogenesis in vivo has yet to be demonstrated. However, it is reasonable to propose that OPN does play a role in tumor biology both directly, by affecting signal transduction and gene expression in tumor cells themselves, and indirectly, perhaps through effects on endothelial cells which thereby promote tumor angiogenesis.

Finally, however, the effects of OPN in cancer may not be straightforward. OPN can be chemotactic for host inflammatory cells (Giachelli et al, 1994), as well as tumor and vascular cells. Thus, an OPN-producing tumor may attract increased numbers of infiltrating host cells such as macrophages and lymphocytes, and these cells themselves also can produce OPN when they are activated. The cells could react against the cancer, or conversely they may also promote angiogenesis in the tumor, by virtue of angiogenic factors they secrete (Leek et al., 2000). These potentially conflicting effects of OPN may well occur simultaneously. Studies with OPN knockout mice have supported this complex picture of the possible roles for OPN in malignancy (e.g. Crawford et al., 1998). Because of the complex role that OPN plays in various physiological contexts, it is likely that its role in cancer is similarly complex, with much still remaining to be learned.

OPN IN HUMAN CANCER AND OPN AS A TUMOR MARKER IN CANCER PATIENTS

While OPN had been shown to be produced by many transformed cell lines in culture (review, Senger et al., 1989), the first assessment of OPN expression in human tumors was conducted by Brown et al. (1994). In that study, OPN RNA was assessed in a variety of human cancer specimens, and most tumors were positive for

OPN RNA However, an interesting feature of this study was the apparent localization of most OPN RNA to macrophages rather than to tumor cells themselves, in spite of the early identification of OPN as a protein produced by tumor cells in culture. In addition, whether OPN expression was important to the clinical course of cancer remained an important question. Distribution of OPN RNA and protein in normal human tissues was also examined, and was found localized predominantly along the lumenal surfaces of secretory epithelial tissues (Brown et al., 1992).

Other studies have demonstrated that OPN can be detected in various human tumor specimens. Bellahcene and Castronovo (1995) reported that OPN immunopositivity was present in the majority of 51 breast tumors studied, while less OPN was detected in 28 benign breast lesions. The data suggested that OPN was present in association with microcalcifications in breast cancer. Over-expression of OPN RNA and protein has also been detected in a series of 25 human esophageal cancers, relative to paired normal tissue (Casson et al., 1997). OPN RNA and protein expression also have been detected in prostate cancers (Thalmann et al., 1999). Overall, the findings to date with regard to OPN expression in human tumors suggest that some, but not all, tumors are OPN-positive. Within tumors, both tumor cells themselves, as well as various host cells (immune, vascular), may be OPN-positive. This variability in OPN expression, both within and between tumors, raised the possibility that OPN levels may contain information important to clinical status and patient prognosis, and thus that this question was worth addressing.

A first indication that OPN over-expression in tumors may be predictive of poor patient prognosis came from a study of twenty-five patients with lung cancer (Chambers et al., 1996). In that study, we examined OPN expression in a series of tumor samples from patients undergoing surgery for lung cancer, along with paired adjacent normal tissue. OPN levels were assessed at the RNA level by northern blotting, and at the protein level in fixed, embedded tissue sections using immunohistochemistry and a high-avidity monoclonal antibody to OPN. We found that normal lung tissue uniformly expressed negligible OPN RNA or protein, while the majority of lung tumor samples were OPN-positive, with variability between tumors in the degree of OPN over-expression. Tumor-associated OPN immunopositivity was found both in lung tumor cells as well as in tumor-associated macrophages. We found no association between tumor OPN positivity and clinico-pathological findings (age, sex, stage or histology). However, even though this was a small study, there was a statistically significant association ($p = 0.0135$) between OPN-immunopositivity in the tumor and poor patient survival. The findings from this study suggested that OPN over-expression in lung tumors may have potential to predict for patient outcome. This suggestion was supported by recent findings of Shijubo et al. (1999). In that study, a significant association between positivity for both OPN and VEGF, and poor patient survival, was seen in a series of 55 lung adenocarcinomas, while this association was not found for 32 squamous cell lung cancers.

Interestingly, however, in a series of ovarian cancers from 30 patients, higher OPN expression was found in cancers of better prognosis (Tiniakos et al., 1998). OPN=s role in cancer may vary depending on the mechanism of spread for different types of cancer. This study does suggest that OPN expression must be examined in the context of specific tumor types, and that the role of OPN in various cancers may be complex.

We had a unique opportunity to study the association of OPN with breast tumor progression in a patient who had bilateral mammary carcinomas of similar histology, and who later developed local and metastatic recurrence (Tuck et al., 1997). In this case report, we described a patient who presented initially with synchronous, bilateral infiltrating mammary carcinomas of the same histological type and grade. The right tumor was associated with positive lymph nodes, while the left-side tumor had not spread to the lymph nodes. The patient later developed right-side local recurrence, and then widespread metastatic disease. We used immunohistochemistry to examine expression of OPN and p53, in both primary tumors as well as in the local recurrence and metastases. We found that the right-side primary tumor was immunopositive for both OPN and p53, while the left-side primary tumor was negative for both markers. The locally recurrent tumor, as well as metastases to lymph nodes and bone, were highly positive for both OPN and p53. We also measured plasma OPN level at the time of last admission, a few months prior to the death of the patient. Plasma OPN levels at this time were significantly higher than the baseline plasma OPN range that we had previously established for healthy women (Bautista et al., 1996; Singhal et al., 1997). Thus, both OPN expression and p53 immunopositivity (likely indicative of p53 mutation), were found to be associated with the more aggressive of the tumors, in association with tumor progression. The findings from this case are consistent with the idea that OPN, both in tumor cells and in plasma, may be a marker for tumor aggressiveness in breast cancer, and elevated levels in a primary tumor may predict for future development of metastases.

This suggestion was tested directly in a study of 154 women with lymph node negative (LNN) breast cancer (Tuck et al., 1998). Additional prognostic markers are needed especially in LNN breast cancer, in order to identify and treat the ~30 per cent of such women who will develop metastases in the future vs. the ~70 per cent who will not develop metastases and who can be spared additional treatment. In this study, we used immunohistochemistry to determine levels and cellular localization (tumor cell vs. infiltrating host cells) of OPN protein in the tumors. We also used in situ hybridization to identify cellular localization of OPN mRNA in a subset of the tumors, to identify possible sources of the OPN protein detected. We found that OPN protein was present in infiltrating macrophages and lymphocytes in 70% of the tumors, too high a proportion to be predictive of patient survival. OPN immunopositivity localized to the tumor cells themselves, in 26% of the tumors. OPN mRNA was detected in tumor cells and in infiltrating inflammatory cells, indicating that both cell types can be a source for OPN detected within the tumors. When the OPN immuno-positivity *within the tumor cells* was scored semi-quantitatively, univariate analysis revealed a statistically significant association between elevated OPN positivity in tumor cells and decreased disease-free ($p = 0.0025$) and overall ($p = 0.0294$) survival, in this group of 154 LNN breast cancer patients. In multivariate analysis (including patient age, menopausal status, tumor size, grade, hormone receptor status, and p53 positivity), OPN remained a significant predictor of overall survival ($p = 0.0138$). This study thus supports the idea that OPN levels within tumor cells may be a useful predictor of patient outcome in breast cancer, and may play a role in tumor progression and aggressiveness.

In addition to the presence of OPN in tumor tissues, OPN was also found to be present in blood (Senger et al., 1988). In that study, blood OPN levels were qualitatively assessed, using Western blotting with a polyclonal antiserum, of barium citrate-precipitated blood samples from 14 patients with various metastatic cancers. Ten of the 14 blood samples were found to have elevated OPN levels.

We developed an antigen-capture ELISA to permit quantitative, rapid and reproducible measurement of OPN levels in blood plasma or other fluids (Bautista et al., 1996; Singhal et al., 1997), using our high-avidity anti-OPN monoclonal antibody (Bautista et al., 1994). We first used this assay to establish quantitative basal plasma OPN values for a cohort of 35 healthy women volunteers (Bautista et al., 1996). Blood samples were taken weekly over a 4-week period. We found that OPN blood levels did not differ between pre- and post-menopausal women, and there was no significant change an individual woman=s OPN blood values over the 4-week period. Levels in pre-menopausal women were thus unaffected by cyclic hormone changes over the menstrual cycle, and levels were also stable for individual post-menopausal women. (Thus, although OPN can be hormonally regulated in some tissues [e.g. Craig and Denhardt, 1991; Vanacker et al., 1999], OPN blood levels do not reflect hormonal changes over the menstrual cycle). The baseline OPN plasma level in healthy women was calculated to be 47 μg/liter (median; range, 22-122 μg/liter), using conditions to increase sensitivity of the assay (i.e., antigen capture at 37 degrees C) (Bautista et al., 1996; Singhal et al., 1997). Preliminary studies suggest that normal OPN blood levels in men do not differ from these values (unpublished data).

Using this ELISA, we then tested for an association between elevated blood OPN levels and patient outcome in breast cancer (Singhal et al., 1997). We measured OPN plasma levels in a series of 70 patients with metastatic breast cancer. We compared these values with those measured for 44 patient controls (women on well follow-up following treatment for primary breast cancer, with no evidence of disease), as well as the normal women volunteers mentioned above. We found that blood OPN levels in women with metastatic breast cancer were 142 μg/liter (median; range, 38-1312 μg/liter). In the patient control group, OPN blood levels were 60 μg/liter (median; range, 15-117). OPN blood levels in the metastatic patients were significantly different from both the patient controls and the healthy women ($p = 0.001$).

In that study (Singhal et al., 1997), we found a statistically significant association between elevated plasma OPN levels and shorter survival, when patients were divided into upper, middle and lower thirds of OPN levels ($p < 0.001$). In addition, using a Cox proportional hazards model, with OPN as a continuous variable, we found an association between increasing OPN and decreasing survival ($p < 0.0001$). Furthermore, we found that increased OPN was associated with larger numbers of sites of metastatic involvement. Elevated plasma OPN was thus significantly associated with shorter survival for women with breast cancer, by several statistical tests.

SUMMARY

Table 1 summarizes the clinical and experimental evidence which we have discussed here, and which supports the idea that OPN plays a role in tumor progression and metastasis. OPN research has progressed from experimental studies of a phosphoprotein detected in cancer cells in culture, and important in development and maintenance of bone and other mineralized tissues, through studies showing that OPN is not only associated with cancer but can contribute functionally, via signal transduction and subsequent effects on gene expression, to malignant properties of cells. Much remains to be learned about the role of OPN in cancer, and in other pathologies and normal situations as well. Recent studies are now suggesting that OPN may be a promising tumor marker, both in tissues and in patients= blood, and may play an important contributing role in tumor biology and metastasis.

Table 1. Clinical and Experimental Evidence of a Role for OPN in Tumor Progression and Metastasis.

CLINICAL	EXPERIMENTAL
! Increased blood OPN in 10/14 patients with metastatic carcinomas (Senger et al., 1988) ! Increased OPN in tumor vs. benign tissue (breast, esophagus, prostate; Bellahcene & Castronovo, 1995; Casson et al., 1997; Thalmann et al., 1999) ! High tumor cell OPN predictive of poor prognosis (lung, breast [LNN]; Chambers et al., 1996; Tuck et al., 1998) ! High plasma OPN associated with poor survival and increased tumor burden in patients with metastatic breast cancer (Singhal et al., 1997)	! Increased OPN in transformed cells of various backgrounds (review Senger et al., 1989) ! Involvement of OPN in cell adhesion (integrins, CD44; Xuan et al., 1994; Weber et al., 1996; Katagiri et al., 1999) ! Induction of cell migration by OPN (Xuan et al., 1995; Tuck et al., 2000) ! Induction of cell invasion by OPN (Tuck et al., 1999) ! Activation of growth factor/receptor (HGF/Met) pathway by OPN (Tuck et al., 2000) ! Induction of secreted protease (eg. uPA) expression/activity by OPN (Tuck et al, 1999; Teti et al., 1998) ! Promotion of metastasis by OPN (reviews Denhardt et al., 1995; Oates et al., 1996, 1997) ! Possible role of OPN in angiogenesis (Liaw et al., 1994, 1995; Prols et al., 1998) ! Inhibition of iNOS by OPN (review Denhardt and Chambers, 1994)

ACKNOWLEDGMENTS

The authors acknowledge the important contributions from members of our research group as well as our collaborators to the research summarized here. This research has been supported by the Canadian Breast Cancer Research Initiative (grants #8426 and 11300) and the U.S. Army Breast Cancer Research Program (DOD DAMD17-96-1-6075, Career Development Award to ABT; the content of this article does not necessarily reflect the position or policy of the U.S. government, and no official endorsement should be inferred).

REFERENCES

Band V, Zajchowski D, Swisshelm K, Trask D, Kulesa V, Cohen C, Connolly J, Sager R. Tumor progression in four mammary epithelial cell lines derived from the same patient.

Cancer Res 1990; 50: 7351-7357.
Bautista DS, Xuan JW, Hota C, Chambers AF, Harris JF. Inhibition of Arg-Gly-Asp (RGD)-mediated cell adhesion to osteopontin by a monoclonal antibody against osteopontin. J Biol Chem 1994; 269: 23280-23285.
Bautista DS, Saad Z, Chambers AF, Tonkin KS, O'Malley FP, Singhal H, Tokmakejian S, Bramwell V, Harris JF. Quantification of osteopontin in human plasma with an ELISA: basal levels in pre- and postmenopausal women. Clin Biochem 1996; 29: 231-239.
Behrend EI, Chambers AF, Wilson SM, Denhardt DT. Comparative analysis of two alternative first exons reported for the mouse osteopontin gene. J Biol Chem 1993; 268: 11172-11175.
Behrend EI, Craig AM, Wilson SM, Denhardt DT, Chambers AF. Reduced malignancy of ras-transformed NIH 3T3 cells expressing antisense osteopontin RNA. Cancer Res 1994; 54: 832-837.
Bellahcene A, Castronovo V. Increased expression of osteonectin and osteopontin, two bone matrix proteins, in human breast cancer. Am J Pathol 1995; 146: 95-100.
Bondy GP, Wilson S, Chambers AF. Experimental metastatic ability of H-ras-transformed NIH3T3 cells. Cancer Res 1985; 45: 6005-6009.
Brooks PC, Stromblad S, Sanders LC, von Schalscha TL, Aimes RT, Stetler-Stevenson WG, Quigley JP, Cheresh DA. Localization of matrix metalloproteinase MMP-2 to the surface of invasive cells by interaction with integrin alpha v beta 3. Cell 1996; 85: 683-693.
Brown LF, Berse B, Van de Water L, Papadopoulos-Sergiou A, Perruzzi CA, Manseau EJ, Dvorak HF, Senger DR. Expression and distribution of osteopontin in human tissues: widespread association with luminal epithelial surfaces. Mol Biol Cell 1992; 3: 1169-1180.
Brown LF, Papadopoulos-Sergiou A, Berse B, Manseau EJ, Tognazzi K, Perruzzi CA, Dvorak HF, Senger DR. Osteopontin expression and distribution in human carcinomas. Am J Pathol 1994; 145: 610-623.
Butler WT. The nature and significance of osteopontin. Connective Tissue Res 1989; 23: 123-136.
Casson AG, Wilson SM, McCart JA, O'Malley F, Ozcelik H, Tsao M-S, Chambers AF. ras mutation, and expression of the ras-regulated genes osteopontin and cathepsin L, in human esophageal cancer. Int J Cancer 1997; 72: 739-745.
Chambers AF, Tuck AB. Oncogene transformation and the metastatic phenotype. Anticancer Res 1988; 8: 861-872.
Chambers AF, Tuck AB. Ras-responsive genes and tumor metastasis. Crit Rev Oncogen 1993; 4: 95-114.
Chambers AF, Hota C, Prince CW. Adhesion of metastatic, ras-transformed NIH 3T3 cells to osteopontin (OPN), fibronectin and laminin. Cancer Res 1993; 53: 701-706.
Chambers AF, Wilson SM, Kerkvliet N, O'Malley FP, Harris JF, Casson AG. Osteopontin expression in lung cancer. Lung Cancer 1996; 15: 311-323.
Craig AM, Denhardt DT. The murine gene encoding secreted phosphoprotein 1 (osteopontin): promoter structure, activity, and induction in vivo by estrogen and progesterone. Gene 1991; 100: 163-171.
Craig AM, Nemir M, Mukherjee BB, Chambers AF, Denhardt DT. Identification of the major

phosphoprotein secreted by many rodent cell lines as 2ar/osteopontin: Enhanced expression in H-ras-transformed 3T3 cells. Biochem Biophys Res Comm 1988; 157: 166-173.

Craig AM, Bowden GT, Chambers AF, Spearman MA, Greenberg AH, Wright JA, McLeod M, Denhardt DT. Secreted phosphoprotein mRNA is induced during multi-stage carcinogenesis in mouse skin and correlates with the metastatic potential of murine fibroblasts. Int J Cancer 1990; 46: 133-137.

Crawford HC, Matrisian LM, Liaw L. Distinct roles of osteopontin in host defense activity and tumor survival during squamous cell carcinoma progression in vivo. Cancer Res 1998; 58: 5206-5215.

Denhardt DT, Guo X. Osteopontin: a protein with diverse functions. FASEB J 1993; 7: 1475-1482.

Denhardt DT, Chambers AF. Overcoming obstacles to metastasis - defenses against host defenses: osteopontin (OPN) as a shield against attack by cytotoxic host cells. J Cell Biochem 1994; 56: 48-51.

Denhardt DT, WT Butler, AF Chambers, DR Senger, eds. *Osteopontin: Role in Cell Signalling and Adhesion.* Annals of the NY Academy of Sciences vol. 760, 1995.

Feng B, Rollo EE, Denhardt DT. Osteopontin (OPN) may facilitate metastasis by protecting cells from macrophage NO-mediated cytotoxicity: evidence from cell lines down-regulated for OPN expression by a targeted ribozyme. Clin Exp Metastasis 1995; 13: 453-462.

Fisher LW, Hawkins GR, Tuross N, Termine JD. Purification and partial characterization of small proteoglycans I and II, bone sialoproteins I and II, and osteonectin from the mineral compartment of developing human bone. J Biol Chem 1987; 262: 9702-9708.

Franzen A, Heinegard D. Isolation and characterization of two sialoproteins present only in bone calcified matrix. Biochem J 1985; 232: 715-724.

Gardner HA, Berse B, Senger DR. Specific reduction in osteopontin synthesis by antisense RNA inhibits the tumorigenicity of transformed Rat1 fibroblasts. Oncogene 1994; 9: 2321-2326.

Giachelli CM, Lombardi D, Johnson RJ, Murry CE, Almeida M. Evidence for a role of osteopontin in macrophage infiltration in response to pathological stimuli in vivo. Am J Pathol 1998; 152: 353-358.

Greenberg AH, Egan SE, Wright JA. Oncogenes and metastatic progression. Invasion Metastasis 1989; 9: 360-378.

Guo X, Zhang YP, Mitchell DA, Denhardt DT, Chambers AF. Identification of a ras-activated enhancer in the mouse osteopontin promoter and its interaction with a putative ETS-related transcription factor whose activity correlates with the metastatic potential of the cell. Mol Cell Biol 1995; 15: 476-487.

Hill SA, Wilson S, Chambers AF. Clonal heterogeneity, experimental metastatic ability, and p21 expression in H-ras-transformed NIH 3T3 cells. J Natl Cancer Inst 1988; 80: 484-490.

Hwang SM, Lopez CA, Heck DE, Gardner CR, Laskin DL, Laskin JD, Denhardt DT. Osteopontin inhibits induction of nitric oxide synthase gene expression by inflammatory mediators in mouse kidney epithelial cells. J Biol Chem 1994; 269: 711-715.

Katagiri YU, Sleemen J, Fujii H, Herrlich P, Hotta H, Tanaka K, Chikuma S, Yagita H,

Okumura K, Murakami M, Saiki I, Chambers AF, Uede T. CD44 variants but not CD44s cooperate with β1-containing integrins to permit cells to bind to osteopontin independently of RGD, thereby stimulating cell motility and chemotaxis. Cancer Res 1999; 59: 219-226.

Koistinen P, Pulli T, Uitto VJ, Nissinen L, Hyypia T, Heino J. Depletion of alphaV integrins from osteosarcoma cells by intracellular antibody expression induces bone differentiation marker genes and suppresses gelatinase (MMP-2) synthesis. Matrix Biol 1999; 18: 239-251.

Leek RD, Hunt NC, Landers RJ, Lewis CE, Royds JA, Harris AL. Macrophage infiltration is associated with VEGF and EGFR expression in breast cancer. J Pathol 2000; 190: 430-436.

Liapis H, Flath A, Kitazawa S. Integrin alpha V beta 3 expression by bone-residing breast cancer metastases. Diagn Mol Pathol 1996; 5: 127-135.

Liaw L, Almeida M, Hart CE, Schwartz SM, Giachelli CM. Osteopontin promotes vascular cell adhesion and spreading and is chemotactic for smooth muscle cells in vitro. Circ Res 1994; 74: 214-224.

Liaw L, Lindner V, Schwartz SM, Chambers AF, Giachelli CM. Osteopontin and β3 integrin are coordinately expressed in regenerating endothelium in vivo and stimulate Arg-Gly-Asp-dependent endothelial migration in vitro. Circ Res 1995; 77: 665-672.

Malyankar UM, Scatena M, Suchland KL, Yun TJ, Clark EA, Giachelli CM. Osteoprotegerin is an αvβ3-induced, NF-κB-dependent survival factor for endothelial cells. J Biol Chem 2000; 275: 20959-20962.

Miyazaki Y, Setoguchi M, Yoshida S, Higuchi Y, Akizuki S, Yamamoto S. The mouse osteopontin gene. Expression in monocytic lineages and complete nucleotide sequence. J Biol Chem 1990; 265: 14432-14438.

Muschel RJ, McKenna WG. Oncogenes and tumor progression. Anticancer Res 1989; 9: 1395-1406.

Oates AJ, Barraclough R, Rudland PS. The identification of osteopontin as a metastasis-related gene product in a rodent mammary tumour model. Oncogene 1996; 13: 97-104.

Oates AJ, Barraclough R, Rudland PS. The role of osteopontin in tumorigenesis and metastasis. Invasion Metastasis 1997; 17: 1-15.

Oldberg A, Franzen A, Heinegard D. Cloning and sequence analysis of rat bone sialoprotein (osteopontin) cDNA reveals an Arg-Gly-Asp cell-binding sequence. Proc Natl Acad Sci 1986; 83: 8819-8823.

Patarca R, Saavedra RA, Cantor H. Molecular and cellular basis of genetic resistance to bacterial infection: the role of the early T-lymphocyte activation-1/osteopontin gene. Crit Rev Immunol 1993; 13:225-246.

Prince CW, Oosawa T, Butler WT, Tomana M, Bhown AS, Bhown M, Schrohenloher RE. Isolation, characterization, and biosynthesis of a phosphorylated glycoprotein from rat bone. J Biol Chem 1987; 262: 2900-2907.

Prince CW. Secondary structure predictions for rat osteopontin. Connect Tissue Res 1989; 21: 15-20.

Prols F, Loser B, Marx M. Differential expression of osteopontin, PC4, and CEC5, a novel mRNA species, during in vitro angiogenesis. Exp Cell Res 1998; 239: 1-10.

Rittling SR, Denhardt DT. Osteopontin function in pathology: lessons from osteopontin-deficient mice. Exp Nephrol 1999; 7: 103-113.

Rollo EE, Laskin DL, Denhardt DT. Osteopontin inhibits nitric oxide production and cytotoxicity by activated RAW264.7 macrophages. J Leukoc Biol 1996; 60: 397-404.

Scatena M, Almeida M, Chaisson ML, Fausto N, Nicosia RF, Giachelli CM. NF-κB mediates αvβ3 integrin-induced endothelial cell survival. J Cell Biol 1998; 141: 1083-1093.

Scott JA, Weir ML, Wilson SM, Xuan JW, Chambers AF, McCormack DG. Osteopontin inhibits inducible nitric oxide synthase activity in rat vascular tissue. Am J Physiol 1998; 275: 2258-2265.

Senger DR, Wirth DF, Hynes RO. Transformed mammalian cells secrete specific proteins and phosphoproteins. Cell 1979; 16: 885-893.

Senger DR, Wirth DF, Hynes RO. Transformation-specific secreted phosophoproteins. Nature 1980; 286: 619-621.

Senger DR, Perruzzi CA, Gracey CF, Papadopoulos A, Tenen DG. Secreted phosphoproteins associated with neoplastic transformation: close homology with plasma proteins cleaved during blood coagulation. Cancer Res 1988; 48: 5770-5774.

Senger DR, Perruzzi CA, Papadopoulos A. Elevated expression of secreted phosphoprotein I (osteopontin, 2ar) as a consequence of neoplastic transformation. Anticancer Res 1989; 9: 1291-1299.

Shijubo N, Uede T, Kon S, Maeda M, Segawa T, Imada A, Hirasawa M, Abe S. Vascular endothelial growth factor and osteopontin in stage I lung adenocarcinoma. Am J Respir Crit Care Med 1999; 160: 1269-1273.

Singhal H, Bautista DS, Tonkin KS, O'Malley FP, Tuck AB, Chambers AF, Harris JF. Elevated plasma osteopontin in metastatic breast cancer associated with increased tumor burden and decreased survival. Clin Cancer Res 1997; 3: 605-611.

Smith JH, Denhardt DT. Molecular cloning of a tumor promoter-inducible mRNA found in JB6 mouse epidermal cells: induction is stable at high, but not at low, cell densities. J Cell Biochem 1987; 34: 13-22.

Sodek J, Ganss B, McKee MD. Osteopontin. Crit Rev Oral Biol Med 2000; 11: 279-303.

Su L, Mukherjee AB, Mukherjee BB. Expression of antisense osteopontin RNA inhibits tumor promoter-induced neoplastic transformation of mouse JB6 epidermal cells. Oncogene 1995; 10: 2163-2169.

Teti A, Farina AR, Villanova I, Tiberio A, Tacconelli A, Sciotino G, Chambers AF, Gulino A, Mackay AR. Activation of MMP-2 by human GCT23 giant cell tumour cells induced by osteopontin, bone sialoprotein and GRGDSP peptide is RGD and cell shape change dependent. Int J Cancer 1998; 77: 82-93.

Thalmann GN, Sikes RA, Devoll RE, Kiefer JA, Markwalder R, Klima I, Farach-Carson CM, Studer UE, Chung LW. Osteopontin: possible role in prostate cancer progression. Clin Cancer Res 1999; 5: 2271-2277.

Tiniakos DG, Yu H, Liapis H. Osteopontin expression in ovarian carcinomas and tumors of low malignant potential (LMP). Hum Pathol 1998; 29: 1250-1254.

Tuck AB, Wilson SM, Khokha R, Chambers AF. Different patterns of gene expression in ras-resistant and ras-sensitive cells. J Natl Cancer Inst 1991; 83: 485-491.

Tuck AB, O'Malley FP, Singhal H, Tonkin KS, Harris JF, Bautista D, Chambers AF. Osteopontin and p53 expression are associated with tumor progression in a case of synchronous, bilateral, invasive mammary carcinomas. Arch Pathol Lab Med 1997;

121: 578-584.
Tuck AB, O'Malley FP, Singhal H, Harris JF, Tonkin KS, Kerkvliet N, Saad Z, Doig GS, Chambers AF. Osteopontin expression in a group of lymph node negative breast cancer patients. Int J Cancer 1998; 79: 502-508.
Tuck AB, Arsenault DM, O=Malley FP, Hota C, Ling M, Wilson SM, Chambers AF. Osteopontin induces increased invasiveness and plasminogen activator expression of human mammary epithelial cells. Oncogene 1999; 18: 4237-4246.
Tuck AB, Elliott BE, Hota C, Tremblay E, Chambers AF. Osteopontin-induced, integrin-dependent migration of human mammary epithelial cells involves activation of hepatocyte growth factor receptor (Met). J Cell Biochem 2000; 78: 465-475.
Van der Pluijm, Vloedgraven H, Papapoulos S, Lowick C, Grzesik W, Kerr J, Robey PG. Attachment characteristics and involvement of integrins in adhesion of breast cancer cell lines to extracellular bone matrix components. Lab Invest 1997; 77: 665-675.
Vanacker JM, Pettersson K, Gustafsson JA, Laudet V. Transcriptional targets shared by estrogen receptor-related receptors (ERRs) and estrogen receptor (ER) α, but not by EBβ. EMBO J 1999; 18: 4270-4279.
Weber GF, Ashkar S, Glimcher MJ, Cantor H. Receptor-ligand interaction between CD44 and osteopontin (Eta-1). Science 1996; 271: 509-512.
Weber GF, Cantor H. The immunology of Eta-1/osteopontin. Cytokine Growth Factor Rev 1996; 7: 241-248.
Wong NC, Mueller BM, Barbas CF, Ruminski P, Quaranta V, Lin EC, Smith JW. AlphaV integrins mediate adhesion and migration of breast carcinoma cell lines. Clin Exp Metastasis 1998; 16: 50-61.
Wu Y, Denhardt DT, Rittling SR. Osteopontin is required for full expression of the transformed phenotype by the ras oncogene. Br J Cancer 2000; 83: 156-163.
Xuan JW, Hota C, Chambers AF. Recombinant GST-human osteopontin fusion protein is functional in RGD-dependent cell adhesion. J Cell Biochem 1994; 54: 247-255.
Xuan JW, Hota C, Shigeyama Y, D'Errico JA, Somerman MJ, Chambers AF. Site-directed mutagenesis of the arginine-glycine-aspartic acid sequence in osteopontin destroys cell adhesion and migration functions. J Cell Biochem 1995; 57: 680-690.

Chapter 14

THE EMERGING ROLE FOR THE MRNA CAP-BINDING PROTEIN, EIF-4E, IN METASTATIC PROGRESSION

Stephen G. Zimmer[1], Jeremy R. Graff[2]

[1]*L.P. Markey Cancer Center, University of Kentucky, Dept. of Microbiology and Immunology, Lexington, KY* [2]*Lilly Research Labs, Cancer Research Division, Eli Lilly and Company, Indianapolis, IN*

INTRODUCTION

The evolution of metastases is a complex process that begins with the establishment of a primary tumor and the subsequent selection of tumor cells for the ability to break from the primary site, invade surrounding normal tissue, enter and survive within the circulation, exit the circulation and colonize the distal tissue site. The ability of the emerging tumor to progress beyond a minimum size and, ultimately, to disseminate to distal tissue sites (i.e. to metastasize) requires a myriad of changes in the expression of key genes- genes that govern angiogenesis (e.g. VEGF, bFGF), invasion (e.g. matrix metalloproteases), autocrine growth stimulation (e.g. TGFs, HER-2/neu) and metastasis (e.g. nm23). It is well-established that the accumulation of multiple genetic "hits" activates dominant oncogenes (e.g. ras) and inactivates tumor suppressor genes (e.g. p16, pRb), thereby facilitating the development of the primary tumor (1, 2). However, the vast majority of the gene products that drive metastatic progression are not genetically altered but, rather, are inappropriately expressed (2-4). Expression of these gene products (i.e. VEGF) has largely been studied at the level of transcriptional control. However, a number of recent studies have implicated translational control as a key element in regulating expression of these critical genes. Moreover, the major growth factor-induced intracellular signaling pathways implicated in cancer converge to hyperactivate protein synthesis, primarily by activating the mRNA cap-binding protein eIF-4E. The central tenet of this review is that the enhanced expression of eIF-4E, which characterizes many human and experimental cancers, contributes to metastatic progression by selectively upregulating the translation of a diverse array of proteins that mediate angiogenesis, invasion and metastasis.

D.R. Welch (ed.), Cancer Metastasis – Related Genes, 247–264.

THE METASTATIC PROCESS

Tumors ultimately arise from a single cell that has overcome negative growth regulatory signals, supercedes apoptosis and evades immune surveillance. Eventually these cellular clusters reach a critical size beyond which continued growth would cease if not for the development of a vascular network that feeds the growing tumor and facilitates continued expansion of the tumor. The tumor stimulates neovascularization by elaborating potent angiogenesis factors, such as vascular endothelial growth factor (VEGF). Individual cells within the tumor subsequently invade the surrounding normal tissue and intravasate through the vascular wall, a process which involves diminished homotypic cell: cell adhesion, enhanced heterotypic cell: cell adhesion and the secretion of degradative enzymes to break down the extracellular matrix components (e.g. the metalloproteases). Once in the circulation, the tumor cells must evade the immune system, lodge or adhere to endothelial cells within the vasculature of a distal tissue, extravasate from the circulation and establish a colony within the tissue of the metastatic site. These cells must again survive immune surveillance, overcome dormancy, evade apoptosis, generate new blood vessels, and remodel the local microenvironment to enable growth of the metastatic colony. To survive and grow within the local microenvironment of the metastatic site, which may differ substantially from that of the primary tumor, the metastatic colonies often must rely upon autocrine growth stimulation of key intracellular signaling pathways (e.g. ras-ERK pathway and PI-3 kinase pathways) (2, 3).

The gene products that mediate each of these key steps during metastatic progression are rarely mutated but are instead aberrantly expressed (4). Moreover, a myriad of diverse proteins must be expressed to mediate the numerous phenotypic alterations necessary for the formation of metastases. While genetic mutations undeniably contribute to tumor formation by enhancing genetic instability and undermining normal growth controls (1), the processes that govern metastatic progression involve quantitative rather than qualitative alterations in the expression of key malignancy related genes (2-4), such as the metalloproteases and VEGF. As such, factors that dramatically alter the "quantitative" expression of key malignancy-related genes may be critical determinants of metastasis (3). Certainly, the transcription of these key malignancy-related genes is quantitatively altered in cancer (4). However, recent studies have now demonstrated that the expression of many of these gene products is also tightly controlled at the level of translation initiation (5-9). Therefore, enhanced translation initiation, driven by enhanced eIF-4E activity, would contribute directly to the evolution of metastases by selectively modulating the translation of key malignancy-related genes (e.g. VEGF, c-myc).

EUKARYOTIC INITIATION FACTOR 4E (eIF-4E) AND CELLULAR TRANSLATION

Eukaryotic initiation factor 4E, eIF-4E, is the 25kDa mRNA cap-binding phosphoprotein that forms a macromolecular complex termed eIF-4F with the RNA helicase, eIF-4A, and the scaffolding protein, eIF-4G. eIF-4E binds the m^7-GpppX cap structure at the 5N terminus of eukaryotic mRNAs, recruiting the mRNA into the eIF-4F complex and, subsequently, into the 48S ribosomal complex (5, 6). The rate-limiting step for eIF-4F is the binding of the $5Nm^7$-GpppX cap structure of eukaryotic cellular mRNAs by eIF-4E (5-9). Subsequently, the secondary structure of an mRNA is melted via the helicase activity of eIF-4A, thereby enabling the 5N-3N scanning of the mRNA and the identification of the translation initiation codon, usually AUG (8, 9).

Under normal cellular conditions, the molar concentration of eIF-4E is lower than that of cellular mRNA. Therefore, to be translated, mRNAs must "compete" for eIF-4E to access the translation machinery. The "competitiveness" of an mRNA is determined largely by the sequence context around the translation initiation codon (i.e. similarity to the Kozak consensus sequence), the position of the translation initiation codon relative to the cap structure, and the secondary structure of the mRNA, particularly within the 5N untranslated region (5N UTR). The majority of cellular mRNAs (e.g. β-actin) may be classified as "competitive" and are typified by short, unstructured, GC-poor, 5N UTRs (Figure 1). These 5N UTRs do not interfere with cap accessibility, processing of the mRNA within the 48S complex or the identification of the initiation codon. In contrast, a select

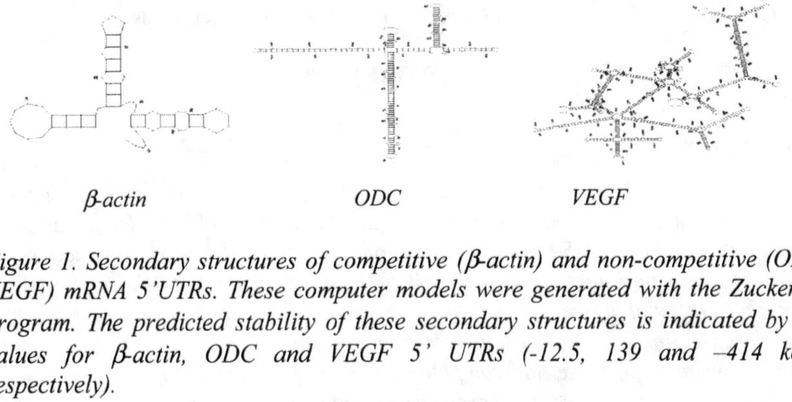

β-actin *ODC* *VEGF*

Figure 1. Secondary structures of competitive (β-actin) and non-competitive (ODC and VEGF) mRNA 5'UTRs. These computer models were generated with the Zucker m-fold program. The predicted stability of these secondary structures is indicated by the ΔG values for β-actin, ODC and VEGF 5' UTRs (-12.5, 139 and –414 kcal/mol, respectively).

group of mRNAs are encumbered by lengthy, GC-rich, highly-structured 5N UTRs (Figure 1), containing one or more upstream initiation codons, that diminish cap accessibility, interfere with 5N-3N scanning, and hinder the identification of the "true" translation initiation codon. These mRNAs, which generally code for growth regulatory proteins such as Ornithine Decarboxylase (ODC) or VEGF, are therefore difficult to translate or "non-competitive" (5-9).

Under normal growth conditions, where the concentrations of active eIF-4E are low, only competitive mRNAs like β-actin would be efficiently translated. With increased active eIF-4E, protein synthesis rates would increase, with a disproportionate increase in the translation of "non-competitive" mRNAs (5, 10). This model then predicts that, under conditions where eIF-4E activity is enhanced, the translation of the non-competitive mRNAs would be "constitutively" increased. Therefore, the enhanced expression and/or activity of eIF-4E, which typifies many human and experimental tumors, would "constitutively" and selectively increase the translation of "non-competitive" mRNAs. By upregulating the translation of these mRNAs, which code for potent mediators of malignancy (e.g. c-myc, cyclin D1, ODC and VEGF), the enhanced activity of eIF-4E would drive metastatic progression (Figure 2).

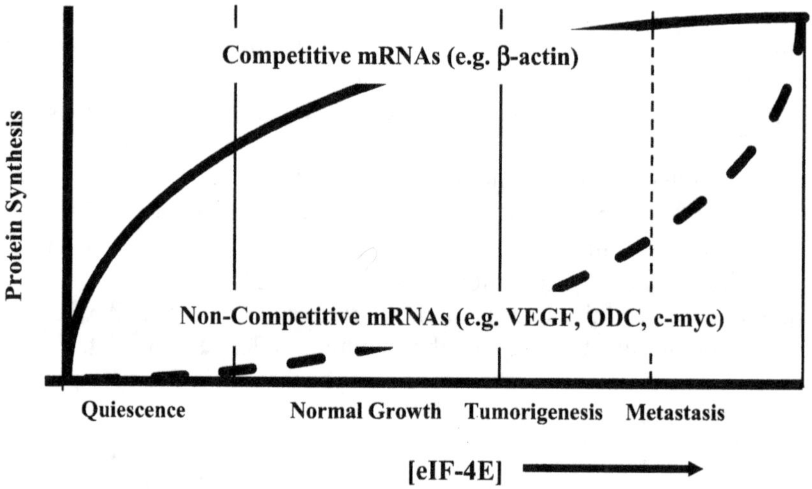

Figure 2. Enhanced activation/ expression of eIF-4E drives metastatic progression by disproportionately increasing the translation of non-competitive mRNAs. Under normal cellular growth conditions, the concentration of activated eIF-4E is low and only competitive mRNAs (solid line) are translated efficiently. As active eIF-4E increases, as in human cancers, global protein synthesis increases with a disproportionate and selective increase in the translation of non-competitive mRNAs (dashed line). Constitutive translation of these non-competitive mRNAs provides the necessary, potent protein mediators that drive metastatic progression (adapted from reference 5).

WHAT FACTORS AFFECT eIF-4E ACTIVITY?

As the rate-limiting member of the eIF-4F complex, eIF-4E governs the flow of mRNAs into polysomes, thereby facilitating the formation of the 48S ribosomal subunit and initiating translation. The activity of eIF-4E is regulated at multiple levels. In normal cells, eIF-4E is complexed with an inhibitory binding protein, 4EBP1 (or PHAS-I). Stimulation of the PI-3 Kinase pathway via insulin or growth factors activates the AKT/PKB kinase which then activates the kinase mTOR (mammalian target of rapamycin). mTOR phosphorylates 4EBP1 at residues Thr-37 and Thr-46, which primes the phosphorylation of three additional sites on 4EBP1 by an as yet unidentified kinase. Phosphorylation of these 5 sites then facilitates dissociation of 4EBP1 from eIF-4E (11-19). eIF-4E can then in turn interact with the scaffolding protein eIF-4G, to which the RNA helicase eIF-4A is also bound. Thus, the release of eIF-4E enables the formation of the eIF-4F complex (ref. 19; see signaling diagram in Figure 3).

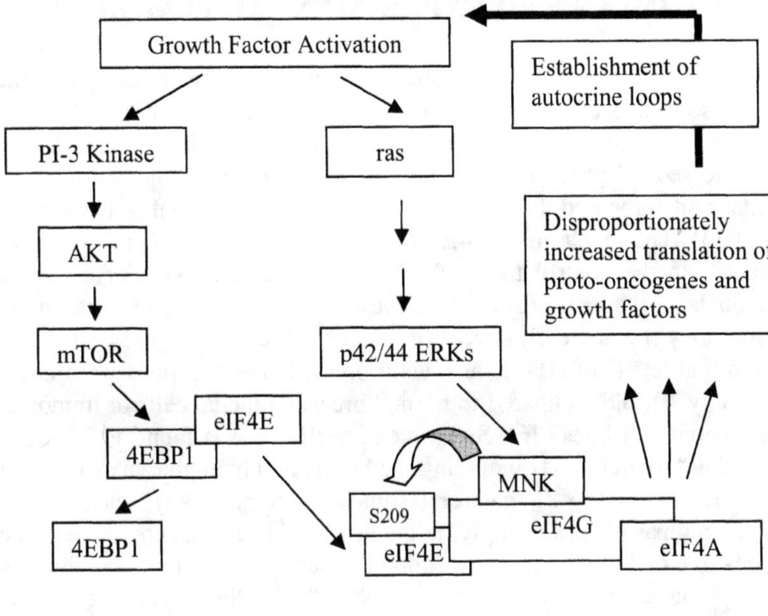

Figure 3. Pathways leading to activation of eIF-4E and the eIF-4F complex. Growth factor activation stimulates both the ras and PI-3 kinase pathways. Stimulation of PI-3 kinase leads to the activation of mTOR downstream of AKT. Subsequently, mTOR phosphorylates 4EBP1 and enables the phosphorylation of additional sites on 4EBP1, causing release of 4EBP1 from eIF-4E. eIF-4E can then bind to eIF-4G, where it is phosphorylated at Ser-209 by MNK, which is activated by the ERKs downstream of ras. As a result of enhanced free eIF-4E and assembly of the functional eIF-4F complex, translation is enhanced. Selective enhancement in the translation of proto-oncoproteins, such as c-myc, and growth factors, such as TGF-β, may drive the formation of autocrine stimulatory loops which facilitate the successful establishment of metastases.

In the eIF-4F complex, the cap-binding activity of eIF-4E is enhanced. Further, the affinity of eIF-4E for the 7-methyl-guanosine cap structure of mRNA is increased via ser-209 phosphorylation of eIF-4E by the kinase MNK1/2, which is activated by the ERK and p38 MAP kinases (20-24) and is docked on eIF-4G

(24). Hence, eIF-4E activity is initiated by release from 4EBP1 as a downstream consequence of PI-3 kinase/AKT/mTOR activation and is further enhanced by MNK1/2-mediated phosphorylation, a downstream consequence of MAP kinase signaling (Figure 3). As such, eIF-4E phosphorylation and activity can be increased by stimuli that activate the MAP kinase and PI-3 kinase signaling pathways, including serum, insulin, phorbol esters, cell stress, cytokines, growth factors (e.g. PDGF, EGF, NGF) and the pp60^{src} and p21^{ras} oncogenes (5, 25-28). Conversely, mitosis, heat shock and adenoviral infection result in reduced cellular protein synthesis and are associated with decreased eIF-4E phosphorylation (5, 28, 29).

eIF-4E IN HUMAN AND EXPERIMENTAL TUMORS

Expression of eIF-4E has been evaluated in a number of human and experimental tumors. These studies have revealed that activation or overexpression of eIF-4E is common to many tumor types. The first study implicating eIF-4E in human cancer assessed expression by immunohistochemistry in human breast carcinomas and compared these levels to those found in normal or benign disease tissue (30-35). The levels of eIF-4E in the cancer were elevated 3-30 fold over that detected in the control tissue. Moreover, increased levels of eIF-4E were found in ductal carcinoma *in situ* (DCIS) lesions, indicating that increased eIF-4E expression may be an early event in tumor development (35). It was also discovered that levels of eIF-4E in several breast tumor cell lines were increased approximately 10-fold compared to normal breast epithelial cells or immortalized breast epithelial cell lines (36). Subsequently, eIF-4E was found to be increased in head and neck carcinomas including the larynx and hypopharynx (37-41), non-Hodgkin's lymphomas (42), gastrointestinal carcinomas (43), and in invasive bladder carcinomas (44). Strikingly, in the head and neck cancers, the presence of eIF-4E positive cells in the margins following surgical resection was found to be a reliable predictor of tumor recurrence (38, 40). Furthermore, in colon cancer cells, eIF-4E expression was increased and appears to be an early event in colon tumor development (45). In our own analyses of eIF-4E expression levels by western blotting, up to 80% of cultured tumor cell lines from neuroblastomas (46) and carcinomas of the breast, prostate and thyroid exhibited elevated eIF-4E levels (Zimmer, et al., unpublished observations). These studies indicate that increased eIF-4E expression may be an integral part of tumor development in a wide variety of tumor cell types. Though human metastases have not been systematically evaluated for elevated eIF-4E expression, the fact that eIF-4E expression is increased in invasive bladder cancer, advanced gastric cancers and is predictive of recurrence in head and neck cancers, strongly suggests that increased eIF-4E expression is involved in malignant progression as well as tumor development. Direct evidence that eIF-4E plays a role in metastasis, as well as tumorigenesis, is derived primarily from studies with animal and human tumor xenografts.

eIF-4E AND TRANSLATIONALLY REGULATED mRNAs IMPORTANT FOR MALIGNANCY

The effect of increased translational efficiency on tumor growth could occur by a myriad of mechanisms. As mentioned, the activity of eIF-4E is rate-limiting for translation initiation, which results in competition by mRNAs for binding to eIF-4E and subsequent flow into polyribosomes. Under cellular conditions wherein the concentration of activated eIF-4E is low, only "competitive" mRNAs will be translated efficiently. As the concentration of activated eIF-4E increases, as a consequence of mitogenic stimulation, overexpression of eIF-4E, or oncogenic transformation (by ras or src), most mRNAs will be translated with increasing efficiency with translation of the "non-competitive" mRNAs being disproportionately increased (see Figure 2). As such, the increased eIF-4E activity that accompanies oncogenesis will result in selectively enhanced translation of certain proteins (i.e. those encoded by non-competitive mRNAs) with only mild enhancement of others (i.e. those encoded by "competitive" mRNAs). Since many of these non-competitive mRNAs code for the growth factors and proto-oncoproteins, constitutively high levels of activated eIF-4E can result in a marked increase in the production of select proteins that profoundly influence cell growth and malignancy.

The list of gene products that can be quantitatively affected by increased eIF-4E activity and increased translational efficiency is expanding and includes gene products that work at multiple points of regulation within the cell (Table 1). Transcription factors, such as c-fos or c-myc, serve to activate transcription of genes that mediate malignancy (e.g. MMP-9). Genes that regulate the cell cycle such as cyclin D1, $p27^{Kip1}$, or mdm2 can regulate the growth of tumor cells as well as influence apoptosis. Increased PDGF or IGF expression may play a key role in stimulating the autocrine growth factor loops necessary for metastases to be established. VEGF and bFGF are critical in stimulating neoangiogenesis. ODC, the rate-limiting enzyme in polyamine biosynthesis, drives transformation and tumor growth in a number of systems. Additionally, gene products that interact with the extracellular environment, such as NMDA, EGFR, MMP-9 or CD44, have been shown to be translationally regulated or affected by changes in eIF-4E levels. All of these gene products impact upon key regulatory pathways and ultimately enable tumor growth and malignancy. All of these are susceptible to translational control. Thus, by specifically influencing the expression of these key molecules, eIF-4E profoundly influences virtually every aspect of tumor growth, invasion, and metastasis (47).

Table 1. mRNAs regulated at the translational level.

GROWTH FACTORS	Reference #
PDGF	48, 49
FGF-2	50, 51
TGF-β 2,3	52, 53
IGF-II	54
VEGF	55
TRANSCRIPTION FACTORS	
c-myc	56, 57, 58
c-fos	59
Spi-1	60
C/EBP	61, 62
CELL CYCLE COMPONENTS	
Mdm-2	63, 64, 65, 66
Cyclin D1	67, 68, 69
$p27^{kip1}$	70, 71
RECEPTORS	
NMDA (NR2A subunit)	72
Her2/Neu	73
INTERLEUKINS	
IL-1B	74
IL-15	75
GROWTH REGULATORY GENES,ONCOGENES	
Ornithine Decarboxylase (ODC)	76, 77, 78, 79, 80
Ornithine Amino Transferase	81
S-adenosylmethionine decarboxylase	82
Lck	83
Pim-1	84
Ribonucleotide Reductase (small subunit)	85
P23	86

eIF-4E AND TUMOR GROWTH IN EXPERIMENTAL TUMOR MODELS

eIF-4E has been used successfully to transform cells in a number of rodent systems. Initial reports from the laboratory of Dr. Nahum Sonenberg revealed that overexpression of eIF-4E alone was able to induce transformation and tumor formation in NIH 3T3, and Rat 1 cells as well as to cooperate with the myc or ras oncogenes to transform primary rat fibroblasts (87, 88). Subsequently, we showed that CHO cells and CREF cells could also be transformed by overexpression of eIF-4E (58). The CHO cells exhibited a large increase in the expression of the myc oncogene and were transformed by morphological criteria and by growth in soft agar but were unable to form tumors in nude mice. Co-transformation with Max, the heterodimeric partner for c-myc, not only induced tumorigenesis but spontaneous metastasis as well (58). Overexpression of eIF-4E alone in CREF cells induced soft agar growth and tumor formation as well as spontaneous and experimental metastases (47, 58). The clones with the highest levels of eIF-4E were most aggressive. Furthermore, examination of cells from these tumors revealed selection for cells with further increased eIF-4E levels. When these tumor-derived cells were re-injected into nude mice, tumors formed much more quickly and metastases were evident at significantly earlier timepoints. Thus, in the CREF system, levels of eIF-4E expression were specifically associated with a more aggressive, metastatic phenotype. Indeed, metastatic progression seemed to select for increased eIF-4E expression (47, Zimmer et al., unpublished).

As the ability of cells to successfully form metastases is often dependent upon the establishment of autocrine feedback loops, we evaluated the CREF-4E cells for MAP kinase activity using an in-the-gel-kinase assay. These analyses revealed a marked stimulation of ERK 1 and 2 that corresponded directly with eIF-4E levels (Figure 4).

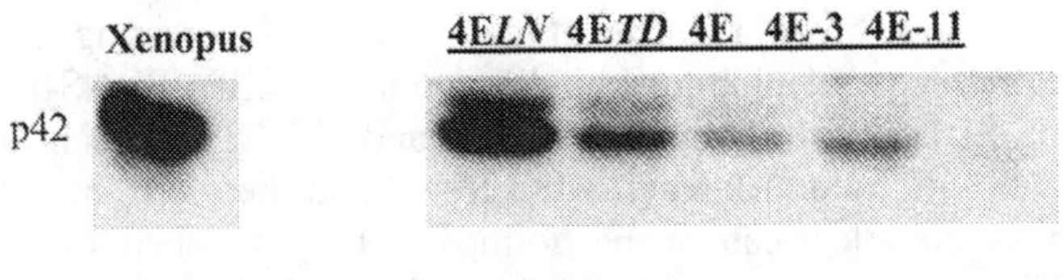

Figure 4. MAP Kinase-in-the-gel assay in CREF cell lines transfected with eIF-4E. 4E-11 is a clone transfected with eIF-4E that did not overexpress eIF-4E compared to the parental CREF control whereas 4E and 4E-3 are clones with defined overexpression of eIF-4E (data not shown). 4E-TD is the tumor-derived version of the 4E clone, whereas 4E-LN is derived from a lung nodule. Expression of eIF-4E increased in the 4E-TD and further in the 4E-LN cell lines (data not shown), corresponding to increased stimulation of the p42 and p44 MAP Kinases. The stimulated Xenopus oocyte control is also shown.

An alternative approach to investigating the role of eIF-4E in the tumorigenic or metastatic phenotype is to reduce the levels or function of eIF-4E in tumorigenic or metastatic cell systems. In the first study of this kind, an antisense construct against the translation initiation region of eIF-4E was transfected into a ras-transformed fibroblast, CREF T24, which is highly tumorigenic and metastatic (89). Antisense RNA expression reduced the level of eIF-4E by approximately 60% (90, 91). These antisense cells reverted morphologically to a more epithelioid shape and exhibited a 95% reduction in the capacity to colonize soft agar. Interestingly, though the levels of eIF-4E were reduced by 60%, these antisense cells grew at the same rate in monolayer culture as the CREF T24 parental or vector only controls (90, 91). When injected into animals, the antisense cells showed a 2-3 fold delay in tumor formation. More significantly, these antisense cell lines also showed a marked reduction in the ability to invade kidney parenchyma when implanted under the renal capsule. Perhaps most striking, the reduction of eIF-4E levels suppressed the ability of these cells to form experimental, pulmonary metastases by as much as 70% following injection into the tail vein (90). Thus, in this ras-transformed CREF system, suppressing eIF-4E levels dampened all aspects of malignancy, including invasion and metastasis.

More recently, a similar approach was used in the MDA-MB-435 human breast carcinoma cell system (32). Similar to the CREF T24 system, reduction of eIF-4E by approximately 70% did not result in dramatic alterations to monolayer growth but had profound consequences on the capacity of the antisense cells to form tumors. The antisense cells only formed tumors in 14% of the animals (2/14) compared to 100% tumor formation in the parental cell line. This reduction in tumor formation correlated with the reduced expression of bFGF protein levels and a shift in the bFGF isoforms produced and correlated with decreased angiogenesis (32). Similar data was obtained with a panel of head and neck squamous cell carcinomas following reduction of eIF-4E levels using antisense methodology (92).

Other studies have taken a different approach to suppressing eIF-4E activity by overexpressing the inhibitory binding protein, 4EBP1/PHAS-1. The binding site of eIF-4E is identical for both 4EBP1 and eIF-4G. As such, when bound to 4EBP1, eIF-4E cannot interact with eIF-4G and therefore the eIF-4F translation initiation complex cannot be formed (11, 19). Therefore, cap-dependent translation initiation is blocked when 4EBP1 is bound to eIF-4E. Overexpression of 4EBP1 in NIH 3T3 cells transformed by eIF-4E reduces tumor formation and cellular growth rate (93). In the highly aggressive breast cancer cell line, MDA-MB-231, 4EBP1 overexpression suppresses tumor formation and yields a 60% reduction in VEGF production (Zimmer et. al., in preparation).

eIF-4E GOVERNS METASTATIC PROGRESSION BY REGULATING EXPRESSION OF MULTIPLE GENE PRODUCTS

Increased eIF-4E expression induces cellular transformation, tumorigenesis and metastatic progression in a number of experimental models. The gene products affected in each model differ however, reinforcing the concept that there are multiple paths to malignancy that can be flavored by the unique genetic and epigenetic composition of the different cell models. For example, the NIH 3T3 cells which overexpress eIF-4E showed translational upregulation of ornithine decarboxylase (ODC) and cyclin D1 but not c-myc (77, 78, 67-69). Conversely, overexpression of eIF-4E in CREF and CHO cells showed enhanced c-myc translation without effects on cyclin D1 expression (47, 58). Furthermore, forced overexpression of eIF-4E was alone sufficient to induce tumorigenesis in NIH3T3 (87) as well as metastasis in CREF cells (47), but did not elicit tumorigenesis in CHO cells (58). These CHO cells required additional transfection with Max. Therefore, the effects of eIF-4E on gene expression and malignancy can differ depending on cell type.

The studies wherein eIF-4E has been suppressed, either by antisense RNA mediated reduction of eIF-4E levels or by overexpression of the inhibitory binding protein 4EBP1, have revealed that eIF-4E can also have profound influence on expression of malignancy-related genes that do not share the lengthy, GC-rich 5N UTRs of genes known to be translationally regulated. In the highly malignant ras-transformed CREF cells (CREFT24), the metastasis-specific spliced variant CD44v6 (94) and the metastasis-associated metalloprotease, MMP-9 (95, 96), are dramatically upregulated compared to non-transformed CREF cells. Reduction of eIF-4E in CREFT24 suppresses expression of both molecules (90). Moreover, selection of these antisense cells *in vivo* during tumor formation for increased eIF-4E expression restored high level expression of both CD44v6 and MMP-9 (90). In three highly tumorigenic, metastatic neuroblastoma cell lines, upregulation of eIF-4E expression also tracked precisely with increased expression of MMP-9 and CD44v6 (46, Fowler et al. unpublished). Therefore the expression of these metastasis-associated genes parallels the expression of eIF-4E in multiple cancer models. Yet, the mRNA for these molecules is not characterized by the lengthy, GC-rich 5N UTRs that typify most translationally controlled molecules (90).

Unlike the metastasis-associated CD44v6 and MMP-9, the metastasis suppressor gene nm23 was specifically upregulated when eIF-4E levels were reduced. Furthermore, when eIF-4E levels were restored following selection for tumor growth *in vivo*, nm23 expression plummeted (90). These data therefore suggest that eIF-4E expression is inversely related to expression of the metastasis suppressor gene, nm23. Recent studies in a human neuroblastoma cell line, which spontaneously progressed to a highly tumorigenic, metastatic phenotype, showed a marked elevation in eIF-4E levels that corresponded to a virtual disappearance of NM23 expression (46). Currently, the mechanistic relationship between eIF-4E and MMP-9, CD44v6 and nm23 are unclear. It is entirely possible that MMP-9 and CD44v6 are regulated by eIF-4E indirectly, by translationally controlled transcription factors (such as AP-1 transcription factors). It is also possible that the splicing of CD44 is directly affected by eIF-4E. Indeed, eIF-4E has been found in the nucleus, where splicing occurs (97, 98), and has recently been co-localized with splicing factors in the nucleus (99). The negative correlation with nm23 expression may reflect regulation similar to that for $p27^{Kip1}$, wherein the mRNA may be translationally inhibited by the binding of an mRNA binding protein that is itself translationally regulated (70, 71).

The activity of eIF-4E influences metastasis by affecting not only proto-oncoprotein expression (e.g. c-myc, ODC, cyclin D1) and expression of the metastasis-associated genes (nm23, CD44v6 and MMP-9), but also by directly regulating expression of critical angiogenesis factors. Both bFGF and VEGF have been shown to be translationally controlled and influenced directly by eIF-4E (30, 50, 51, 55). The upregulation of these potent angiogenesis factors would not only enhance primary tumor growth, but would also facilitate the formation and growth of metastases.

Enhanced eIF-4E activity has also been shown to facilitate the establishment of autocrine growth stimulatory loops. Establishing autocrine stimulation is key to the growth and survival of the metastatic cell as the foreign microenvironment of the metastatic site may not supply the same growth factors as the microenvironment of the primary site (100-102). Overexpression of eIF-4E has been shown to facilitate autocrine growth stimulation in Xenopus oocytes (103), NIH3T3 cells (104), CHO (58) and CREF cells (see Figure 4). Though the autocrine growth factors have yet to be fully defined in these systems, it is clear that eIF-4E can elicit stimulation of the ras-ERK pathway (see figure 4). As PDGF, TGF-β, IGF, VEGF and Her-2/neu are all translationally controlled, it is intriguing to speculate that eIF-4E may upregulate one or more of these potent growth factors and thereby feed the establishment of autocrine stimulatory loops.

Finally, recent reports have begun to implicate increased eIF-4E expression and/or activity in the suppression of apoptosis (105-109). Indeed, the increased resistance to apoptosis mediated by the ras oncogene seems to be dependent upon eIF-4E. It is unclear just how eIF-4E may suppress apoptosis though it is certainly plausible to posit that eIF-4E upregulates an array of anti-apoptotic growth factors (e.g. IGF-II).

SUMMARY

The successful evolution of metastases requires the interplay of numerous and diverse gene products-metalloproteases (MMP-9), adhesion factors (CD44v6), apoptosis suppressors (bcl-2), angiogenesis factors (VEGF), and proto-oncoproteins (c-myc, cyclin D1). Expression of each of these potent proteins is profoundly influenced by the activity of the mRNA cap-binding protein, eIF-4E. Indeed, eIF-4E overexpression has been shown to selectively influence a wide array of key proteins including c-myc, cyclin D1, VEGF, bFGF, CD44v6, MMP-9 and nm23. By selectively and coordinately upregulating expression of these potent proteins, enhanced eIF-4E activity, which characterizes a variety of human tumors and has been associated with malignant progression, can drive metastatic progression. As such, eIF-4E represents a potential convergence point whereby many key metastasis genes are regulated and may therefore represent a novel, compelling target for therapeutic intervention. Indeed, blocking eIF-4E activity dramatically suppresses malignancy by specifically suppressing the expression of key malignancy-related genes like cyclin D1, bFGF, VEGF and ODC (47 and references therein). Blocking eIF-4E activity has also recently been shown to induce apoptosis (109). Indeed, eIF-4E may be an attractive new therapeutic target precisely because each and every function necessary for the evolution and growth of a metastatic nodule is mediated or influenced by a gene product that is regulated translationally. The development of novel reagents to interfere with eIF-4E and translation initiation may provide a very promising alternative to current anti-cancer therapies.

ACKNOWLEDGMENTS

This work was supported in part by a grant from the National Cancer Institute (CA 77614) to SGZ. The authors are indebted to Dr. Chin-Tin Chen for assistance with the MAP kinase in-the-gel assay.

REFERENCES

1. Vogelstein B, Kinzler KW. The multistep nature of cancer. Trends Genet. 9:138-141, 1993.
2. Liteplo RG, Frost P, Kerbel RS. Genetic and epigenetic aspects of tumor progression. Basic Life Sci. 33:285-305, 1985.
3. Nicholson GL. Tumor cell instability, diversification, and progression to the metastatic phenotype: from oncogene to oncofetal expression. Cancer Res. 47:1473-1487, 1987.
4. Sager R. Expression genetics in cancer: shifting the focus from DNA to RNA. Proc. Natl. Acad. Sci. 94:952-955, 1997.
5. Rhoads RE, Joshi-Barve S, Rinker-Schaeffer C. Mechanism of action and regulation of protein synthesis initiation factor 4E: effects on mRNA discrimination, cellular growth rate, and oncogenesis. Prog. In Nucl. Acid Res. and Mol. Biol. 46:183-219, 1993.
6. Sonenberg N, Gingras AC. The mRNA 5' cap-binding protein eIF-4E and the control of cell growth. Curr. Opin. Cell. Biol. 10:268-75, 1998.
7. Kozak M. Regulation of translation in eukaryotic systems. Ann. Rev. Cell. Biol. 8:197-225, 1992.
8. Kozak M. An analysis of 5'-noncoding sequences from 699 vertebrate messenger RNAs. Nucleic Acid Res. 15:8125-8147, 1989.
9. Kozak M. An analysis of vertebrate mRNA sequences: intimations of translational control. J. Cell Biol. 115:887-903, 1991.
10. Lodish HF. Model for regulation of mRNA translation applied to haemoglobin synthesis. Nature (London) 251:385-388, 1974.
11. Raught R, Gingras AC. eIF4E activity is regulated at multiple levels. Int. J. Biochem. Cell Biol. 31:43-57, 1999.
12. Lin TA, Kong X, Haystead TAJ, Pause A, Belsham G, Sonenberg N, Lawrence JC Jr. PHAS-1 as a link between mitogen activated protein kinase and translation initiation. Science 266:653-656, 1994.
13. Pause A, Gelsham GJ, Gingras AC, Donze O, Lin TA, Lawrence JC, Sonenberg N. Insulin-dependent stimulation of protein synthesis by phosphorylation of a regulator of 5'-cap function. Nature 371:762-767, 1994.
14. Gingras AC, Kennedy SG, O'Leary MA, Sonenberg N, Hay N. 4E-BP1, a repressor of the mRNA translation, is phosphorylated and inactivated by the Akt (PKB) signaling pathway. Genes Dev. 12:502-513, 1998.
15. Brunn GJ, Hudson CC, Sekulic A, Williams JM, Hosoi J, Houghton PJ, Lawrence

JC Jr., Abraham RT. Phosphorylation of the translational repressor PHAS-1 by the mammalian target of rapamycin. Science 277: 99-101, 1997.
16. Hara K, Yonezawa K, Kozlowoski MT, Sugimoto T, Andrabi K, Weng QP, Kasuga M, Nishimoto I, Avruch J. Regulation of eIF-4E BP1 phosphorylation by mTOR. J. Biol. Chem. 272:26457-26463, 1997.
17. Fadden P, Haystead TAJ, Lawrence JC. Identification of phosphorylation site in the translational regulator, PHAS-1, that are controlled by insulin and rapamycin in rat adipocytes. J. Biol. Chem. 272:10240-10247, 1997.
18. Gingras AC, Gygi SP, Raught B, Polakiewicz RD, Abraham RT, Hoekstra MF, Aebersold R, Sonenberg N. Regulation of 4E-BP1 phosphorylation: a novel two-step mechanism. Genes Dev. 13:1422-1437, 1999.
19. Haghighat A, Mader S, Pause A, Sonenberg N. Repression of cap-dependent translation by 4E-binding protein 1: competition with p220 for binding to eukaryotic initiation factor-4E. EMBO J. 14:5701-5709, 1995.
20. Joshi B, Cai AL, Keiper BD, Minich WB, Menlez R, Beach C, Stepinski J, Stolarski R, Darzykiewicz E, Rhoads, RE. Phosphorylation of eukaryotic protein synthesis initiation factor 4E at ser-209. J. Biol. Chem. 270:14597-14603, 1995.
21. Makkinje A, Xiong HS, Li M, Damuni Z. Phosphorylation of eukaryotic protein synthesis initiation factor 4E by insulin-stimulated protamine kinase. J. Biol. Chem. 270:14824-14828, 1995.
22. Waskiewicz AJ, Flynn A, Proud CG, Cooper JA. Mitogen-activated protein kinases activate the serine/threonine kinases Mnk1 and Mnk2. EMBO J. 16:1909-1920, 1997.
23. Waskiewicz AJ, Johnson JC, Penn B, Mahalingam M, Kimball SR, Cooper JA. Phosphorylation of the cap-binding protein eukaryotic translation initiation factor 4E by protein kinase Mnk1 in vivo. Mol. Cell. Biol. 19:1871-80, 1999.
24. Pyronnet S, Imataka H, Gingras AC, Fukunaga R, Hunter T, Sonenberg N. Human eukaryotic initiation factor 4G (eIF4G) recruits mnk1 to phosphorylate eIF-4E. EMBO J 18: 270-279, 1999.
25. Frederickson RM, Montine KS, Sonenberg N. Phosphorylation of eukaryotic translation initiation factor 4E is increased in Src-transformed cell lines. Mol. Cell. Biol. 11: 2896-2900, 1991.
26. Frederickson RM, Mushynski WE, Sonenberg N. Phosphorylation of translation factor eIF-4E is induced in a ras-dependent manner during nerve growth factor-mediated PC12 cell differentiation. Mol. Cell. Biol. 12:1239-1247, 1992.
27. Rinker-Schaeffer CW, Austin V, Zimmer S, Rhoads RE. Ras transformation of cloned rat embryo fibroblasts results in increased rates of protein synthesis and phosphorylation of eukaryotic initiation factor 4E. J. Biol. Chem. 267:10659-10664, 1992.
28. Wang X, Flynn A, Waskiewicz AJ, Webb BLJ, Vries RG, Baines IA, Cooper JA, Proud CG. The phosphorylation of eukaryotic initiation factor eIF4E in response to phorbol esters, cell stresses, and cytokines is mediated by distinct MAP kinase pathways. J. Biol. Chem. 273:9373-9377, 1998.
29. Vries RG, Flynn A, Patel JC, Wang X, Denton RM, Proud CG. Heat shock increases the association of binding protein-1 with initiation factor 4E. J. Biol. Chem. 272: 32779-32784, 1997.
30. DeBenedetti A, Harris A. eIF4E expression in tumours. Int. J. Biochem. Cell Biol. 31:59-72, 1999.
31. Kerekatte V, Smiley K, Hu B, Smith A, Gelder F, DeBenedetti A. The

protooncogene/translation factor eIF-4E - A survey of its expression in breast carcinomas. Int. J. Cancer 65:27-31, 1995.

32. Nathan CA, Carter P, Liu L, Li BD, Abreo F, Tudor A, Zimmer SG, DeBenedetti A. Elevated expression of eIF-4E and FGF-2 isoforms during vascularization of breast carcinomas. Oncogene 15:1087-1095, 1997.
33. Li BDL, Liu L, Dawson M, DeBenedetti A. Overexpression of eukaryotic initiation factor 4E (eIF4E) in breast carcinoma. Cancer 79:2385-2389, 1997.
34. Li BDL, McDonald J, Nassar R, DeBenedetti A. Clinical outcome in stage 1 to 3 breast carcinomas and eIF4E overexpression. An. Surg. Soc. 227:756-762, 1998.
35. DeFatta RJ, Turbat-Herrera EA, Li BD, Anderson W, DeBenedetti A. Elevated expression of eIF-4E in confined early breast cancer lesions: possible role of hypoxia. Int. J. Cancer 80:516-522, 1999.
36. Anthony B, Carter P, DeBenedetti A. Overexpression of the proto-oncogene/translation factor 4E in breast-carcinoma cell lines. Int. J. Cancer 65:858-863, 1996.
37. Haydon MS, Googe JD, Sorrells DS, Ghali GE, Li BD. Progression of eIF4E gene amplification and overexpression in benign and malignant tumors of the head and neck. Cancer 88:2803-2810, 2000.
38. Nathan CA, Sanders K, Abreo FW, Nassar R, Glass J. Correlation of p53 and the proto-oncogene eIF4E in larynx cancers: prognostic implications. Cancer Res, 60:3599-3604, 2000.
39. Nathan CA, Liu L, Li B, Nandy I, Abreo F, DeBenedetti A. Detection of the proto-oncogene eIF4E in surgical margins may predict recurrence in head and neck cancer. Oncogene 15:579-584, 1997.
40. Franklin S, Pho T, Abreo FW, Nassar R, DeBenetti A, Stucker FJ, Nathan CA. Detection of the proto-oncogene eIF4E in larynx and hypopharynx cancers. Arch Otolaryngol Head Neck Surg 125:177-182, 1999.
41. Sorrells DL Jr, Ghali GE, DeBenedetti A, Nathan CA, Li BD. Progressive amplification and overexpression of the eukaryotic initiation factor 4E gene in different zones of head and neck cancers. J Oral Maxillofac Surg, 57:294-299,1999.
42. Wang S, Rosenwald IB, Hutzler MJ, Pihan GA, Savas L, Chen JJ, Woda BA. Expression of the eukaryotic translation initiation factors 4E and 2alpha in non-Hodgkin's lymphomas. Am J Pathol 155: 247-255, 1999.
43. Martin ME, Perez MI, Redondo C, Alvarez MI, Salinas M, Fando JL. 4E binding protein 1 expression is inversely correlated to the progression of gastrointestinal cancers. Int J Biochem Cell Biol 32: 633-642, 2000.
44. Crew JP, Fuggle S, Bicknell R, Cranston DW, DeBenedetti A, Harris AL. Eukaryotic initiation factor-4E in superficial and muscle invasive bladder cancer and its correlation with vascular endothelial growth factor expression and tumor progression. Br J Cancer 82: 161-166, 2000.
45. Rosenwald IB, Chen JJ, Wang S, Savas L, London IM, Pullman J. Upregulation of protein synthesis initiation factor eIF-4E is an early event during colon carcinogenesis. Oncogene 18:2507-17, 1999.
46. Fowler CL, Zimmer CC, Zimmer SG. Spontaneous progression of a stage IV-S human neuroblastoma cell line involves the increased expression of the protooncogenes ras and eukaryotic initiation factor 4E. Pediatr Pathol Mol Med 19: 433-447, 2000.
47. Zimmer, SG, DeBenedetti A, Graff JR. Translational control of malignancy: the mRNA cap-binding protein,eIF-4E, as a central regulator of tumor formation,

growth, invasion, and metastasis. Anticancer Res. 20: 1343-1351,2000.

48. Rao CD, Pech M, Robbins KC, Aaronson SA. The 5' untranslated sequence of the c-sis/PDGF 2 transcript is a potent translational inhibitor. Mol. Cell. Biol. 8: 284-292, 1988.
49. Wang C, Stiles CD. Regulation of platelet-derived growth factor A messenger RNA translation in differentiating F9 teratocarcinoma cells. Cell Growth Diff. 4: 871-877, 1993.
50. Prats H, Kaghad M, Prats AP, Klagsbrun M, Lelias JM, Liazun P, Chalon P, Tauber JP, Amalric F, Smith JA, Caput D. High molecular mass forms of basic fibroblast growth factor are initiated by alternative CUG codons. Proc. Natl. Acad. Sci. USA 86: 1836-1840, 1989.
51. Kevil C, Carter P, Hu B, DeBenedetti A. Translational enhancement of FGF-2 by eIF-4 factors, and alternate utilization of CUG and AUG codons for translation initiation. Oncogene 11: 2339-2348, 1995.
52. Noma T, Glick AB, Geiser AG, O'Reilly MA, Miller J, Roberts AB, Sporn MB. Molecular cloning and structure of the human transforming growth factor-beta 2 gene promoter. Growth Factors 4:247-255, 1991.
53. Arrick BA, Lee AL, Grendell RL, Derynck R. Inhibition of translation of transforming growth factor-beta 3 mRNA by its 5' untranslated region. Mol. Cell. Biol. 11: 4306-4313, 1991.
54. Nilsson FC, Gammeltoft S, Christiansen J. Translational discrimination of mRNAs coding for human insulin-like growth factor II. J. Biol. Chem. 265: 13431-13434, 1990.
55. Kevil C, DeBenedetti A, Payne KD, Coe LL, Laroux S, Alexander S. Translational regulation for vascular permeability factor by eukaryotic initiation factor 4E: Implications for tumor angiogenesis. Int. J. Cancer 65: 785-790, 1996.
56. Saito H, Hayday AC, Wiman K, Hayward WS, Tonegawa S. Activation of the c-myc gene by translocation: a model for translational control. Proc. Natl. Acad. Sci. USA 80: 7476-7480, 1983.
57. Darveau A, Pelletier J, Sonenberg N. Differential efficiencies of in vitro translation of mouse c-myc transcripts differing in the 5' untranslated region. Proc. Natl. Acad. Sci. USA 82: 2315-2319, 1985.
58. DeBenedetti A, Joshi B, Graff JR, Zimmer SG. CHO cells transformed by the translation factor eIF-4E display increased c-myc expression, but require overexpression of Max for tumorigenicity. Mol. Cell. Diff. 2: 347-371, 1994.
59. Zerial M, Toschi L, Ryseek RP, Schuermann M, Mueller R, Bravo R. The product of a novel growth factor activated gene, fosB, interacts with Jun proteins enhancing their DNA binding activity. EMBO J. 8: 805-813, 1989.
60. Hensold JO, Stratton CA, Barth C. The conserved 5'-untranslated leader of Spi-1 (PU.1) mRNA is highly structured and potently inhibits translation in vitro but not in vivo. Nucl. Acids Res. 25: 2869-2876, 1997.
61. Raught B, Gingras AC, James A, Medina D, Sonenberg N, Rosen JM. Expression of a translationally regulated, dominant-negative CCAAT/enhancer-binding protein beta isoform and up-regulation of the eukaryotic translation initiation factor 2alpha are correlated with neoplastic transformation of mammary epithelial cells. Cancer Res. 56: 4382-4386, 1996.
62. Calkhoven CF, Muller C, Leutz A. Translational control of C/EBPalpha and C/EBPbeta isoform expression. Genes Dev. 14: 1920-1932, 2000.
63. Landers JE, Haines DS, Strauss JF 3rd, George DL. Enhanced translation: a novel

mechanism of mdm2 oncogene overexpression identified in human tumor cells. Oncogene 9: 2745-2750, 1994.

64. Landers JE, Cassel SL, George DL. Translational enhancement of mdm2 oncogene expression in human tumor cells containing a stabilized wild-type p53 protein. Cancer Res. 57:3562-3568, 1997.
65. Capoulade C, Bressac-de Paillerets B, Lefrere I, Ronsin M, Feunteun J, Tursz T, Wiels J. Overexpression of MDM2, due to enhanced translation, results in inactivation of wild-type p53 in Burkitt's lymphoma cells. Oncogene 16: 1603-1610, 1998.
66. Brown CY, Mize GJ, Pineda M, George DL, Morris DR. Role of two upstream open reading frames in the translational control of oncogene mdm2. Oncogene 18: 5631-5637, 1999.
67. Rosenwald IB, Karatzas A, Sonenberg N, Schmidt EV. Elevated levels of cyclin D1 protein in response to increased expression of eukaryotic initiation factor eIF4E. Mol. Cell. Biol. 13: 7358-7363, 1993.
68. Rosenwald EB, Kaspar R, Rosseau D, Gehrke L, Leboulch P, Chen JJ, Schmidt EV, Sonenberg N, London IM. Eukaryotic translation initiation factor 4E regulates expression of cyclin D at transcriptional and posttranscriptional levels. J. Biol. Chem. 270: 21176-21180, 1995.
69. Rousseau D, Kaspar R, Rosenwald I, Gehrke L, Sonenberg N. Translation initiation of ornithine decarboxylase and nucleocytoplasmic transport of cyclin D1 mRNA are increased in cells overexpressing eukaryotic initiation factor 4E. Proc. Natl. Acad. Sci. U.S.A. 93: 1065-70, 1996.
70. Agrawal D, Hauser P, McPherson F, Dong F, Garcia A, Pledger WJ. Repression of $p27^{kip1}$ synthesis by platelet-derived growth factor in BALB/c 3T3 cells. Mol. Cell. Biol. 16: 4327-4336, 1996.
71. Hengst L, Reed SI. Translational control of $p27^{kip1}$ accumulation during the cell cycle. Science 271: 1861-1864, 1996.
72. Wood MW, Van Dongen HM, Van Dongen AM. The 5' untranslated region of the N-methyl-D-aspartate receptor NR2A subunit controls efficiency of translation. J. Biol. Chem. 271: 8115-20, 1996.
73. Child SJ, Miller MK, Geballe AP. Cell type-dependent and –independent control of HER-2/ *neu* translation. Int. J. Biochem. Cell Biol. 31: 201-213, 1999.
74. Kaspar R, Gerke L. *In*: Translational Control, CSH Press, pp. 152, 1992.
75. Bamford RN, Battiata AP, Burton JD, Sharma H, Waldmann TA. Interleukin (IL) 15/IL-T production by the adult T-cell leukemia cell line HuT-102 is associated with a human T-cell lymphotrophic virus type IR region/IL-15 fusion message that lacks many upstream AUGs that normally attenuate IL-15 mRNA translation. Proc. Natl. Acad. Sci. USA 93: 2897-2902, 1996.
76. Shantz LM, Pegg AE. Overproduction of ornithine decarboxylase caused by relief of translational repression is associated with neoplastic transformation. Cancer Res. 54: 2313-2316, 1994.
77. Shantz LM, Hu RH, Pegg AE. Regulation of ornithine decarboxylase in a transformed cell line that overexpresses translation initiation factor eIF-4E. Cancer Research 56: 3265-3269, 1996.
78. Graff JR, DeBenedetti A, Olson JW, Tamez P, Casero RA, Zimmer SG. Translation of ODC mRNA and polyamine transport are suppressed in ras-transformed CREF cells by depleting translation initiation factor 4E. BBRC 240: 15-20, 1997.
79. Manzella JM, Rychlik W, Rhoads RE, Hershey JW, Blackshear PJ. Insulin induction

of ornithine decarboxylase. Importance of mRNA secondary structure and phosphorylation of eukaryotic initiation factors eIF-4B and cIF4E. J. Biol. Chem. 266: 2383-2389, 1991.

80. Grens A, Scheffler IE. The 5'- and 3'-untranslated regions of ornithine decarboxylase mRNA affect the translational efficiency. J. Biol. Chem. 265: 11810-11816, 1990.
81. Fagan RJ, Lazaris-Karatzas A, Sonenberg N, Rozan R. Translational control of ornithine aminotransferase. Modulation by initiation factor eIF4E. J. Biol. Chem. 266: 16518-16523, 1991.
82. Shantz LM, Viswanath R, Pegg AE. Role of the 5'-untranslated region of the mRNA in the synthesis of S-adenosylmethionine decarboxylase and its regulation by spermine. Biochem J. 302:765-772, 1994.
83. Marth JD, Overell RW, Meier KE, Krebs EG, Permutter RM. Translational activation of the lck proto-oncogene. Nature 332: 171-173, 1988.
84. Hoover DS, Wingett DG, Zhang J, Reeves R, Magnuson NS. Pim-1 protein expression is regulated by its 5'-untranslated region and translation initiation factor eIF4E. Cell Growth Diff. 8:1371-1380, 1997.
85. Abid MR, Li Y, Anthony C, De Benedetti A. Translational regulation of ribonucleotide reductase by eukaryotic initiation factor 4E links protein synthesis to the control of DNA replication. J. Biol. Chem. 274: 35991-35998, 1999.
86. Bommer UA, Lazaris-Karatzas A, De Benedetti A, Nurnberg P, Benndorf R, Bielka H, Sonenberg N. Translational regulation of the mammalian growth-related protein P23: involvement of eIF-4E. Cell Mol Biol Res. 40: 633-641, 1994.
87. Lazaris-Karatzas A, Montine KS, Sonenberg N. Malignant transformation by a eukaryotic initiation factor subunit that binds to mRNA cap. Nature 345: 544-547, 1990.
88. Lazaris-Karatzas A, Sonenberg N. The mRNA 5' cap-binding protein, eIF-4E, cooperates with v-myc or E1A in the transforming of primary rodent fibroblasts. Mol. Cell. Biol. 12: 1234-1238, 1992.
89. Boylan JF, Jackson J, Steiner MR, Shih TY, Duigou GJ, Roszman T, Fisher PB, Zimmer SG. Role of the Ha-ras (RasH) oncogene in mediating progression of the tumor cell phenotype (review). Anticancer Res. 10: 717-724, 1990.
90. Graff JR, Boghaert ER, DeBenedetti A, Tudor DM, Zimmer SG. Reduction of translation initiation factor 4E reduces tumor growth, invasion and metastasis of ras-transformed cloned rat embryo fibroblast. Int. J. Cancer 60:255-263, 1995.
91. Rinker-Schaeffer CW, Graff JR, DeBenedetti A, Zimmer SG, Rhoads RE. Decreasing the level of translation initiation factor 4E with antisense RNA causes reversal of ras-mediated transformation and tumorigenesis of cloned rat embryo fibroblasts. Int. J. Cancer 55:841-847, 1993.
92. DeFatta RJ, Nathan,CA, DeBenedetti A. Antisense RNA to eIF4E suppresses oncogenic properties of a head and neck squamous cell carcinoma cell line. Laryngoscope 110: 928-933, 2000.
93. Rousseau, D., A-C Gingras, A. Pause and N. Sonenberg. (1996) The eIF-4E-binding proteins 1 and 2 are negative regulators of cell growth. Oncogene 13: 2415-2420.
94. Gunthert U, Stauder R, Mayer B, Terpe HJ, Finke L, Friedrichs K. Are CD44 variant isoforms involved in human tumor progression? Cancer Surv. 24: 19-42, 1995.
95. Bernhard EJ, Graber SB, Muschel RJ. Direct evidence linking expression of matrix metalloproteinase 9 (92-kDa gelatinase/collagenase) to the metastatic phenotype in transformed rat embryo cells. Proc. Natl. Acad. Sci. 91:4293-4297, 1994.

96. Su ZZ, Leon JA, Austin VA, Zimmer SG, Fisher PB. Wild type 5 adenovirus transforming genes function as trans-dominant suppressors of oncogenesis in mutant type 5 adenovirus transformed rat embryo fibroblast cells. Cancer Res. 53: 1929-1938, 1993.
97. Dostie J, Ferraiuolo M, Pause A, Adam SA, Sonenberg N. A novel shuttling protein, 4E-T, mediates the nuclear import of the mRNA 5' cap-binding protein, eIF-4E. EMBO J. 19: 3142-3156, 2000.
98. Lejbkowicz F, Goyer C, Darveau A, Neron S, Lemieux R, Sonenberg N. A fraction of the mRNA 5' cap-binding protein, eukaryotic initiation factor 4E, localizes to the nucleus. Proc. Nat. Acad. Sci. USA 89: 612-9616, 1992.
99. Dostie J, Lejbkowicz F, Sonenberg N. Nuclear eukaryotic initiation factor 4E (eIF4E) colocalizes with splicing factors in speckles. J. Cell. Biol. 148: 239-247, 2000.
100. Wright JA, Turley EA, Greenberg AH. Transforming growth factor beta and fibroblast growth factor as promoters of tumor progression to malignancy. Crit. Rev. Oncog. 4: 473-492, 1993.
101. Herlyn M, and Malkowicz SB. Regulatory pathways in tumor growth and invasion. Lab. Invest. 64: 262-271, 1991.
102. Liotta LA, Wewer U, Rao NC, Schiffman E, Sracke M, Guirguis R, Thorgeeirsson U, Muschel R, Sobel M. Biochemical mechanisms of tumor invasion and metastases. Adv. Exp. Med. Biol. 233: 161-169, 1988.
103. Klein PS, Melton DA. Induction of mesoderm in *Xenopus laevis* embryos by translation initiation factor 4E. Science 265: 803-805, 1994.
104. Lazaris-Karatzas A, Smith MR, Frederickson RM, Jaramillo ML, Liu Y-L, Kung H-F, Sonenberg N. Ras mediates translation initiation factor 4E-induced malignant transformation. Genes and Development 6: 1631-1642, 1992.
105. Polunovsky VA, Gingras AC, Sonenberg N, Peterson M, Tan A, Rubins JB, Manivel JC, Bitterman PB. Translational control of the antiapoptotic function of Ras. J. Biol. Chem. 275: 24776-24780, 2000.
106. Tan A, Bitterman P, Sonenberg N, Peterson M, Polunovsky V. Inhibition of Myc-dependent apoptosis by eukaryotic translation initiation factor 4E requires cyclin D1. Oncogene 19: 1437-1447, 2000.
107. Polunovsky VA, Rosenwald IB, Tan AT, White J, Chiang L, Sonenberg N, Bitterman PB. Translational control of programmed cell death: eukaryotic translation initiation factor 4E blocks apoptosis in growth-factor-restricted fibroblasts with physiologically expressed or deregulated Myc. Mol. Cell. Biol. 16: 6573-81, 1996.
108. Bushell M, Wood W, Clemens MJ, Morley SJ. Changes in integrity and association of eukaryotic protein synthesis factors during apoptosis. Eur. J. Biochem. 267: 1083-1091, 2000.
109. Herbert TP, Fahraeus R, Prescott A, Lane DP, Proud CG. Rapid induction of apoptosis mediated by peptides that bind initiation factor 4E. Curr. Biology 10: 793-796, 2000.

INDEX

A375 melanoma
effect of Phe deprivation 193
effect of Tyr deprivation 193
AFT-1
targeting 160
Amino acid deprivation
effect on tumor invasion 194
Amino acid restriction 191
Antisense MTA1 gene
cell growth 56
AP-2
lack of expression 155
metastatic effects 155
regulation of MMP-2 154
role in tumor growth 154
Assessory molecules
cooperative interaction with integrins 74
Astrocytes
contribution to the brain-metastatic specificity of melanoma cells 98
ATF-1
as a survival factor 159
effect on human melanoma cells 157, 158
effect on tumorigenicity 157
role in melanoma progression 157
Autocrine motility factor
clinical utilization 116
identification 110
mechanism of secretion 111
receptor systems 116
regulating metastasis 109
Autocrine motility factor receptor 112
cell surface expression 114
regulation 116
signaling cell motility 114

Brain invasion 94
Brain metastases
associated genes 89
in melanoma 90
molecular determinants 90
role of neurotrophins 91
role of p75 neurotrophin receptor 93
Brain metastatic melanoma cells
heparanase production 96
neurotrophins enhance invasion 96
Brain metastatic phenotype 94
Breast cancer 191
heterochromatic changes 170
role of BRSM1 209
role of heterochromatin-associated protein 1 169
Breast cancer cells
localization of MTA1 protein 55
Breast metastatic suppressor 1
discovery 210
functional characterization 212
mouse homolog 212
role in breast carcinoma 209
suppressing metastasis 211, 213

Cadherins 67
Cancer cells
malignant behavior 233
Cell adhesion molecules 66
Cellular translation
role of eIF-4E 249
Chromosome 17 19
Chromosome 6 21, 22
dormancy 226
identifying the *trans*-regulatory locus 222
lung micro-environment 226
role in KiSS1 222
role in metastasis suppression 219
c-Kit
role in the progresson of melanoma metastasis 148
Cooperative integrin interaction 65
CREB
as a survival factor 159
effect on human melanoma cells 157, 158
effect on tumorigenicity 157
role in melanoma progression 157
targeting 160

Drosophila HP1 171

EIF-4E
- affecting factors 250
- cellular translation 249
- effect of translationally regulated mRNAs 253
- in experimental tumors 252
- in human tumors 252
- metastatic progression 254
- role in experimental tumor models 254
- role in metastatic progression 247
- tumor growth 254

Epithelial cancer cell proliferation 51
ERK pathway 36, 42
Euchromatin 173

Gene changes
- associated with amino acid restriction 191

Gene discovery 23
Gene expression
- in melanoma metastasis 145
- regulation by integrins 72

Gene silencing 171, 178, 184
Genomic organization
- in heterochromatin-associated protein 1 183

Genomic stability
- in heterochromatin-associated protein 1 182

Growth factors
- cooperation with integrins 76

Hematogenous metastases 5
Heparanase
- produced by astrocytes 98

Heparanase functionality
- in melanoma prgression 94

Heterochromatic changes
- in breast cancer cells 170

Heterochromatin 169
- and gene silencing 171

Heterochromatin-associated protein 1
- down regulation 175
- experimental expression 179
- genomic organization 183
- genomic stability 182
- in breast cancer 174
- in breast cancer invasion 169
- in breast cancer invasive metastatic phenotype 175, 180
- in metastasis 169, 177
- reduced 177

Histone deacetylase
- interaction with MTA1 protein 58

Human melanoma cells
- role of ATF-1 157-159
- role of CREB 157-159
- survival factors 159

Human MTA1 gene
- metastasis-associated 51

Ig super family 68
Integrins 69
- cooperation with growth factors 74
- cooperation with tetraspan molecules 76
- coöperative interaction with assessory molecules 74
- cooperative interactions with proteolytic enzyme 78

ECM interactions 70
- general structure 69
- matrix assembly 70
- regulation of gene expression 72
- regulation of proliferation 71
- signaling proliferation 71

Invasion
- during amino acid restriction 191

JNK kinases 39

KAI1 12
KiSS1
- discovery 220
- effects on non-melanoma cell lines 227
- expression 221
- in melanoma and normal tissue 221
- mechanisms 225
- metastasis suppression 222
- regulation 221
- role in melanoma metastasis suppression 219
- role of chromosome 6 222

Malignancy
- role of eIF-4E 253

Mammalian HP1-like proteins 172
- expression patterns 172
- role in gene silencing 172, 180

subcellular localization 172
MAP kinase
promoting metastases 42
suppressing metastases 42
MAP kinase pathways 35, 36
MCAM/MUC18
role in melanoma metastasis 149
MEK family members 35
MEK pathway 36, 42
Melanoma cells
brain-metastatic specificity 98
Melanoma metastasis
gene regulation 145
role of c-Kit 148
role of MCAM/MUC18 149
Melanoma metastasis suppression
role of KiSS1 219
Melanoma progression
role of ATF-1 157
role of CREB 157
Metastasis
during amino acid restriction 191
Metastasis suppression
by chromosome 6 219
Metastasis suppressor activity 7
Metastasis suppressor genes
associated with amino acid restriction 191
chromosomal regions 8
clinical utility 23, 133
identification 7, 10, 13
influence of Phe deprivation 196
influence of Tyr deprivation 196
regulation of metastatic growth 19
Metastatic colonization 6
Metastatic process 248
Metastatic progression
governed by eIF-4E 254
Metastatic propensity
prediction 1
regulation 4
Methionine 193
MKK4
in A375 melanoma cells 193
MMP-2
regulated by AP-2 154
Molecular determinants
of brain metastases 90
Mouse homolog
in breast metastatic suppressor 1 212
MTA1 antisense oligonucleotides 56
MTA1 gene
expression in human cell lines 55
expression in human tumors 55
localization in breast cancer cells 55
sequence analysis 53
MTA1 protein
interaction with histone deacetylase 58
levels 56
possible functions 58
multiple gene products
role of EIF-4E 254

NDP kinase 131
Neurotrophin production 94
Neurotrophins
role in brain metastatic events 91
Nm23 12
NM23 metastasis suppressor gene 123
biochemical activities 129
cloning 123
differentiation 127
expression 123
structure 131
transfection experiments 126
translational research 133
Nm23-H1
re-expression 134

Osteopontin 231
as a ras oncogene-induced gene 232
as a tumor marker 236
contribution to the malignant behavior of cancer cells 233
in human cancer 236

p38 MAP kinase 39, 44
p75 neurotrophin receptor
role in brain metastases 93
role in tumor invasion 93
Phe deprivation
in A375 melanoma 193
on differential modulation of gene expression 200
on metastasis-suppressor gene expression 196
Phenylalanine, *see* Phe

Prostate cancer 191
Proteolytic enzyme
cooperative interaction with integrins 78
Putative biochemical activities 129

Ras-regulated gene 231
Reciprocal cytokine regulation 99

SAPK kinase 39, 44
SEK1
in A375 melanoma cells 193
Selectins 67
Single chain versus fragment
antiATF-1 160
Stress-response protein kinases 39

Tetraspan molecules
cooperation with integrins 76
Translational research
of Nm23 133
Translationally regulated mRNAs
effect on eIF-4E 253
Trans-regulatory locus
identification 6 222
Tumor growth
role of eIF-4E 254
Tumor marker
osteopontin 236
Tumor metastases
role of cooperative integrin interactions 65
Tumor models
experimental 254
Tumorigenicity
role of ATF-1 158
role of CREB 157
Tyr deprivation
differential modulation of gene expression 200
in A375 melanoma 193
on metastasis-suppressor gene expression 196
Tyrosine, *see* Tyr

GPSR Compliance
The European Union's (EU) General Product Safety Regulation (GPSR) is a set of rules that requires consumer products to be safe and our obligations to ensure this.

If you have any concerns about our products, you can contact us on

ProductSafety@springernature.com

In case Publisher is established outside the EU, the EU authorized representative is:

Springer Nature Customer Service Center GmbH
Europaplatz 3
69115 Heidelberg, Germany

www.ingramcontent.com/pod-product-compliance
Lightning Source LLC
LaVergne TN
LVHW011751200926
848484LV00003B/669
* 9 7 8 1 4 0 2 0 0 5 2 2 0 *